Plant Breeding: Theory and Methods

Plant Breeding: Theory and Methods

Brittany Hilbert

SYRAWOOD
PUBLISHING HOUSE
New York

Published by Syrawood Publishing House,
750 Third Avenue, 9th Floor,
New York, NY 10017, USA
www.syrawoodpublishinghouse.com

Plant Breeding: Theory and Methods
Brittany Hilbert

International Standard Book Number: 978-1-64740-354-6 (Hardback)

Cataloging-in-Publication Data

Plant breeding : theory and methods / edited by Brittany Hilbert.
p. cm.
Includes bibliographical references and index.
ISBN 978-1-64740-354-6
1. Plant breeding. 2. Breeding. 3. Agriculture. I. Hilbert, Brittany.
SB123 .P53 2023
631.52--dc23

TABLE OF CONTENTS

PREFACE

The world is advancing at a fast pace like never before. Therefore, the need is to keep up with the latest developments. This book was an idea that came to fruition when the specialists in the area realized the need to coordinate together and document essential themes in the subject. That's when I was requested to be the editor. Editing this book has been an honour as it brings together diverse authors researching on different streams of the field. The book collates essential materials contributed by veterans in the area which can be utilized by students and researchers alike.

Plant breeding refers to the science of modifying the traits of plants to generate desired qualities. It has been utilized to improve the quality of nutrition in the food products for animals and humans. Plant breeding aims to create crop varieties with superior and unique features that can be used for a range of agricultural purposes. The most commonly studied traits are those associated with end-use quality attributes such as flavor or concentrations of certain biological molecules, biotic and abiotic stress tolerance, processing ease, and biomass or grain production. The type of quantitative or qualitative traits a plant will have is determined by its genes. Plant breeding can be done in a variety of ways, such as selecting plants with desirable features for propagation and using approaches that rely on the knowledge of genetics. This book presents the complex subject of plant breeding in the most comprehensible and easy to understand language. The readers would gain knowledge that would broaden their perspective about the theories and methods of plant breeding.

Each chapter is a sole-standing publication that reflects each author´s interpretation. Thus, the book displays a multi-facetted picture of our current understanding of application, resources and aspects of the field. I would like to thank the contributors of this book and my family for their endless support.

Brittany Hilbert

1

Linear Selection Indices: An Overview

Abstract We describe the main characteristics of two approaches to the linear selection indices theory. The first approach is called *standard linear selection indices* whereas the second of them is called *eigen selection index methods*. In the first approach, the economic weights are fixed and known, whereas in the second approach the economic weights are fixed but unknown. This is the main difference between both approaches and implies that the eigen selection index methods include to the standard linear selection indices because they do not require that the economic weights be known. Both types of indices predict the net genetic merit and maximize the selection response, and they give the breeder an objective criterion to select individuals as parents for the next selection cycle. In addition, in the prediction they can use phenotypic, markers, and genomic information. In both approaches, the indices can be unrestricted, null restricted or predetermined proportional gains and can be used in the context of single-stage or multistage breeding selection schemes. We describe the main characteristics of the two approaches to the linear selection indices theory and we finish this chapter describing the Lagrange multiplier method, which is the main tool to maximize the selection index responses.

Linear selection indices that assume that economic weights are fixed and known to predict the net genetic merit are based on the linear selection index theory originally developed by Smith (1936), Hazel and Lush (1942), and Hazel (1943). They are called *standard linear selection indices* in this introduction. Linear selection indices that assume that economic weights are fixed but unknown are based on the linear selection index theory developed by Cerón-Rojas et al. (2008a, 2016) and are called *Eigen selection index methods*. The Eigen selection index methods include the standard linear selection indices as a particular case because they do not require the economic weights to be known. To understand the Eigen selection index methods theory, the point is to see that this is an application of the canonical correlation theory to the standard linear selection index context. The multistage linear selection index theory will be described only in the context of the standard linear selection indices. As we shall see, there are three main types of LSI: phenotypic, marker, and genomic. Each can be unrestricted, null restricted or predetermined proportional gains and can be used in the context of single-stage or multistage breeding selection schemes.

For each specific selection index described in this book, we have used an acronym. For example, the Smith (1936), Hazel and Lush (1942), and Hazel (1943) index was denoted LPSI (linear phenotypic selection index), whereas the Cerón-Rojas et al. (2008a) index was denoted ESIM (Eigen selection index method), etc. For additional details, see Table 1.1 and the Preface of this book. We think that such notation gives the reader a more general point of view of the relationship that exists among all the indices described in this book.

Table 1.1 Chapter where the index was described, authors who developed the selection index, acronym of the index used in this book, and description of the acronym

Chapter	Authors who developed the index	Acronym	Description
2	Smith(1936), Hazel and Lush (1942), Hazel (1943)	LPSI[a]	Linear phenotypic selection index
	Williams (1962a)	BLPSI[a]	Base linear phenotypic selection index
3	Kempthorne and Nordskog (1959)	RLPSI[a]	Restricted linear phenotypic selection index
	Mallard (1972), Harville (1975), Tallis (1985), Itoh and Yamada (1987)	PPG-LPSI[a]	Predetermined proportional gain linear phenotypic selection index
	Pesek and Baker (1969), Yamada et al. (1975), Itoh and Yamada (1986)	DG-LPSI[a]	Desired gains linear phenotypic selection index
4	Lande and Thompson (1990)	LMSI[b]	Linear marker selection index
	Lange and Whittaker (2001)	GW-LMSI[b]	Genome-wide linear marker selection index
5	Togashi et al. (2011), Ceron-Rojas et al. (2015)	LGSI[c]	Linear genomic selection index
	Dekkers (2007)	CLGSI[d]	Combined linear genomic selection index
6	Kempthorne and Nordskog (1959), Ceron-Rojas et al. (2015)	RLGSI[c]	Restricted linear genomic selection index
	Tallis(1985), Ceron-Rojas et al. (2015)	PPG-LGSI[c]	Predetermined proportional gain linear genomic selection index
	Kempthorne and Nordskog (1959), Dekker (2007)	CRLGSI[d]	Combined restricted linear genomic selection index
	Tallis (1985), Dekker (2007)	PPG-CLGSI[d]	Predetermined proportional gain combined linear genomic selection index
7	Cerón-Rojas et al. (2008a)	ESIM[a]	Eigen selection index method
	Cerón-Rojas et al. (2008a)	RESIM[a]	Restricted eigen selection index method
	Cerón-Rojas et al. (2016)	PPG-ESIM[a]	Predetermined proportional gain eigen selection index method

(continued)

Table 1.1 (continued)

Chapter	Authors who developed the index	Acronym	Description
8	Cerón-Rojas et al. (2008b)	MESIM[b]	Molecular eigen selection index method
	Crossa and Cerón-Rojas (2011)	GW-ESIM[b]	Genome-wide eigen selection index method
	Dekkers (2007), Cerón-Rojas et al. (2008b)	GESIM[d]	Genomic eigen selection index method
	Dekkers (2007), Cerón-Rojas et al. (2008a)	RGESIM[d]	Restricted genomic eigen selection index method
	Dekkers (2007), Cerón-Rojas et al. (2016)	PPG-GESIM[d]	Predetermined proportional gain genomic eigen selection index method
9	Cochran (1951), Young (1964)	MLPSI[a]	Multistage linear phenotypic selection index
	Cochran (1951), Young (1964), Kempthorne and Nordskog (1959)	MRLPSI[a]	Multistage restricted linear phenotypic selection index
	Cochran (1951), Young (1964), Tallis (1985)	MPPG-LPSI[a]	Multistage predetermined proportional gain linear phenotypic selection index
	Cochran (1951), Young (1964), Ceron-Rojas et al. (2015)	MLGSI[c]	Multistage linear genomic selection index
	Cochran (1951), Young (1964), Kempthorne and Nordskog (1959), Ceron-Rojas et al. (2015)	MRLGSI[c]	Multistage restricted linear genomic selection index
	Cochran (1951), Young (1964), Tallis (1985), Ceron-Rojas et al. (2015)	MPPG-LGSI[c]	Multistage predetermined proportional gain linear genomic selection index

[a]Indices that use only phenotypic information
[b]Indices that use marker and phenotypic information jointly
[c]Indices that use only genomic information
[d]Indices that use genomic and phenotypic information jointly in the prediction of the net genetic merit

1.1 Standard Linear Selection Indices

1.1.1 Linear Phenotypic Selection Indices

Three main linear phenotypic selection indices used to predict the net genetic merit and select parents for the next selection cycle are the LPSI, the null restricted LPSI (RLPSI), and the predetermined proportional gains LPSI (PPG-LPSI). The LPSI is an unrestricted index, whereas the RLPSI and the PPG-LPSI allow restrictions to be imposed equal to zero and predetermined proportional gain restrictions respectively, on the trait expected genetic gain per trait values to make some traits change their mean values based on a predetermined level while the rest of the trait means remain without restrictions. All these indices are linear combinations of several observable and optimally weighted phenotypic trait values.

The simplest linear phenotypic selection index (LPSI) can be written as $I_B = \mathbf{w}'\mathbf{y}$, where $\mathbf{w}$ is a known vector of economic values and $\mathbf{y}$ is a vector of phenotypic values. We called this index the *base linear phenotypic selection index* (BLPSI). In this case, the breeder does not need to estimate any parameters, and some authors have indicated that the BLPSI is a good predictor of the net genetic merit ($H = \mathbf{w}'\mathbf{g}$, where $\mathbf{g}$ is a vector of true unobservable breeding values) when no data are available for estimating the phenotypic ($\mathbf{P}$) and genotypic ($\mathbf{G}$) covariance matrices. When the traits are independent and the economic weights are also known, the LPSI can be written as $I = \sum_{i=1}^{t} w_i h_i^2 y_i$, and when the economic weights are not known, the LPSI is $I = \sum_{i=1}^{t} h_i^2 y_i$, where w_i is the ith economic weight and h_i^2 is the heritability of trait y_i. In Chap. 2 (Sects. 2.5.1 and 2.5.2), we will show that the foregoing three indices are particular cases of the more general LPSI, i.e., $I = \mathbf{b}'\mathbf{y}$, where $\mathbf{b}$ is the I vector of coefficients and $\mathbf{y}$ is the vector of observable trait phenotypic values. In the latter case, we need to estimate matrices $\mathbf{P}$ and $\mathbf{G}$.

The LPSI was originally proposed by Smith (1936) in the plant breeding context; later Hazel and Lush (1942) and Hazel (1943) extended the LPSI to the context of animal breeding. These authors made a clear distinction between the LPSI and the net genetic merit. The net genetic merit was defined as a linear combination of the unobservable true breeding values of the traits weighted by their respective economic values. In the LPSI theory, the main assumptions are: the genotypic values that make up the net genetic merit are composed entirely of the additive effects of genes, the LPSI and the net genetic merit have a joint normal distribution, and the regression of the net genetic merit on LPSI values is linear. Two of the main parameters of this index are the selection response and the expected genetic gain per trait or multi-trait selection response. The LPSI selection response is associated with the mean of the net genetic merit and was defined as the mean of the progeny of the selected parents or the mean of the future population (Cochran 1951). The selection response enables breeders to estimate the expected selection progress before carrying it out. This information gives improvement programs a clearer orientation and helps to predict the success of the adopted selection method and choose the option that is technically most effective on a scientific basis (Costa et al. 2008). On the other hand, the LPSI expected genetic gain per trait, or multi-trait selection response, is the population mean of each trait under selection of the progeny of the selected parents. Thus, although the LPSI selection response is associated with the mean of the net genetic merit, the LPSI expected genetic gain per trait is associated with the mean of each trait under selection. The foregoing definition of selection response and the expected genetic gain per trait are valid for all selection indices described in this book.

One of the main problems of the LPSI is that when used to select individuals as parents for the next selection cycle, the expected mean of the traits can increase or decrease in a positive or negative direction without control. This was the main reason why Kempthorne and Nordskog (1959) developed the basics of the restricted LPSI (RLPSI), which allows restrictions to be imposed equal to zero on the expected genetic gain of some traits whereas the expected genetic gain of other traits increases (or decreases) without any restrictions being imposed. Based on the results of the RLPSI, Tallis (1962) and James (1968) proposed a selection index called predetermined proportional gains LPSI (PPG-LPSI), which attempts to make some traits change their expected genetic gain values based on a predetermined level, while the rest of the traits remain without restrictions. Mallard (1972) pointed out that the PPG-LPSI proposed by Tallis (1962) and James (1968) does not provide optimal genetic gains and was the first to propose an optimal PPG-LPSI based on a slight modification of the RLPSI. Other optimal PPG-LPSIs were proposed by Harville (1975) and Tallis (1985). Itoh and Yamada (1987) showed that the Mallard (1972) index is equal to the Tallis (1985) index and that, except for a proportional constant, the Tallis (1985) index is equal to the Harville (1975) index. Thus, in reality, there is only one optimal PPG-LPSI.

In Chap. 3 (Sect. 3.1.1 and 3.2.1), we show that $\mathbf{b}_R = \mathbf{Kb}$ and $\mathbf{b}_P = \mathbf{K}_P\mathbf{b}$ are the vectors of coefficients of the RLPSI and PPG-LPSI, respectively, where $\mathbf{b}$ is the LPSI vector of coefficients. Matrices $\mathbf{K}$ and $\mathbf{K}_P$ are idempotent ($\mathbf{K} = \mathbf{K}^2$ and $\mathbf{K}_P = \mathbf{K}_P^2$), that is, they are projectors. Matrix $\mathbf{K}$ projects $\mathbf{b}$ into a space smaller than the original space of $\mathbf{b}$ because the restrictions imposed on the expected genetic gains per trait are equal to zero (Sect. 3.1.1). The reduction of the space into which matrix $\mathbf{K}$ projects $\mathbf{b}$ will be equal to the number of null restrictions imposed by the breeder on the expected genetic gain per trait, or multi-trait selection response. In the PPG-LPSI context, matrix $\mathbf{K}_P$ has the same function as $\mathbf{K}$ (see Sect. 3.2.1 for details).

The aims of the LPSI, RLPSI, and PPG-LPSI are to:

1. Predict the unobservable net genetic merit values of the candidates for selection.
2. Maximize the selection response and the expected genetic gain for each trait.
3. Provide the breeder with an objective rule for evaluating and selecting several traits simultaneously (Baker 1974).

The LPSI is described in Chap. 2, and the RLPSI and PPG-LPSI are described in Chap. 3. As we will be see in this book, the RLPSI and PPG-LPSI theories can be extended to all selection indices described in this book. Also, the main objectives of all selection indices described in this book are the same as those of the LPSI, RLPSI, and PPG-LPSI.

1.1.2 Linear Marker Selection Indices

The linear marker selection index (LMSI) and the genome-wide LMSI (GW-LMSI) are employed in marker-assisted selection (MAS) and are useful in training populations when there is phenotypic and marker information; both are a direct application of the LPSI theory to the MAS context. The LMSI was originally proposed by Lande and Thompson (1990), and the GW-LMSI was proposed by Lange and Whittaker (2001). The fundamental idea of these authors is based on the fact that crossing two inbred lines generates linkage disequilibrium between markers and quantitative trait loci (QTL), which is useful for identifying markers correlated with the traits of interest and estimating the correlation between each of the selected markers and the trait; the selection criteria are then based upon this marker information (Moreau et al. 2007). The LMSI combines information on markers linked to QTL and the phenotypic values of the traits to predict the net genetic merit of the candidates for selection because it is not possible to identify all QTL affecting the economically important traits (Li 1998). That is, unless all QTL affecting the traits of interest can be identified, phenotypic values should be combined with the marker scores to increase LMSI efficiency (Dekkers and Settar 2004).

Moreau et al. (2000) and Whittaker (2003) found that the LMSI is more effective than LPSI only in early generation testing and that LMSI increased costs because of molecular marker evaluation. The LMSI assumes that favorable alleles are known, as are their average effects on phenotype (Lande and Thompson 1990; Hospital et al. 1997). This assumption is valid for major gene traits but not for quantitative traits that are influenced by the environment and many QTLs with small effects interacting among them and with the environment. The LMSI requires regressing phenotypic values on marker-coded values and, with this information, constructing the marker score for each individual candidate for selection, and then combining the marker score with phenotypic information using the LMSI to obtain a final prediction of the net genetic merit. Several authors (Lange and Whittaker 2001; Meuwissen et al. 2001; Dekkers 2007; Heffner et al. 2009) have criticized the LMSI approach because it makes inefficient use of the available data. It would be preferable to use all the available data in a single step to achieve maximally accurate estimates of marker effects. In addition, because the LMSI is based on only a few large QTL effects, it violates the selection index assumptions of multivariate normality and small changes in allele frequencies.

Lange and Whittaker (2001) proposed the genome-wide LMSI (GW-LMSI) as a possible solution to LMSI problems. The GW-LMSI is a single-stage procedure that treats information at each individual marker as a separate trait. Thus, all marker information can be entered together with phenotypic information into the GW-LMSI, which is then used to predict the net genetic merit and select candidates. Both selection indices are described in Chap. 4.

1.1.3 Linear Genomic Selection Indices

The linear genomic selection index (LGSI) is a linear combination of genomic estimated breeding values (GEBVs) and was originally proposed by Togashi et al. (2011); however, Ceron-Rojas et al. (2015) developed the LGSI theory completely. The advantage of the LGSI over the other indices lies in the possibility of reducing the intervals between selection cycles by more than two thirds. A 4-year breeding cycle (including 3 years of field testing) is thus reduced to only 4 months, i.e., the time required to grow and cross a plant. As a result, thousands of candidates for selection can be evaluated without ever taking them out to the field (Lorenz et al. 2011).

In the LGSI, phenotypic and marker data from the training population are fitted in a statistical model to estimate all available marker effects; these estimates are then used to obtain GEBVs that are predictors of breeding values in a testing population for which there is only marker information. The GEBV can be obtained by multiplying the genomic best linear unbiased predictor (GBLUP) of the estimated marker effects in the training population (Van Raden 2008) by the coded marker values obtained in the testing population in each selection cycle. Applying the LGSI in plant or animal breeding requires genotyping the candidates for selection to obtain the GEBV, and predicting and ranking the net genetic merit of the candidates for selection using the LGSI. An additional genomic selection index was given by Dekkers (2007); however, this index can only be used in training populations because GEBV and phenotypic information are jointly used to predict the net genetic merit. Both indices are described in Chap. 5 and in Chap. 6, we describe both indices in the context of the restricted selection indices.

1.2 Eigen Selection Index Methods

The eigen selection index methods are described in Chaps. 7 and 8. As we shall see, these indices are only used in training populations and can be unrestricted, restricted, and predetermined proportional gains selection indices; they can also use phenotypic and/or marker information to predict the net genetic merit. In the context of this linear selection index theory, it is assumed that economic weights are fixed but unknown. The eigen selection index methods is based on the canonical correlation theory and applied to the LPSI, RLSPI, etc., selection indices's context.

1.2.1 Linear Phenotypic Eigen Selection Index Method

Cerón-Rojas and Sahagún-Castellanos (2005) and Cerón-Rojas et al. (2006) proposed a phenotypic selection index in the principal component context that has low accuracy; later, Cerón-Rojas et al. (2008a, 2016) developed the eigen selection index method (ESIM), the restricted ESIM (RESIM) and the predetermined proportional gain ESIM (PPG-ESIM) in the canonical correlations context (Hotelling 1935, 1936). The ESIM is an unrestricted index, but the RESIM and PPG-ESIM allow null and predetermined restrictions respectively to be imposed on the expected genetic gains of some traits, whereas the rest remain without restrictions. The latter three indices use only phenotypic information to predict the individual net genetic merit of the candidate for selection and use the elements of the first eigenvector of the multi-trait heritability as the index vector of coefficients and the first eigenvalue of the multi-trait heritability in their selection response. The main objectives of the three indices are to predict the unobservable net genetic merit values of the candidates for selection, maximize the selection response and the expected genetic gain per trait, and provide the breeder with an objective rule for evaluating and selecting several traits simultaneously. Their main characteristics are:

1. They do not require the economic weights to be known.
2. The first eigenvector of the multi-trait heritability is used as their vector of coefficients, and the first eigenvalue of the multi-trait heritability is used in the selection response.
3. Owing to the properties associated with eigen analysis, it is possible to use the theory of similar matrices (Harville 1997) to change the direction and proportion of the expected genetic gain values without affecting the accuracy.
4. The sampling statistical properties of ESIM are known.
5. The PPG-ESIM does not require a proportional constant.

Finally, the main theory describe in Chapter 7 was developed by Cerón-Rojas et al.(2008a, 2016) based on the canonical correlation framework. That is, ESIM and its variants (RESIM, MESIM, PPG-ESIM) are applications of the canonical correlation theory to the LPSI context.

1.2.2 Linear Marker and Genomic Eigen Selection Index Methods

Cerón-Rojas et al. (2008b) and Crossa and Cerón-Rojas (2011) extended the ESIM to a molecular ESIM (MESIM) and to a genome-wide ESIM (GW-ESIM), respectively, similar to the linear molecular selection index (LMSI) and to the genome-wide LMSI (GW-LMSI). The MESIM and GW-ESIM have problems similar to those associated with the LMSI and GW-LMSI respectively (Chap. 4 for details). The MESIM and GW-ESIM use phenotypic information and markers linked to QTL to predict the net genetic merit, but the GW-ESIM omits the molecular selection step in the prediction. The main difference among the MESIM, the GW-ESIM, the LMSI, and the GW-LMSI is how they obtain the vector of coefficients: while the LMSI and GW-LMSI obtain the vector of coefficients according to the LPSI theory, the MESIM and the GW-ESIM obtain the vector of coefficients based on canonical correlation analysis and the singular value decomposition theory.

It is possible to extend the ESIM to a genomic ESIM (GESIM), and the restricted RESIM and the PPG-ESIM can be extended to a restricted genomic ESIM (RGESIM) and to a predetermined proportional gain genomic ESIM (PPG-GESIM) that use phenotypic and GEBV information jointly to predict the net genetic merit of the candidates for selection, maximizing the selection response and optimizing the expected genetic gain per trait; but although the GESIM is not constrained, the RGESIM and the PPG-GESIM allow null and predetermined restrictions respectively to be imposed on the expected genetic gain to make some traits change their mean values based on a predetermined level, while the rest of the traits remain without any restriction.

1.3 Multistage Linear Selection Indices

Multistage linear selection indices are methods of selecting one or more individual traits available at different times or stages and are applied mainly in animals and tree breeding where the traits under consideration become evident at different ages. The theory of these indices is based on the independent culling level method and the standard linear selection index theory. There are two main approaches associated with these indices:

1. The *optimal* multistage linear selection index, which takes into consideration the correlation among indices at different stages when makes selection.
2. The *selection index updating* or *decorrelated* multistage linear selection index, in which the correlation among indices at different stages is zero when makes selection.

These indices can use phenotypic or GEBV information to predict the net genetic merit or combine phenotypic and GEBV in the prediction. These indices can also be unrestricted, null restricted or predetermined proportional gains. In this book, we describe only the *optimal* multistage linear selection index in Chap. 9 and, in this book, it is called simply multistage linear selection index.

Multistage linear selection indices are a cost-saving strategy for improving multiple traits, because not all traits need to be measured at each stage. Thus, when traits have a developmental sequence in ontogeny or there are large differences in the costs of measuring several traits, the efficiency of this index over LPSI efficiency can be substantial (Xu et al. 1995). Xu and Muir (1992) have indicated that the optimal multistage linear phenotypic selection index (MLPSI) increases selection intensity on traits measured at an earlier age, and, with fixed facilities, a greater number of individuals can be selected at an earlier age. For example, if some individuals can be culled before final traits are measured (e.g., weaning weights in swine and beef cattle breeding), savings are realized in terms of feed, labor, and facilities. With the LPSI, the same individuals must be measured for each trait; thus, the number of traits measured per mature individual is the same as that for an immature individual.

The original MLPSI was developed by Cochran (1951) in the two-stage context and later, Young (1964) and Cunningham (1975) combined the LPSI theory with the independent culling method to simultaneously select more than one trait in the multistage selection context. This selection method was called *multistage selection* by Cochran (1951) and Young (1964) and *multistage index selection* by Cunningham (1975).

The MLPSI theory can also be adapted to the genomic selection context, where it is possible to develop an optimal multistage unrestricted, restricted, and predetermined proportional gains linear genomic selection index. The latter indices are linear combinations of estimated breeding values (GEBV) used to predict the individual net genetic merit and select individual traits available at different stages in a non-phenotyped testing population and are called *multistage linear genomic selection indices*. The advantage of these indices over the other selection indices lies in the possibility of reducing the intervals between selection cycles or stages by more than two thirds.

One of the main problems of all the multistage selection indices is that after the first selection stage their values could be non-normally distributed. In addition, for more than two stages, those indices require computationally sophisticated multiple integration techniques to derive selection intensities, and there are problems of convergence when the traits and the index values of successive stages are highly correlated. Furthermore, the computational time could be unacceptable if the number of selection stages becomes too high (Börner and Reinsch 2012). One possible solution to these problems was given by Xu and Muir (1992) in the *selection index updating* or *decorrelated* multistage linear phenotypic selection index context. However, one problem with the *decorrelated* multistage selection index is that its accuracy and selection response is generally lower than the accuracy and selection response of the multistage selection index described in this book.

1.4 Stochastic Simulation of Four Linear Phenotypic Selection Indices

Chapter 10 describes a stochastic simulation of four linear indices: LPSI, ESIM, RLPSI, and RESIM. We think that stochastic simulation can contribute to a better understanding of the relationship between these indices and their accuracies to predict the net genetic merit.

1.5 RIndSel: Selection Indices with R

Chapter 11 describes how RIndSel can be used to determine individual candidates as parents for the next cycle of improvement. RIndSel is a graphical unit interface that uses the selection index theory to make selection. The index can be a linear combination of phenotypic values, genomic estimated breeding values or a linear combination of phenotypic values and marker scores.

1.6 The Lagrange Multiplier Method

To obtain the constrained linear selection indices (e.g., RLPSI, PPG-LPSI, RESIM) described in Chaps. 3, 6, 7, 8, and 9, we used the method of Lagrange multipliers. This is a powerful method for finding extreme values (maxima or minima) of constrained functions. For example, the covariance between the breeding value vector ($\mathbf{g}$) and the LPSI ($I = \mathbf{b}'\mathbf{y}$) is $Cov(I, \mathbf{g}) = \mathbf{Gb}$. In the LPSI context, the $\mathbf{Gb}$ vector can take any value (positive or negative) which could be a problem for some breeding objectives. That is, the breeder could be interested in improving only $(t - r)$ of t $(r < t)$ traits, leaving r of them fixed; that is, the expected genetic gains of r traits will be equal to zero for a specific selection cycle. In such cases, we want r covariances between the linear combinations of $\mathbf{g}$ ($\mathbf{U}'\mathbf{g}$) and the $I = \mathbf{b}'\mathbf{y}$ to be zero, i.e., $Cov(I, \mathbf{U}'\mathbf{g}) = \mathbf{U}'\mathbf{G}\mathbf{b} = \mathbf{0}$, where $\mathbf{U}'$ is a matrix with r 1's and $(t - r)$ 0's; 1 indicates that the trait is restricted and 0 that the trait is not restricted. This is the main problem of the RLPSI, and the method of Lagrange multipliers is useful for solving that problem.

In the constrained linear selection indices context, the method of Lagrange multipliers involves maximizing (or minimizing) the Lagrange function: $L[H, I, \mathbf{g}, \mathbf{v}] = f(H, I) + \mathbf{v}'g(\mathbf{g}, I)$, where the elements of vector $\mathbf{v}'$ are called Lagrange multipliers. In the RLPSI context, $f(H, I) = E[(H - I)^2] = \mathbf{w}'\mathbf{G}\mathbf{w} + \mathbf{b}'\mathbf{P}\mathbf{b} - 2\mathbf{w}'\mathbf{G}\mathbf{b}$ is the mean squared difference between I and H. Let $g(\mathbf{g}, I) = Cov(I, \mathbf{U}'\mathbf{g}) = \mathbf{U}'\mathbf{G}\mathbf{b}$ be the covariances between the linear combinations of $\mathbf{g}$ ($\mathbf{U}'\mathbf{g}$), and $I = \mathbf{b}'\mathbf{y}$, the LPSI. Then, to find the RLPSI vector of coefficients $\mathbf{b}_R = \mathbf{K}\mathbf{b}$, we need to minimize the Lagrange function: $\mathbf{b}'\mathbf{P}\mathbf{b} + \mathbf{w}'\mathbf{G}\mathbf{w} - 2\mathbf{w}'\mathbf{G}\mathbf{b} + 2\mathbf{v}'\mathbf{C}'\mathbf{b}$, with respect to vectors $\mathbf{b}$ and $\mathbf{v}' = [v_1 \; v_2 \; \cdots \; v_{r-1}]$, where $\mathbf{v}$ is a vector of Lagrange multipliers (see Chap. 3, Sect. 3.1.1 for details). Schott (2005) has given additional details associated with the method of Lagrange multipliers.

References

Baker RJ (1974) Selection indexes without economic weights for animal breeding. Can J Anim Sci 54:1–8

Börner V, Reinsch N (2012) Optimising multistage dairy cattle breeding schemes including genomic selection using decorrelated or optimum selection indices. Genet Sel Evol 44(1):11

Cerón-Rojas JJ, Sahagún-Castellanos J (2005) A selection index based on principal components. Agrociencia 39:667–677

Cerón-Rojas JJ, Crossa J, Sahagún-Castellanos J, Castillo-González F, Santacruz-Varela A (2006) A selection index method based on eigenanalysis. Crop Sci 46:1711–1721

Cerón-Rojas JJ, Sahagún-Castellanos J, Castillo-González F, Santacruz-Varela A, Crossa J (2008a) A restricted selection index method based on eigenanalysis. J Agric Biol Environ Stat 13 (4):421–438

Cerón-Rojas JJ, Sahagún-Castellanos J, Castillo-González F, Santacruz-Varela A, Benítez-Riquelme I, Crossa J (2008b) A molecular selection index method based on eigenanalysis. Genetics 180:547–557

Ceron-Rojas JJ, Crossa J, Arief VN, Basford K, Rutkoski J, Jarquín D, Alvarado G, Beyene Y, Semagn K, DeLacy I (2015) A genomic selection index applied to simulated and real data. G3 (Bethesda) 5:2155–2164

Cerón-Rojas JJ, Crossa J, Toledo FH, Sahagún-Castellanos J (2016) A predetermined proportional gains eigen selection index method. Crop Sci 56:2436–2447

Cochran WG (1951) Improvement by means of selection. In: Neyman J (ed) Proceedings of the second Berkeley symposium on mathematical statistics and probability. University of California Press, Berkeley, CA, pp 449–470

Costa MM, Di Mauro AO, Unêda-Trevisoli SH, Castro Arriel NH, Bárbaro IM, Dias da Silveira G, Silva Muniz FR (2008) Analysis of direct and indirect selection and indices in soybean segregating populations. Crop Breed Appl Biotechnol 8:47–55

Crossa J, Cerón-Rojas JJ (2011) Multi-trait multi-environment genome-wide molecular marker selection indices. J Indian Soc Agric Stat 62(2):125–142

Cunningham EP (1975) Multi-stage index selection. Theor Appl Genet 46:55–61

Dekkers JCM (2007) Prediction of response to marker-assisted and genomic selection using selection index theory. J Anim Breed Genet 124:331–341

Dekkers JCM, Settar P (2004) Long-term selection with known quantitative trait loci. Plant Breed Rev 24:311–335

Harville DA (1975) Index selection with proportionality constraints. Biometrics 31(1):223–225

Harville DA (1997) Matrix Algebra from a statistician's perspective. Springer, New York

Hazel LN (1943) The genetic basis for constructing selection indexes. Genetics 8:476–490

Hazel LN, Lush JL (1942) The efficiency of three methods of selection. J Hered 33:393–399

Heffner EL, Sorrells ME, Jannink JL (2009) Genomic selection for crop improvement. Crop Sci 49 (1):12

Hospital F, Moreau L, Lacoudre F, Charcosset A, Gallais A (1997) More on the efficiency of marker-assisted selection. Theor Appl Genet 95:1181–1189

Hotelling H (1935) The most predictable criterion. J Educ Psychol 26:139–142

Hotelling H (1936) Relations between two sets of variables. Biometrika 28:321–377

Itoh Y, Yamada Y (1986) Re-examination of selection index for desired gains. Genet Sel Evol 18 (4):499–504

Itoh Y, Yamada Y (1987) Comparisons of selection indices achieving predetermined proportional gains. Genet Sel Evol 19(1):69–82

James JW (1968) Index selection with restriction. Biometrics 24:1015–1018

Kempthorne O, Nordskog AW (1959) Restricted selection indices. Biometrics 15:10–19

Lande R, Thompson R (1990) Efficiency of marker-assisted selection in the improvement of quantitative traits. Genetics 124:743–756

Lange C, Whittaker JC (2001) On prediction of genetic values in marker-assisted selection. Genetics 159:1375–1381

Li Z (1998) Molecular analysis of epistasis affecting complex traits. In: Paterson AH (ed) Molecular dissection of complex traits. CRC Press, Boca Raton, New York, pp 119–1130

Lorenz AJ, Chao S, Asoro FG, Heffner EL, Hayashi T et al (2011) Genomic selection in plant breeding: knowledge and prospects. Adv Agron 110:77–123

Mallard J (1972) The theory and computation of selection indices with constraints: a critical synthesis. Biometrics 28:713–735

Meuwissen THE, Hayes BJ, Goddard ME (2001) Prediction of total genetic value using genome-wide dense marker maps. Genetics 157:1819–1829

Moreau L, Lemarie S, Charcosset A, Gallais A (2000) Economic efficiency of one cycle of marker-assisted selection efficiency. Crop Sci 40:329–337

Moreau L, Hospital F, Whittaker J (2007) Marker-assisted selection and introgression. In: Balding DJ, Bishop M, Cannings C (eds) Handbook of statistical genetics, vol 1, 3rd edn. Wiley, New York, pp 718–751

Pesek J, Baker RJ (1969) Desired improvement in relation to selection indices. Can J Plant Sci 49:803–804

Schott JR (2005) Matrix analysis for statistics, 2nd edn. Wiley, Hoboken, NJ

Smith HF (1936) A discriminant function for plant selection. In: Papers on quantitative genetics and related topics. Department of Genetics, North Carolina State College, Raleigh, NC, pp 466–476

Tallis GM (1962) A selection index for optimum genotype. Biometrics 18:120–122

Tallis GM (1985) Constrained selection. Jpn J Genet 60(2):151–155

Togashi K, Lin CY, Yamazaki T (2011) The efficiency of genome-wide selection for genetic improvement of net merit. J Anim Sci 89:2972–2980

Van Raden PM (2008) Efficient methods to compute genomic predictions. J Dairy Sci 91:4414–4423

Whittaker JC (2003) Marker-assisted selection and introgression. In: Balding DJ, Bishop M, Cannings C (eds) Handbook of statistical genetics, vol 1, 2nd edn. Wiley, New York, pp 554–574

Xu S, Muir WM (1992) Selection index updating. Theor Appl Genet 83:451–458

Xu S, Martin TG, Muir WM (1995) Multistage selection for maximum economic return with an application to beef cattle breeding. J Anim Sci 73:699–710

Yamada Y, Yokouchi K, Nishida A (1975) Selection index when genetic gains of individual traits are of primary concern. Jpn J Genet 50(1):33–41

Young SSY (1964) Multi-stage selection for genetic gain. Heredity 19:131–143

2

LPSI: Construction, Parameters and Properties

Abstract The main distinction in the linear phenotypic selection index (LPSI) theory is between the net genetic merit and the LPSI. The net genetic merit is a linear combination of the true unobservable breeding values of the traits weighted by their respective economic values, whereas the LPSI is a linear combination of several observable and optimally weighted phenotypic trait values. It is assumed that the net genetic merit and the LPSI have bivariate normal distribution; thus, the regression of the net genetic merit on the LPSI is linear. The aims of the LPSI theory are to predict the net genetic merit, maximize the selection response and the expected genetic gains per trait (or multi-trait selection response), and provide the breeder with an objective rule for evaluating and selecting parents for the next selection cycle based on several traits. The selection response is the mean of the progeny of the selected parents, whereas the expected genetic gain per trait, or multi-trait selection response, is the population means of each trait under selection of the progeny of the selected parents. The LPSI allows extra merit in one trait to offset slight defects in another; thus, with its use, individuals with very high merit in one trait are saved for breeding even when they are slightly inferior in other traits. This chapter describes the LPSI theory and practice. We illustrate the theoretical results of the LPSI using real and simulated data. We end this chapter with a brief description of the quadratic selection index and its relationship with the LPSI.

2.1 Bases for Construction of the Linear Phenotypic Selection Index

The study of quantitative traits (QTs) in plants and animals is based on the mean and variance of phenotypic values of QTs. Quantitative traits are phenotypic expressions of plant and animal characteristics that show continuous variability and are the result of many gene effects interacting among them and with the environment. That is, QTs are the result of unobservable gene effects distributed across plant or animal genomes that interact among themselves and with the environment to produce the observable characteristic plant and animal phenotypes (Mather and Jinks 1971; Falconer and Mackay 1996).

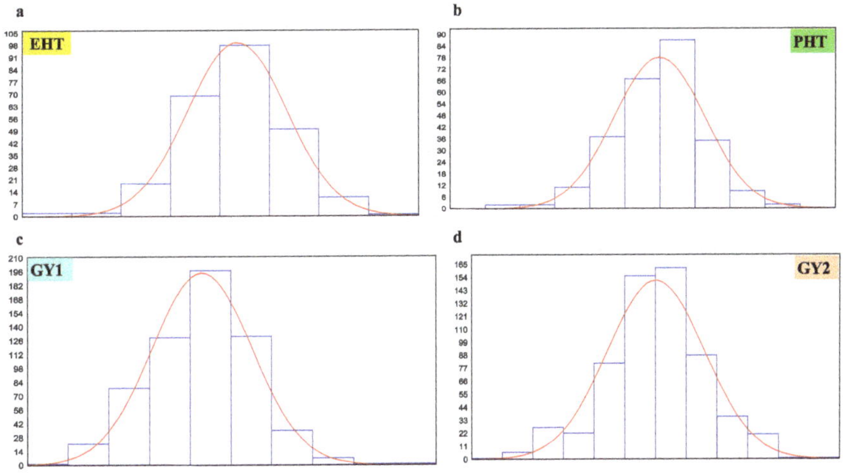

Fig. 2.1 Distribution of 252 phenotypic means of two maize (*Zea mays*) F_2 population traits: plant height (*PHT*, cm; **a**) and ear height (*EHT*, cm; **b**), evaluated in one environment, and of 599 - phenotypic means of the grain yield (*GY1* and *GY2*, ton ha^{-1}; **c** and **d** respectively) of one double haploid wheat (*Triticum aestivum* L.) population evaluated in two environments

The QTs are the traits that concern plant and animal breeders the most. They are particularly difficult to analyze because heritable variations of QTs are masked by larger nonheritable variations that make it difficult to determine the genotypic values of individual plants or animals (Smith 1936). However, as QTs usually have normal distribution (Fig. 2.1), it is possible to apply normal distribution theory when analyzing this type of data.

Any phenotypic value of QTs (y) can be divided into two main parts: one related to the genes and the interactions (g) among them (called genotype), and the other related to the environmental conditions (e) that affect genetic expression (called environment effects). Thus, the genotype is the particular assemblage of genes possessed by the plant or animal, whereas the environment consists of all the nongenetic circumstances that influence the phenotypic value of the plant or animal (Cochran 1951; Bulmer 1980; Falconer and Mackay 1996). In the context of only one environment, the phenotypic value of QTs (y) can be written as

$$y = g + e, \tag{2.1}$$

where g denotes the genotypic values that include all types of gene and interaction values, and e denotes the deviations from the mean of g values. For two or more environments, Eq. (2.1) can be written as $y = g + e + ge$, where ge denotes the interaction between genotype and environment. Assumptions regarding Eq. (2.1) are:

1. The expectation of e is zero, $E(e) = 0$.
2. Across several environments, the expectation of y is equal to the expectation of g, i.e., $E(g) = \mu_g = E(y) = \mu_y$.
3. The covariance between g and e is equal to 0.

The g value can be partitioned into three additional components: additive genetic (a) effects (or intra-locus additive allelic interaction), dominant genetic (d) effects (or intra-locus dominance allelic interaction), and epistasis (ι) effects (or inter-loci allelic interaction) such that $g = a + d + \iota$. In this book, we have assumed that $g = a$.

According to Kempthorne and Nordskog (1959), the following four theoretical conditions are necessary to construct a valid LPSI:

1. The phenotypic value (Eq. 2.1) shall be additively made up of two parts: a genotypic value (g) (defined as the average of the phenotypic values possible across a population of environments), and an environmental contribution (e).
2. The genotypic value g is composed entirely of the additive effects of genes and is thus the individual breeding value.
3. The genotypic economic value of an individual is its net genetic merit.
4. The phenotypic values and the net genetic merit are such that the regression of the net genetic merit on any linear function of the phenotypic values is linear.

Under assumptions 1 to 4, the offspring of a mating will have a genotypic value equal to the average of the breeding values of the parents (Kempthorne and Nordskog 1959). Additional conditions for practical objectives are:

5. Selection is practiced at only one stage of the life cycle.
6. The generations do not overlap.
7. All individuals below a certain level of desirability are culled without exception.
8. Selected individuals have equal opportunity to have offspring (Hazel and Lush 1942).
9. The LPSI values in the ith selection cycle and the LPSI values in the $(i + 1)$th selection cycle do not correlate.
10. The correlation between the LPSI and the net genetic merit should be at its maximum in each selection cycle.

Conditions 5 to 10 indicate that the LPSI is applying in a single stage context.

2.2 The Net Genetic Merit and the LPSI

Not all the individual traits under selection are equally important from an economic perspective; thus, the economic value of a trait determines how important that trait is for selection. Economic value is defined as the increase in profit achieved by improving a particular trait by one unit (Tomar 1983; Cartuche et al. 2014). This means that for several traits, the total economic value is a linear combination of the breeding values of the traits weighted by their respective economic values (Smith 1936; Hazel and Lush 1942; Hazel 1943; Kempthorne and Nordskog 1959); this is called *the net genetic merit of one individual* and can be written as

$$H = \mathbf{w}'\mathbf{g}, \tag{2.2}$$

where $\mathbf{g}' = [g_1 \ g_2 \ \dots \ g_t]$ is a vector of true unobservable breeding values and $\mathbf{w}' = [w_1 \quad w_2 \quad \dots \quad w_t]$ is a vector of known and fixed economic weights. Equation (2.2) has several names, e.g., linear aggregate genotype (Hazel 1943), genotypic economic value (Kempthorne and Nordskog 1959), net genetic merit (Akbar et al. 1984; Cotterill and Jackson 1985), breeding objective (Mac Neil et al. 1997), and total economic merit (Cunningham and Tauebert 2009), among others. In this book, we call Eq. (2.2) *net genetic merit* only. The values of $H = \mathbf{w}'\mathbf{g}$ are unobservable but they can be simulated for specific studies, as is seen in the examples included in this chapter and in Chap. 10, where four indices have been simulated for many selection cycles.

In practice, the net genetic merit of an individual is not observable; thus, to select an individual as parent of the next generation, it is necessary to consider its overall merit based on several observable traits; that is, we need to construct an LPSI of observable phenotypic values such that the correlation between the LPSI and $H = \mathbf{w}'\mathbf{g}$ is at a maximum. The LPSI should be a good predictor of $H = \mathbf{w}'\mathbf{g}$ and should be useful for ranking and selecting among individuals with different net genetic merits. The LPSI for one individual can be written as

$$I = \mathbf{b}'\mathbf{y}, \tag{2.3}$$

where $\mathbf{b}' = [b_1 \quad b_2 \quad \cdots \quad b_t]$ is the I vector of coefficients, t is the number of traits on I, and $\mathbf{y}' = [y_1 \quad y_2 \quad \cdots \quad y_t]$ is a vector of observable trait phenotypic values usually centered with respect to its mean. The LPSI allows extra merit in one trait to offset slight defects in another. With its use, individuals with very high merit in some traits are saved for breeding, even when they are slightly inferior in other traits (Hazel and Lush 1942). Only one combination of **b** values allows the correlation of the LPSI with $H = \mathbf{w}'\mathbf{g}$ for a particular set of traits to be maximized.

Figure 2.2 indicates that the regression of the net genetic merit on the LPSI is lineal and that the correlation between the LPSI and the net genetic merit is maximal in each selection cycle. Also, note that the true correlations between the LPSI and the net genetic merit, and the true regression coefficients of the net genetic merit over the LPSI are the same, but the estimated correlation values between the LPSI and the net genetic merit are lower than the true correlation (Fig. 2.2). Table 2.1 indicates that the LPSI in the ith selection cycle and the LPSI in the $(i + 1)$th selection cycle do not correlate. However, in practice, the correlation values between any pair of LPSIs could be different from zero in successive selection cycles.

One fundamental assumption of the LPSI is that $I = \mathbf{b}'\mathbf{y}$ has normal distribution. This assumption is illustrated in Fig. 2.3 for two real datasets: a maize (*Zea mays*) F_2 population with 252 lines and three traits—grain yield (ton ha^{-1}); plant height (cm) and ear height (cm)—evaluated in one environment; and a double haploid wheat (*Triticum aestivum* L.) population with 599 lines and one trait—grain yield (ton ha^{-1})—evaluated in three environments. Figure 2.3 indicates that, in effect, the LPSI values approach normal distribution when the number of lines is very large.

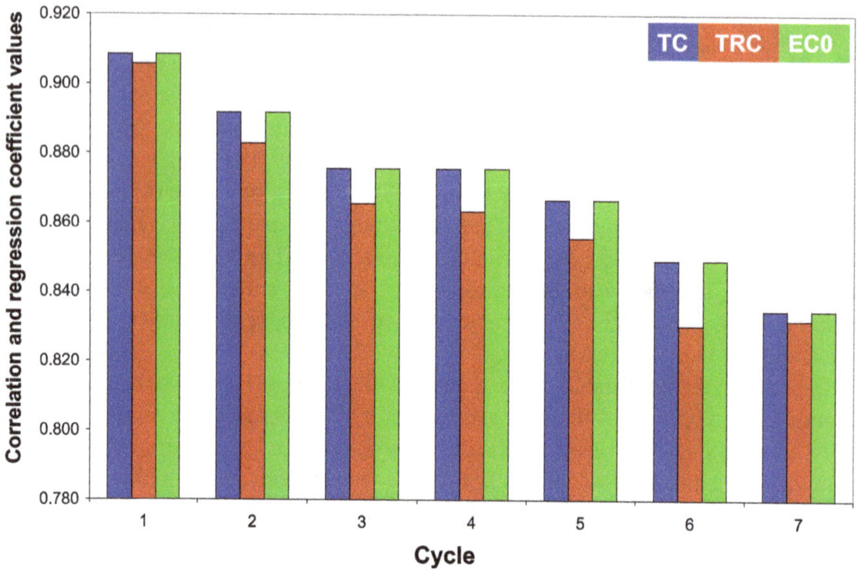

Fig. 2.2 True correlation (*TC*) and estimated correlation (*ECO*) values between the linear phenotypic selection index (*LPSI*) and the net genetic merit for seven selection cycles, and true regression coefficient (*TRC*) of the net genetic merit over the LPSI for four traits and 500 genotypes in one environment simulated for seven selection cycles

Table 2.1 Estimated correlation values between the linear phenotypic selection index (LPSI) values in seven simulated selection cycles

1	2	3	4	5	6	7
1.000	0.199	0.256	0.220	0.168	0.225	0.123
0.199	1.000	0.225	0.252	0.284	0.292	0.362
0.256	0.225	1.000	0.198	0.276	0.267	0.213
0.220	0.252	0.198	1.000	0.258	0.224	0.240
0.168	0.284	0.276	0.258	1.000	0.269	0.195
0.225	0.292	0.267	0.224	0.269	1.000	0.325
0.123	0.362	0.213	0.240	0.195	0.325	1.000

2.3 Fundamental Parameters of the LPSI

There are two fundamental parameters associated with the LPSI theory: the selection response (R) and the expected genetic gain per trait (**E**). In general terms, the selection response is the difference between the mean phenotypic values of the offspring (μ_O) of the selected parents and the mean of the entire parental generation (μ_P) before selection, i.e., $R = \mu_O - \mu_P$ (Hazel and Lush 1942; Falconer and Mackay 1996). The expected genetic gain per trait (or multi-trait selection response) is the covariance between the breeding value vector and the LPSI (I) values weighted by the standard deviation of the variance of $I(\sigma_I)$, i.e., $\frac{Cov(I,\mathbf{g})}{\sigma_I} = \frac{\mathbf{Gb}}{\sigma_I}$, multiplied by the

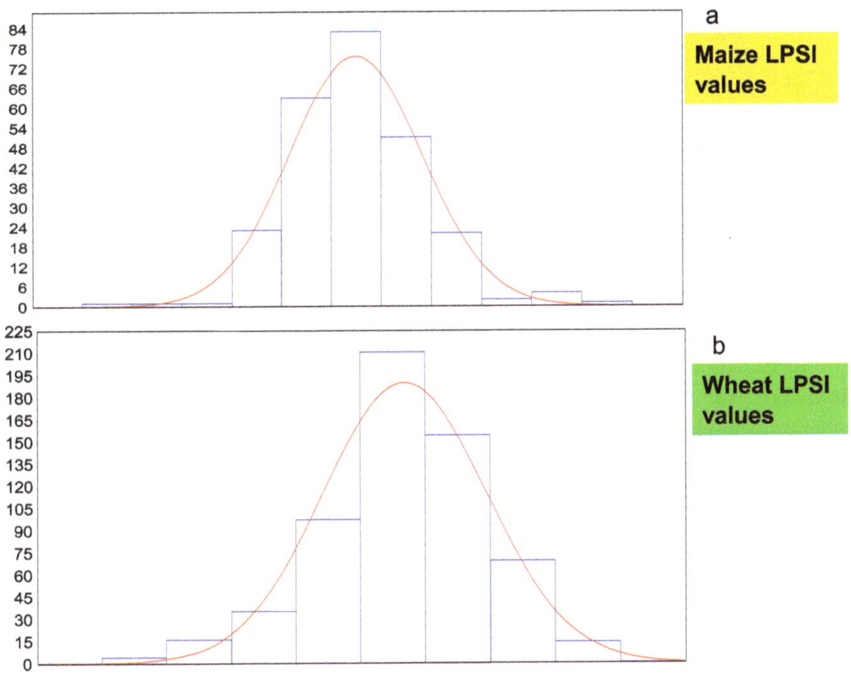

Fig. 2.3 *Maize LPSI* (Fig. 2.3a) is the distribution of 252 values of the LPSI constructed with the phenotypic means of three maize (*Zea mays*) F_2 population traits: grain yield (ton ha^{-1}), PHT (cm) and EHT (cm), evaluated in one environment. *Wheat LPSI* (Fig. 2.3b) is the distribution of 599 LPSI values constructed with the phenotypic means of the grain yield (ton ha^{-1}) of a double haploid wheat (*Triticum aestivum* L.) population evaluated in three environments

selection intensity. This is one form of the LPSI multi-trait selection response. In the univariate context, the expected genetic gain per trait is the same as the selection response.

One additional way of defining the selection response is based on the selection differential (D). The selection differential is the mean phenotypic value of the individuals selected as parents (μ_S) expressed as a deviation from the population mean (μ_P) or parental generation before the selection was made (Falconer and Mackay 1996); that is, $D = \mu_S - \mu_P$. Thus, another way of defining R is as the part of the expected differential of selection ($D = \mu_S - \mu_P$) that is gained when selection is applied (Kempthorne and Nordskog 1959); that is

$$R = \frac{Cov(g, y)}{\sigma_y^2} D = k\sigma_y h^2, \tag{2.4}$$

where $Cov(g, y) = \sigma_g^2$ is the covariance between g and y, g is the individual breeding value associated with trait y, σ_y^2 is the variance of y, $k = \frac{D}{\sigma_y}$ is the standardized selection differential or selection intensity, and $h^2 = \frac{\sigma_g^2}{\sigma_y^2}$ is the heritability of trait y in the base population. Heritability (h^2) appears in Eq. (2.4) as a measure of the accuracy with which animals or plants having the highest genetic values can be chosen by selecting directly for phenotype (Hazel and Lush 1942).

The selection response (Eq. 2.4) is the mean of the progeny of the selected parents or the future population mean of the trait under selection (Cochran 1951). Thus, the selection response enables breeders to estimate the expected progress of the selection before carrying it out. This information gives improvement programs a clearer orientation and helps to predict the success of the selection method adopted and choose the option that is technically most effective on a scientific base (Costa et al. 2008). Equation (2.4) is very powerful but its application requires strong assumptions. For example, Eq. (2.4) assumes that the trait of interest does not correlate with other traits having causal effects on fitness and, in its multivariate form the validity of predicted change rests on the assumption that all such correlated traits have been measured and incorporated into the analysis (Morrissey et al. 2010).

2.3.1 *The LPSI Selection Response*

The univariate selection response (Eq. 2.4) can also be rewritten as

$$R = k\sigma_y h^2 = k\sigma_g \rho_{gy}, \tag{2.5}$$

where σ_g was defined in Eq. (2.4) and ρ_{gy} is the correlation between g and y. Thus, as $H = \mathbf{w}'\mathbf{g}$ and $I = \mathbf{b}'\mathbf{y}$ are univariate random variables, the selection response of the LPSI (R_I) can be written in a similar form as Eq. (2.5), i.e.,

$$R_I = k_I \sigma_H \rho_{HI}, \tag{2.6}$$

where σ_H and σ_I are the standard deviation and ρ_{HI} the correlation between $H = \mathbf{w}'\mathbf{g}$ and $I = \mathbf{b}'\mathbf{y}$ respectively; $k_I = \frac{\mu_{IA} - \mu_{IB}}{\sigma_I}$ is the standardized selection differential or the selection intensity associated with the LPSI; μ_{IA} and μ_{IB} are the means of the LPSI values after and before selection respectively. The second part of Eq. (2.6) ($k_I \sigma_H \rho_{HI}$) indicates that the genetic change due to selection is proportional to k_I, σ_H, and ρ_{HI} (Kempthorne and Nordskog 1959). Thus, the genetic gain that can be achieved by selecting for several traits simultaneously within a population of animals or plants is the product of the selection differential (k_I), the standard deviation of $H = \mathbf{w}'\mathbf{g}$ (σ_H), and the correlation between $H = \mathbf{w}'\mathbf{g}$ and $I = \mathbf{b}'\mathbf{p}$ (ρ_{HI}). Selection intensity k_I is limited by the rate of reproduction of each species, whereas σ_H is relatively beyond man's control; hence, the greatest opportunity for increasing selection progress is by ensuring that ρ_{HI} is as large as possible (Hazel 1943). In general, it is assumed that k_I

and σ_H are fixed and $\mathbf{w}$ known and fixed; hence, R_I is maximized when ρ_{HI} is maximized only with respect to the LPSI vector of coefficients $\mathbf{b}$.

Equation (2.6) is the mean of $H = \mathbf{w}'\mathbf{g}$, whereas $\sigma_H^2\rho_{HI}^2(1-v)$ is its variance and $\rho_{HI}^* = \rho_{HI}\sqrt{\frac{1-v}{1-v\rho_{HI}^2}}$ the correlation between $H = \mathbf{w}'\mathbf{g}$ and $I = \mathbf{b}'\mathbf{p}$ after selection was carried out (Cochran 1951), where $v = k_I(k_I - \tau)$ and τ is the truncation point. For example, if the selection intensity is 5%, $k_I = 2.063$, $\tau = 1.645$, and $v = 0.862$ (Falconer and Mackay 1996, Table A). In *R* (in this case R denotes a platform for data analysis, see Kabakoff 2011 for details), the truncation point and selection intensity can be obtained as v <– *qnorm*(1 – q) and k <– *dnorm*(v)/q, respectively, where q is the proportion retained. Both the variance and the correlation (ρ_{HI}^*) are reduced by selection. If $H = \mathbf{w}'\mathbf{g}$ could be selected directly, the gain in $H = \mathbf{w}'\mathbf{g}$ would be k_I. Thus, the gain due to indirect selection using $I = \mathbf{b}'\mathbf{p}$ is a fraction ρ_{HI} of that due to direct selection using $H = \mathbf{w}'\mathbf{g}$. As k_I increases, R_I increases (Eq. 2.6), $\sqrt{\sigma_H^2\rho_{HI}^2(1-v)}$ and ρ_{HI}^* decrease, and the effects are in the same direction as ρ_{HI}^* increases (Cochran 1951). These results should be valid for all selection indices described in this book.

Smith (1936) gave an additional method to obtain Eq. (2.6). Suppose that we have a large number of plant lines and we select one proportion q for further propagation. In addition, assume that the values of I for each line are normally distributed with variance $\sigma_I^2 = \mathbf{b}'\mathbf{P}\mathbf{b}$; let I be transformed into a variable u, with unit variance and mean at zero, that is, $u = \frac{I-\mu_I}{\sigma_I}$, where μ_I is the mean of I. Assume that all I values higher than I' value are selected; then the value of $u' = \frac{I'-\mu_I}{\sigma_I}$ corresponding to any given value of q may be ascertained from a table of the standard normal probability integral (Fig. 2.4).

Assuming that the expectations of H and I are $E(H) = 0$ and $E(I) = \mu_I$, the conditional expectation of H given I can be written as

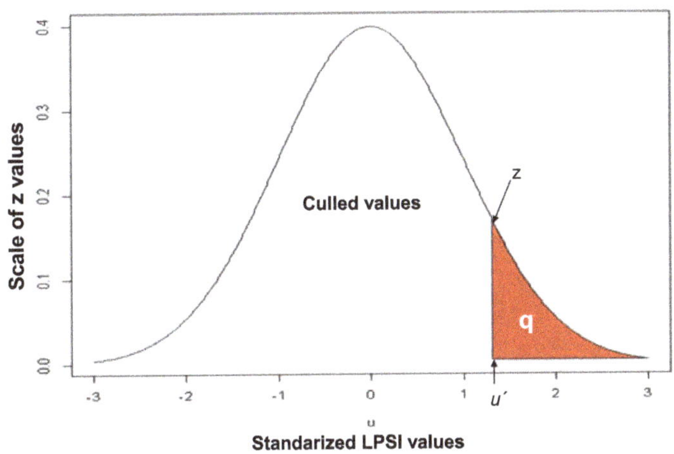

Fig. 2.4 Graph of standardized LPSI values showing how a population can be separated sharply at a given point (u') into a selected fraction (q), denoted by the *red area*, and a remainder that is culled, denoted by the *white area*

$E(H/I)=\frac{\sigma_{HI}}{\sigma_I^2}[I-\mu_I]=\frac{\sigma_{HI}}{\sigma_I^2}\sigma_I u=B\sigma_I u$, where $B=\frac{\sigma_{HI}}{\sigma_I^2}$, $\sigma_{HI}=\mathbf{w}'\mathbf{Gb}$ is the covariance between H and I, and $\sigma_I^2=\mathbf{b}'\mathbf{Pb}$ is the variance of I. Therefore, if σ_I^2 and σ_{HI} are fixed, the LPSI selection response (R_I) can be obtained as the expectation of the selected population, which has univariate left truncated normal distribution. A truncated distribution is a conditional distribution resulting when the domain of the parent distribution is restricted to a smaller region (Hattaway 2010). In the LPSI context, a truncation distribution occurs when a sample of individuals from the parent distribution is selected as parents for the next selection cycle, thus creating a new population of individuals that follow a truncated normal distribution. Thus, we need to find $E[E(H/I)]=q^{-1}B\sigma_I E(u)$, or, using integral calculus,

$$E[E(H/I)]=\frac{B\sigma_I}{q}\int_{u=u'}^{\infty}\frac{u}{\sqrt{2\pi}}\exp\left\{-\frac{1}{2}u^2\right\}du=\frac{z}{q}\sigma_H\rho_{HI}, \tag{2.7}$$

where $z=\frac{\exp\{-0.5u'^2\}}{\sqrt{2\pi}}$ is the height of the ordinate of the normal curve at the lowest value of u' retained and q is the proportion of the population of animal or plant lines that is selected (Fig. 2.4). The proportion q that must be saved depends on the reproductive rate and longevity of the species under consideration and on whether the population is expanding, stationary or declining in numbers. The ordinate (z) of the normal curve is determined by the proportion selected (q) (Fig. 2.4). The amount of progress is expected to be larger as q becomes smaller; that is, as selection becomes more intense (Hazel and Lush 1942). Kempthorne and Nordskog (1959) showed that $\frac{z}{q}=k_I$. Thus, Eqs. (2.6) and (2.7) are the same, that is, $E[E(H/I)]=R_I$.

2.3.2 *The Maximized Selection Response*

The main objective of the LPSI is to maximize the mean of $H=\mathbf{w}'\mathbf{g}$ (Eq. 2.7). Assuming that $\mathbf{P}$, $\mathbf{G}$, $\mathbf{w}$, and k_I are known, to maximize R_I we can either maximize ρ_{HI} or minimize the mean squared difference between I and H, $E[(H-I)^2]=\mathbf{w}'\mathbf{Gw}+\mathbf{b}'\mathbf{Pb}-2\mathbf{w}'\mathbf{Gb}$ with respect to $\mathbf{b}$, that is, $\frac{\partial}{\partial\mathbf{b}}E\left[(H-I)^2\right]=2\mathbf{Pb}-2\mathbf{Gw}=\mathbf{0}$, from where

$$\mathbf{b}=\mathbf{P}^{-1}\mathbf{Gw} \tag{2.8}$$

is the vector that simultaneously minimizes $E[(H-I)^2]$ and maximizes ρ_{HI}, and then $R_I=k_I\sigma_H\rho_{HI}$.

By Eq. (2.8), the maximized LPSI selection response can be written as

$$R_I = k_I\sqrt{\mathbf{b}'\mathbf{P}\mathbf{b}}. \tag{2.9}$$

The maximized LPSI selection response predicts the mean improvement in H due to indirect selection on I only when $\mathbf{b} = \mathbf{P}^{-1}\mathbf{G}\mathbf{w}$ (Harris 1964) and is proportional to the standard deviation of the LPSI variance (σ_I) and the standardized selection differential or the selection intensity (k_I).

The maximized LPSI selection response (Eq. 2.9) it related to the Cauchy–Schwarz inequality (Rao 2002; Cerón-Rojas et al. 2006), which establishes that for any pair of vectors **u** and **v**, if **A** is a positive definite matrix, then the inequality $(\mathbf{u}'\mathbf{v})^2 \le (\mathbf{v}'\mathbf{A}\mathbf{v})(\mathbf{u}'\mathbf{A}^{-1}\mathbf{u})$ holds. Kempthorne and Nordskog (1959) proved that maximizing $\rho_{HI}^2 = \dfrac{(\mathbf{w}'\mathbf{G}\mathbf{b})^2}{(\mathbf{w}'\mathbf{G}\mathbf{w})(\mathbf{b}'\mathbf{P}\mathbf{b})}$ also maximizes R_I. According to Eqs. (2.6) and (2.7), R_I^2 can be written as $R_I^2 = k_I^2\dfrac{(\mathbf{w}'\mathbf{G}\mathbf{b})^2}{(\mathbf{b}'\mathbf{P}\mathbf{b})}$, such that maximizing R_I^2 is equivalent to maximizing $\frac{(\mathbf{w}'\mathbf{G}\mathbf{b})^2}{(\mathbf{b}'\mathbf{P}\mathbf{b})}$. Let $\mathbf{G}\mathbf{w} = \mathbf{u}$, $\mathbf{b} = \mathbf{v}$, and $\mathbf{A} = \mathbf{P}$, by the Cauchy–Schwarz inequality $\frac{(\mathbf{w}'\mathbf{G}\mathbf{b})^2}{(\mathbf{b}'\mathbf{P}\mathbf{b})} \le \mathbf{w}'\mathbf{G}\mathbf{P}^{-1}\mathbf{G}\mathbf{w}$. This implies that the maximum is reached when $\frac{(\mathbf{w}'\mathbf{G}\mathbf{b})^2}{(\mathbf{b}'\mathbf{P}\mathbf{b})} = \mathbf{w}'\mathbf{G}\mathbf{P}^{-1}\mathbf{G}\mathbf{w}$, at which point $R_I = k_I\sqrt{\mathbf{w}'\mathbf{G}\mathbf{P}^{-1}\mathbf{G}\mathbf{w}}$. This latter result is the same as Eq. (2.9) when $\mathbf{b} = \mathbf{P}^{-1}\mathbf{G}\mathbf{w}$.

Result $R_I = k_I\sqrt{\mathbf{w}'\mathbf{G}\mathbf{P}^{-1}\mathbf{G}\mathbf{w}}$ obtained using the Cauchy–Schwarz inequality corroborates that $\mathbf{b} = \mathbf{P}^{-1}\mathbf{G}\mathbf{w}$ (Eq. 2.8) is a global minimum when the mean squared difference between I and H ($E[(H - I)^2]$) is minimized, and a global maximum when the correlation ρ_{HI} between I and H is maximized because $R_I = k_I\sqrt{\mathbf{b}'\mathbf{P}\mathbf{b}} = k_I\sqrt{\mathbf{w}'\mathbf{G}\mathbf{P}^{-1}\mathbf{G}\mathbf{w}}$ only when $\mathbf{b} = \mathbf{P}^{-1}\mathbf{G}\mathbf{w}$.

2.3.3 The LPSI Expected Genetic Gain Per Trait

Whereas $R = \dfrac{Cov(g,y)}{\sigma_y^2}D$ (Eq. 2.4) denotes the selection response in the univariate case, $\mathbf{E} = \frac{Cov(I,\mathbf{g})}{\sigma_I}$ denotes the LPSI expected genetic gain per trait. Also, except by $\frac{D}{\sigma_y}$, $\frac{Cov(g,y)}{\sigma_y}$ and $\frac{Cov(I,\mathbf{g})}{\sigma_I}$ are mathematically equivalent and whereas $\frac{Cov(g,y)}{\sigma_y}$ is the covariance between g and y weighted by the standard deviation of the variance of y, $\frac{Cov(I,\mathbf{g})}{\sigma_I}$ is the covariance between the breeding value vector and the LPSI values weighted by the standard deviation of the variance of LPSI. This means that in effect, **E** is the LPSI multi-trait selection response and can be written as

$$\mathbf{E} = k_I\frac{\mathbf{G}\mathbf{b}}{\sigma_I}, \tag{2.10}$$

where $\mathbf{G}$, σ_I and k_I were defined earlier. As Eq. (2.10) is the covariance between $I = \mathbf{b}'\mathbf{p}$ and $\mathbf{g}' = [g_1 \quad g_2 \quad \dots \quad g_t]$ divided by σ_I, considering g_j and $I = \sum_{j=1}^{t} b_j y_j$, the genetic gain in the *j*th index trait due to selection on I will be

$$\frac{k_I}{\sigma_I} Cov(I, g_j) = \frac{k_I}{\sigma_I} \left[b_1\sigma_{1j} + b_2\sigma_{2j} + \dots + b_j\sigma_j^2 + \dots + b_t\sigma_{tj} \right] = k_I \frac{\mathbf{b}'\boldsymbol{\sigma}_j}{\sigma_I}, \quad (2.11)$$

where $\boldsymbol{\sigma}'_j = \left[\sigma_{1j} \; \cdots \; \sigma_j^2 \; \cdots \; \sigma_{tj} \right]$ is a vector of genotypic covariances of the *j*th index trait with all the index traits (Lin 1978; Brascamp 1984).

If Eq. (2.11) is multiplied by its economic weight, we obtain a measure of the economic value of each trait included in the net genetic merit (Cunningham and Tauebert 2009). In percentage terms, the economic value attributable to genetic change in the *j*th trait can be written as

$$w_j \frac{\mathbf{b}'\boldsymbol{\sigma}_j}{\sigma_I^2} 100. \quad (2.12)$$

In addition, the percentage reduction in the net genetic merit of overall genetic gain if the *j*th trait is omitted from the LPSI (Cunningham and Tauebert 2009) is

$$\left[1 - \sqrt{1 - \frac{b_j^2}{\sigma_I^2 \varphi_j^{-2}}} \right] 100, \quad (2.13)$$

where φ_j^{-2} is the *j*th diagonal element of the inverse of the phenotypic covariance matrix $\mathbf{P}^{-1}$ and b_j^2 the square of the *j*th coefficient of the LPSI. Equations (2.12) and (2.13) are measures of the importance of each trait included in the LPSI when makes selection.

2.3.4 *Heritability of the LPSI*

As the variance of $I = \mathbf{b}'\mathbf{y}$ is equal to $\sigma_I^2 = \mathbf{b}'\mathbf{P}\mathbf{b} = \mathbf{b}'\mathbf{G}\mathbf{b} + \mathbf{b}'\mathbf{R}\mathbf{b}$, where $\mathbf{P} = \mathbf{G} + \mathbf{R}$, $\mathbf{G}$ and $\mathbf{R}$ are the phenotypic, genetic, and residual covariance matrices respectively, then the LPSI heritability (Lin and Allaire 1977; Nordskog 1978) can be written as

$$h_I^2 = \frac{\mathbf{b}'\mathbf{G}\mathbf{b}}{\mathbf{b}'\mathbf{P}\mathbf{b}}. \quad (2.14)$$

When selecting a trait, the correlation between the phenotypic and genotypic values is equal to the square root of the trait's heritability ($\rho_{gy} = h$); however, in the LPSI context, when $\mathbf{b} = \mathbf{P}^{-1}\mathbf{G}\mathbf{w}$, the maximized correlation between H and I is $\rho_{HI} = \sqrt{\frac{\mathbf{b}'\mathbf{P}\mathbf{b}}{\mathbf{w}'\mathbf{G}\mathbf{w}}} = \frac{\sigma_I}{\sigma_H}$, whereas $h_I = \sqrt{\frac{\mathbf{b}'\mathbf{G}\mathbf{b}}{\mathbf{b}'\mathbf{P}\mathbf{b}}}$ is the square root of I heritability; that is,

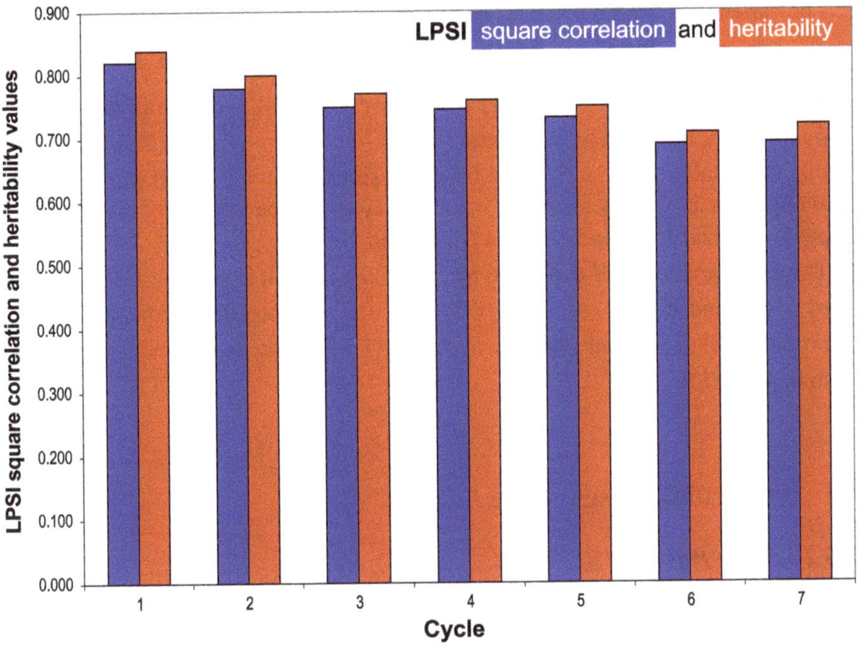

Fig. 2.5 Estimated values of the square correlation between the LPSI and the net genetic merit ($H = \mathbf{w}'\mathbf{g}$) and the LPSI heritability for four traits and 500 genotypes in one environment simulated for seven selection cycles

from a mathematical point of view, $\rho_{HI} \neq h_I$. In practice, h_I^2 and ρ_{HI}^2 give similar results (Fig. 2.5).

2.4 Statistical LPSI Properties

Assuming that H and I have joint bivariate normal distribution, $\mathbf{b} = \mathbf{P}^{-1}\mathbf{G}\mathbf{w}$, and $\mathbf{P}$, $\mathbf{G}$ and $\mathbf{w}$ are known, the statistical LPSI properties (Henderson 1963) are the following:

1. The variance of I (σ_I^2) and the covariance between H and I (σ_{HI}) are equal, i.e., $\sigma_I^2 = \sigma_{HI}$. We can demonstrate this property noting that as $\mathbf{b} = \mathbf{P}^{-1}\mathbf{G}\mathbf{w}$, $\sigma_I^2 = \mathbf{b}'\mathbf{P}\mathbf{b}$, and $\sigma_{HI} = \mathbf{w}'\mathbf{G}\mathbf{b}$, then $\sigma_I^2 = \left(\mathbf{w}'\mathbf{G}\mathbf{P}^{-1}\right)\mathbf{P}\mathbf{P}^{-1}\mathbf{G}\mathbf{w} = \mathbf{w}'\mathbf{G}\mathbf{P}^{-1}\mathbf{G}\mathbf{w}$, and $\sigma_{HI} = \mathbf{w}'\mathbf{G}\mathbf{P}^{-1}\mathbf{G}\mathbf{w}$; i.e., $\sigma_I^2 = \sigma_{HI}$. This last result implies that when $\mu_I = 0$, $E(H/I) = I$.
2. The maximized correlation between H and I is equal to $\rho_{HI} = \frac{\sigma_I}{\sigma_H}$. That is,

$$\rho_{HI} = \frac{\mathbf{w}'\mathbf{G}\mathbf{b}}{\sqrt{\mathbf{w}'\mathbf{G}\mathbf{w}}\sqrt{\mathbf{b}'\mathbf{P}\mathbf{b}}} = \frac{\mathbf{w}'\mathbf{G}\mathbf{P}^{-1}\mathbf{G}\mathbf{w}}{\sqrt{\mathbf{w}'\mathbf{G}\mathbf{w}}\sqrt{\mathbf{w}'\mathbf{G}\mathbf{P}^{-1}\mathbf{G}\mathbf{w}}} = \sqrt{\frac{\mathbf{w}'\mathbf{G}\mathbf{P}^{-1}\mathbf{G}\mathbf{w}}{\mathbf{w}'\mathbf{G}\mathbf{w}}} = \frac{\sigma_I}{\sigma_H}, \text{ thus, } \rho_{HI} = \frac{\sigma_I}{\sigma_H}.$$

3. The variance of the predicted error, $Var(H - I) = (1 - \rho_{HI}^2)\sigma_H^2$, is minimal. Note that $Var(H - I) = E\left[(H - I)^2\right] = \sigma_I^2 + \sigma_H^2 - 2\sigma_{HI}$, and when $\mathbf{b} = \mathbf{P}^{-1}\mathbf{G}\mathbf{w}$, $\sigma_I^2 = \sigma_{HI}$, from where $Var(H - I) = \sigma_H^2 - \sigma_I^2 = (1 - \rho_{HI}^2)\sigma_H^2$ is minimal because by Eq. (2.8), $\mathbf{b} = \mathbf{P}^{-1}\mathbf{G}\mathbf{w}$ minimizes $Var(H - I) = (1 - \rho_{HI}^2)\sigma_H^2$. Thus, the larger ρ_{HI}, the smaller $E[(H - I)^2]$ and the more similar I and H are. If $\rho_{HI} > 0$, I and H tend to be positively related; if $\rho_{HI} < 0$, they tend to be negatively related; and if $\rho_{HI} = 0$, I and H are independent (Anderson 2003).
4. The total variance of H explained by I is $\sigma_I^2 = \rho_{HI}^2\sigma_H^2$. It is evident that if $\rho_{HI} = 1$, $\sigma_I^2 = \sigma_H^2$, and if $\rho_{HI} = 0$, $\sigma_I^2 = 0$. That is, the variance of H explained by I is proportional to ρ_{HI}, and when ρ_{HI} is close to 1, σ_I^2 is close to σ_H^2, and if ρ_{HI} is close to 0, σ_I^2 is close to 0.

2.5 Particular Cases of the LPSI

2.5.1 *The Base LPSI*

To derive the LPSI theory, we assumed that the phenotypic (**P**) and the genotypic (**G**) covariance matrix, and the vector of economic values (**w**) are known. However, **P**, **G**, and **w** are generally unknown and it is necessary to estimate them. There are many methods for estimating **P** and **G** (Lynch and Walsh 1998) and **w** (Cotterill and Jackson 1985; Magnussen 1990). However, when the estimator of $\mathbf{P}(\widehat{\mathbf{P}})$ is not positive definite (all eigenvalues positive) or the estimator of $\mathbf{G}(\widehat{\mathbf{G}})$ is not positive semidefinite (no negative eigenvalues), the estimator of $\mathbf{b} = \mathbf{P}^{-1}\mathbf{G}\mathbf{w}$ ($\widehat{\mathbf{b}} = \widehat{\mathbf{P}}^{-1}\widehat{\mathbf{G}}\mathbf{w}$) could be biased. In this case, the base linear phenotypic selection index (BLPSI):

$$I_B = \mathbf{w}'\mathbf{y} \tag{2.15}$$

may be a better predictor of $H = \mathbf{w}'\mathbf{g}$ than the *estimated* LPSI $\widehat{I} = \widehat{\mathbf{b}}'\mathbf{y}$ (Williams 1962a; Lin 1978) if the vector of economic values **w** is indeed known. Many authors (Williams 1962b; Harris 1964; Hayes and Hill 1980, 1981) have investigated the influence of parameter estimation errors on LPSI accuracy and concluded that those errors affect the accuracy of $\widehat{I} = \widehat{\mathbf{b}}'\mathbf{y}$ when the accuracy of $\widehat{\mathbf{P}}$ and $\widehat{\mathbf{G}}$ is low. If vector **w** values are known, the BLPSI has certain advantages because of its simplicity and its freedom from parameter estimation errors (Lin 1978). Williams (1962a) pointed out that the BLPSI is superior to $\widehat{I} = \widehat{\mathbf{b}}'\mathbf{y}$ unless a large amount of data is available for estimating **P** and **G**.

There are some problems associated with the BLPSI. For example, what is the BLPSI selection response and the BLPSI expected genetic gains per trait when no data are available for estimating **P** and **G**? The BLPSI is a better selection index than the standard LPSI only if the correlation between the BLPSI and the net genetic merit is higher than that between the LPSI and the net genetic merit (Hazel 1943).

However, if estimations of **P** and **G** are not available, how can the correlation between the base index and the net genetic merit be obtained? Williams (1962b) pointed out that the correlation between the BLPSI and $H = \mathbf{w}'\mathbf{g}$ can be written as

$$\rho_{HI_B} = \sqrt{\frac{\mathbf{w}'\mathbf{G}\mathbf{w}}{\mathbf{w}'\mathbf{P}\mathbf{w}}} \tag{2.16}$$

and indicated that the ratio ρ_{HI_B}/ρ_{HI} can be used to compare LPSI efficiency versus BLPSI efficiency; however, in the latter case, at least the estimates of **P** and **G**, i.e., $\widehat{\mathbf{P}}$ and $\widehat{\mathbf{G}}$, need to be known.

In addition, Eq. (2.15) is only an assumption, not a result, and implies that **P** and **G** are the same. That is, $\mathbf{b} = \mathbf{P}^{-1}\mathbf{G}\mathbf{w} = \mathbf{w}$ only when $\mathbf{P} = \mathbf{G}$, which indicates that the BLPSI is a special case of the LPSI. Thus, to obtain the selection response and the expected genetic gains per trait of the BLPSI, we need some information about **P** and **G**. Assuming that the BLPSI is indeed a particular case of the LPSI, the BLPSI selection response and the BLPSI expected genetic gains per trait could be written as

$$R_B = k_I\sqrt{\mathbf{w}'\mathbf{P}\mathbf{w}}, \tag{2.17}$$

and

$$\mathbf{E}_B = k_I\frac{\mathbf{G}\mathbf{w}}{\sqrt{\mathbf{w}'\mathbf{P}\mathbf{w}}}, \tag{2.18}$$

respectively. The parameters of Eqs. (2.17) and (2.18) were defined earlier.

There are additional implications if $\mathbf{b} = \mathbf{P}^{-1}\mathbf{G}\mathbf{w} = \mathbf{w}$. For example, if $\mathbf{P} = \mathbf{G}$, then $\rho_{HI_B} = \sqrt{\frac{\mathbf{w}'\mathbf{G}\mathbf{w}}{\mathbf{w}'\mathbf{P}\mathbf{w}}}$ and BLPSI heritability $h^2_{I_B} = \frac{\mathbf{w}'\mathbf{G}\mathbf{w}}{\mathbf{w}'\mathbf{P}\mathbf{w}}$ are equal to 1. However, in practice, the estimated values of the $\rho_{HI_B}(\widehat{\rho}_{HI_B})$ are usually lower than the estimated values of the $\rho_{HI}(\widehat{\rho}_{HI})$ (Fig. 2.6).

2.5.2 *The LPSI for Independent Traits*

Suppose that the traits under selection are independent, then **P** and **G** are diagonal matrices and $\mathbf{b} = \mathbf{P}^{-1}\mathbf{G}\mathbf{w}$ is a vector of single-trait heritabilities multiplied by the economic weights, because $\mathbf{P}^{-1}\mathbf{G}$ is the matrix of multi-trait heritabilities (Xu and Muir 1992). Based on this result, Hazel and Lush (1942) and Smith et al. (1981) used trait heritabilities multiplied by the economic weights (or heritabilities only) as coefficients of the LPSI. Thus, when the traits are independent and the economic weights are known, the LPSI can be constructed as

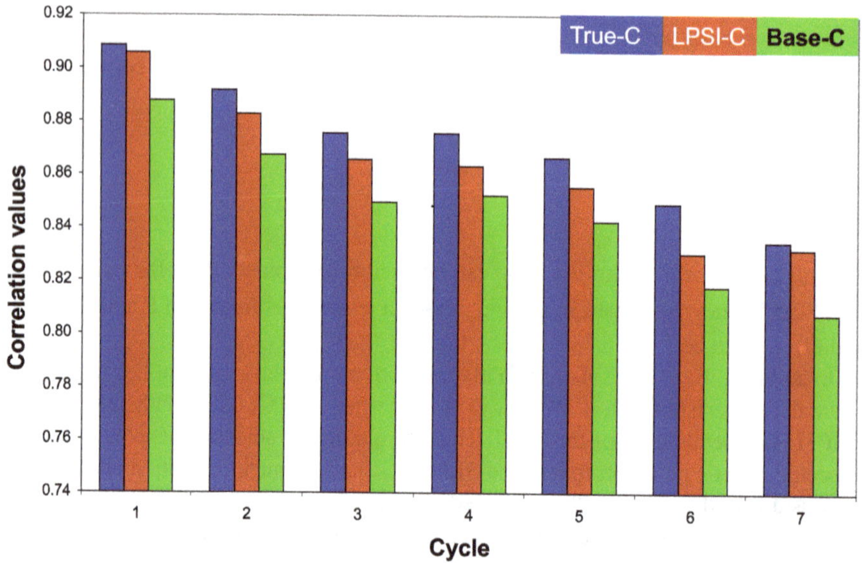

Fig. 2.6 Values of the true correlation between the LPSI and the net genetic merit ($H = \mathbf{w}'\mathbf{g}$) (*True-C*), the estimated correlation between the LPSI and H (*LPSI-C*), and the estimated correlation between the base index and H (*Base-C*) for four traits and 500 genotypes in one environment simulated for seven selection cycles

$$I = \sum_{i=1}^{t} w_i h_i^2 y_i, \tag{2.19}$$

and when the economic weights are unknown, the LPSI can be constructed as

$$I = \sum_{i=1}^{t} h_i^2 y_i. \tag{2.20}$$

The selection response of Eq. (2.19) and (2.20) can be seen in Hazel and Lush (1942).

2.6 Criteria for Comparing LPSI Efficiency

Assuming that the intensity of selection is the same in both indices, we can compare BLPSI ($I_B = \mathbf{w}'\mathbf{y}$) efficiency versus LPSI efficiency to predict the net genetic merit in percentage terms as

$$p = 100(\lambda - 1), \tag{2.21}$$

where $\lambda = \frac{\rho_{HI}}{\rho_{HI_B}}$ (Williams 1962b; Bulmer 1980). Therefore, when $p = 0$, the efficiency of both indices is the same; when $p > 0$, the efficiency of the LPSI is higher than the base index efficiency, and when $p < 0$, the base index efficiency is higher than LPSI efficiency (Fig. 2.6). Equation (2.21) is useful for comparing the efficiency of any linear selection index, as we shall see in this book.

2.7 Estimating Matrices G and P

To derive the LPSI theory we assumed that matrices **P** and **G** are known. In practice, we have to estimate them. Matrices **P** and **G** can be estimated by analysis of variance (ANOVA), maximum likelihood or restricted maximum likelihood (REML) (Baker 1986; Lynch and Walsh 1998; Searle et al. 2006; Hallauer et al. 2010). Equation (2.1) is the simplest model because we only need to estimate two variance components: the genotypic variance (σ_g^2) and the residual variance (σ_e^2), from where the phenotypic variance for trait y is the sum of σ_g^2 and σ_e^2, that is, $\sigma_y^2 = \sigma_g^2 + \sigma_e^2$. However, to construct matrices **P** and **G**, we also need the covariance between any two traits. Thus, if y_i and y_j $(i, j = 1, 2, \cdots, t)$ are any two traits, then the covariance between y_i and y_j ($\sigma_{y_{ij}}$) can be written as $\sigma_{y_{ij}} = \sigma_{g_{ij}} + \sigma_{e_{ij}}$, where $\sigma_{g_{ij}}$ and $\sigma_{e_{ij}}$ denote the genotypic and residual covariance respectively of traits y_i and y_j.

Several authors (Baker 1986; Lynch and Walsh 1998; Hallauer et al. 2010) have described ANOVA methods for estimating matrix **G** using specific design data, for example, half-sib, full-sib, etc., when the sample sizes are well balanced. In the ANOVA method, observed mean squares are equal to their expected values; the expected values are linear functions of the unknown variance components; thus the resulting equations are a set of simultaneous linear equations in the variance components. The expected values of mean squares in the ANOVA method do not need assumptions of normality because the variance component estimators do not depend on normality assumptions (Lynch and Walsh 1998; Hallauer et al. 2010).

In cases where the sample sizes are not well balanced, Lynch and Walsh (1998) and Fry (2004) proposed using the REML method to estimate matrix **G**. The REML estimation method does not require a specific design or balanced data and can be used to estimate genetic and residual variance and covariance in any arbitrary pedigree of individuals. The REML method is based on projecting the data in a subspace free of fixed effects and maximizing the likelihood function in this subspace, and has the advantage of producing the same results as the ANOVA in balanced designs (Blasco 2001).

In the context of the linear mixed model, Lynch and Walsh (1998) have given formulas for estimating variances σ_g^2 and σ_e^2 that can be adapted to estimate covariances $\sigma_{g_{ij}}$ and $\sigma_{e_{ij}}$. Suppose that we want to estimate σ_g^2 and σ_e^2 for the qth trait ($q = 1, 2 \cdots, t$ = number of traits) in the absence of dominance and epistatic effects using the model $\mathbf{y}_q = \mathbf{1}\mu_q + \mathbf{Z}\mathbf{g}_q + \mathbf{e}_q$, where the vector of averages $\mathbf{y}_q$~NMV $(\mathbf{1}\mu_q, \mathbf{V}_q)$ is $g \times 1$ (g = number of genotypes in the population) and has multivariate normal distribution; $\mathbf{1}$ is a $g \times 1$ vector of ones, μ_q is the mean of the qth trait, $\mathbf{Z}$ is an identity matrix $g \times g$, $\mathbf{g}_q$~NMV$(\mathbf{0}, \mathbf{A}\sigma^2_{gq})$ is a vector of true breeding values, and $\mathbf{e}_q$~NMV$(\mathbf{0}, \mathbf{I}\sigma^2_{e_q})$ is a $g \times 1$ vector of residuals, where NMV stands for normal multivariate distribution. Matrix $\mathbf{A}$ denotes the numerical relationship matrix between individuals (Lynch and Walsh 1998; Mrode 2005) and $\mathbf{V}_q = \mathbf{A}\sigma^2_{gq} + \mathbf{I}\sigma^2_{e_q}$.

The expectation–maximization algorithm allows the REML to be computed for the variance components $\sigma^2_{g_q}$ and $\sigma^2_{e_q}$ by iterating the following equations:

$$\sigma^{2(n+1)}_{g_q} = \sigma^{2(n)}_{g_q} + \frac{\left(\sigma^{2(n)}_{g_q}\right)^2}{g}\left[\mathbf{y}'_q\left(\mathbf{T}^{(n)}\mathbf{A}\mathbf{T}^{(n)}\right)\mathbf{y}_q - tr\left(\mathbf{T}^{(n)}\mathbf{A}\right)\right] \tag{2.22}$$

and

$$\sigma^{2(n+1)}_{e_q} = \sigma^{2(n)}_{e_q} + \frac{\left(\sigma^{2(n)}_{e_q}\right)^2}{g}\left[\mathbf{y}'_q\left(\mathbf{T}^{(n)}\mathbf{T}^{(n)}\right)\mathbf{y}_q - tr\left(\mathbf{T}^{(n)}\right)\right], \tag{2.23}$$

where, after n iterations, $\sigma^{2(n+1)}_{g_q}$ and $\sigma^{2(n+1)}_{e_q}$ are the estimated variance components of $\sigma^2_{g_q}$ and $\sigma^2_{e_q}$ respectively; $tr(.)$ denotes the trace of the matrices within brackets; $\mathbf{T} = \mathbf{V}_q^{-1} - \mathbf{V}_q^{-1}\mathbf{1}\left(\mathbf{1}'\mathbf{V}_q^{-1}\mathbf{1}\right)\mathbf{1}'\mathbf{V}_q^{-1}$ and $\mathbf{V}_q^{-1}$ is the inverse of matrix $\mathbf{V}_q = \mathbf{A}\sigma^2_{gq} + \mathbf{I}\sigma^2_{e_q}$. In $\mathbf{T}^{(n)}$, $\mathbf{V}_q^{-1(n)}$ is the inverse of matrix $\mathbf{V}_q^{(n)} = \mathbf{A}\sigma^{2(n)}_{\gamma q} + \mathbf{I}\sigma^{2(n)}_{e_q}$.

The additive genetic and residual covariances between the observations of the qth and ith traits, $\mathbf{y}_q$ and $\mathbf{y}_i$ ($\sigma_{g_{q,i}}$ and $\sigma_{e_{q,i}}$, q, i = 1, 2, …, t), can be estimated using REML by adapting Eqs. (2.22) and (2.23). Note that the variance of the sum of $\mathbf{y}_q$ and $\mathbf{y}_i$ can be written as $Var(\mathbf{y}_i + \mathbf{y}_q) = \mathbf{V}_i + \mathbf{V}_q + 2\mathbf{C}_{iq}$, where $\mathbf{V}_i = \mathbf{A}\sigma^2_{gi} + \mathbf{I}\sigma^2_{e_i}$ is the variance of $\mathbf{y}_i$ and $\mathbf{V}_q = \mathbf{A}\sigma^2_{gq} + \mathbf{I}\sigma^2_{e_q}$ is the variance of $\mathbf{y}_q$; in addition, $2\mathbf{C}_{iq} = 2\mathbf{A}\sigma_{giq} + 2\mathbf{I}\sigma_{eiq} = 2Cov(\mathbf{y}_i, \mathbf{y}_q)$ is the covariance of $\mathbf{y}_q$ and $\mathbf{y}_i$, and σ_{giq} and σ_{eiq} are the additive and residual covariances respectively associated with the covariance of $\mathbf{y}_q$ and $\mathbf{y}_i$. Thus, one way of estimating σ_{giq} and σ_{eiq} is by using the following equation:

$$0.5Var(\mathbf{y}_i + \mathbf{y}_q) - 0.5Var(\mathbf{y}_i) - 0.5Var(\mathbf{y}_q), \tag{2.24}$$

for which Eqs. (2.22) and (2.23) can be used. Equations (2.22) to (2.24) are used to estimate $\mathbf{P}$ and $\mathbf{G}$ in the illustrative examples of this book.

2.8 Numerical Examples

2.8.1 Simulated Data

This data set was simulated by Ceron-Rojas et al. (2015) and can be obtained at http://hdl.handle.net/11529/10199. The data were simulated for eight phenotypic selection cycles (C0 to C7), each with four traits (T_1, T_2, T_3 and T_4), 500 genotypes, and four replicates for each genotype (Fig. 2.7). The LPSI economic weights for T_1,

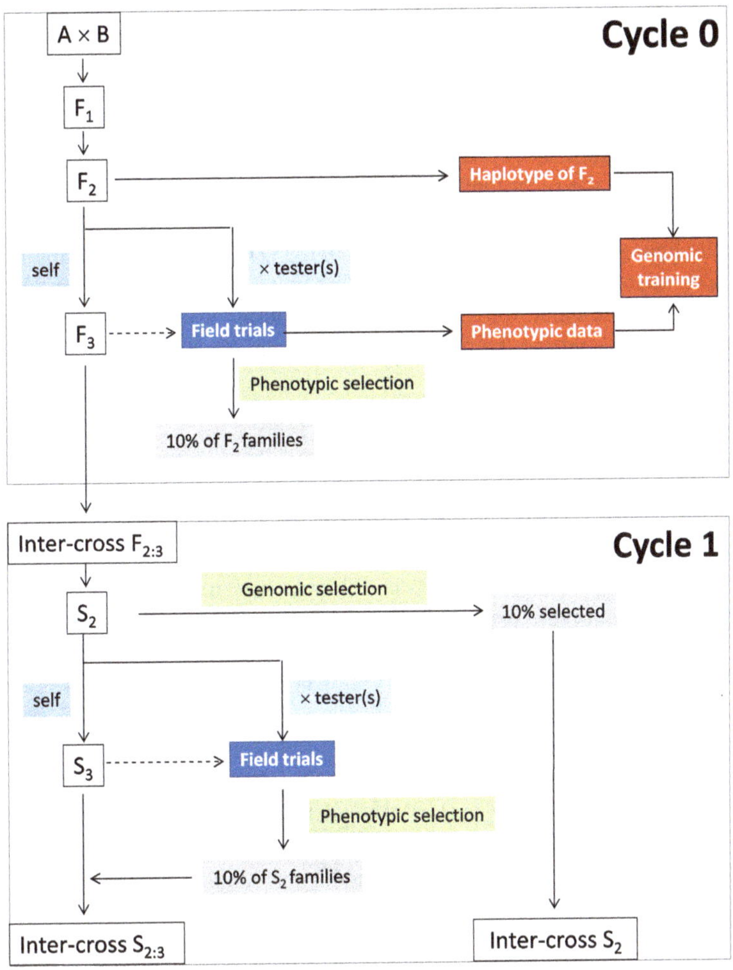

Fig. 2.7 Schematic illustration of the steps followed to generate data sets 1 and 2 for the seven selection cycles using the linear phenotypic selection index and the linear genomic selection index. Dotted lines indicate the process used to simulate the phenotypic data (according to Ceron-Rojas et al. 2015)

T_2, T_3 and T_4 were 1, −1, 1, and 1 respectively. Each of the four traits was affected by a different number of quantitative trait loci (QTLs): 300, 100, 60, and 40, respectively. The common QTLs affecting the traits generated genotypic correlations of −0.5, 0.4, 0.3, −0.3, −0.2, and 0.1 between T_1 and T_2, T_1 and T_3, T_1 and T_4, T_2 and T_3, T_2 and T_4, and T_3 and T_4 respectively. The genotypic value of each plant was generated based on its haplotypes and the QTL effects for each trait.

Simulated data were generated using QU-GENE software (Podlich and Cooper 1998; Wang et al. 2003). A total of 2500 molecular markers were distributed uniformly across 10 chromosomes, whereas 315 QTLs were randomly allocated over the ten chromosomes to simulate one maize (*Zea mays* L.) population. Each QTL and molecular marker was biallelic and the QTL additive values ranged from 0 to 0.5. As QU-GENE uses recombination fraction rather than map distance to calculate the probability of crossover events, recombination between adjacent pairs of markers was set at 0.0906; for two flanking markers, the QTL was either on the first (recombination between the first marker and QTL was equal to 0.0) or the second (recombination between the first marker and QTL was equal to 0.0906) marker; excluding the recombination fraction between 15 random QTLs and their flanking markers, which was set at 0.5, i.e., complete independence (Haldane 1919), to simulate linkage equilibrium between 5% of the QTLs and their flanking markers. In addition, in every case, two adjacent QTLs were in complete linkage. For each trait, the phenotypic value for each of four replications of each plant was obtained from QU-GENE by setting the per-plot heritability of T_1, T_2, T_3, and T_4 at 0.4, 0.6, 0.6, and 0.8 respectively.

2.8.2 *Estimated Matrices, LPSI, and Its Parameters*

For this example, we used only cycle C1 data and traits T_1, T_2, and T_3. The phenotypic and genotypic estimated covariance matrices for traits T_1, T_2, and T_3 were $\widehat{\mathbf{P}} = \begin{bmatrix} 62.50 & -12.74 & 8.53 \\ -12.74 & 17.52 & -3.38 \\ 8.53 & -3.38 & 12.31 \end{bmatrix}$ and $\widehat{\mathbf{G}} = \begin{bmatrix} 36.21 & -12.93 & 8.35 \\ -12.93 & 13.04 & -3.40 \\ 8.35 & -3.40 & 9.96 \end{bmatrix}$ respectively, whereas the inverse of matrix $\widehat{\mathbf{P}}$ was $\widehat{\mathbf{P}}^{-1} = \begin{bmatrix} 0.01997 & 0.01251 & -0.01040 \\ 0.01251 & 0.06809 & 0.01005 \\ -0.01040 & 0.01005 & 0.09123 \end{bmatrix}$. The estimated heritabilities for T_1, T_2, and T_3 were $\widehat{h_1^2} = 0.579$, $\widehat{h_2^2} = 0.744$, and $\widehat{h_2^2} = 0.809$ respectively.

According to matrices $\widehat{\mathbf{P}}^{-1}$ and $\widehat{\mathbf{G}}$, and because $\mathbf{w}' = [1 \quad -1 \quad 1]$, the estimated vector of coefficients was $\widehat{\mathbf{b}}' = \mathbf{w}'\widehat{\mathbf{GP}}^{-1} = [0.555 \quad -1.063 \quad 1.087]$, from which the estimated LPSI can be written as $\widehat{I} = 0.555T_1 - 1.063T_2 + 1.087T_3$. Table 2.2 presents the first 20 genotypes, the means of the three traits (T1, T2 and T3) and the first 20 estimated unranked LPSI values of the 500 simulated genotypes for cycle C1.

Table 2.2 Number of genotypes, means of the trait (T1, T2 and T3) values, and unranked values of the LPSI for part of a simulated data set

	Means of the trait values			Unranked
Number of genotypes	T1	T2	T3	LPSI values
1	164.46	39.63	34.66	86.81
2	144.39	50.77	34.65	63.82
3	157.48	48.04	37.9	77.52
4	167.30	47.98	30.49	74.97
5	164.11	49.89	32.03	72.85
6	166.26	40.44	29.93	81.81
7	154.59	52.22	30.31	63.22
8	160.00	42.91	31.23	77.12
9	158.51	46.32	34.52	76.25
10	163.63	45.43	35.73	81.35
11	156.16	46.75	35.58	75.62
12	171.38	41.17	35.13	89.52
13	153.17	54.18	36.23	66.79
14	149.89	52.33	31.13	61.39
15	159.63	49.01	31.72	70.96
16	160.70	42.51	32.99	79.85
17	157.07	45.49	28.4	69.68
18	167.50	41.69	36.73	88.55
19	159.17	50.6	36.25	73.93
20	161.80	46.58	37.33	80.84

According to the means of the three traits, the first estimated LPSI value was obtained as

$$\widehat{I}_1 = 0.555(164.46) - 1.063(39.63) + 1.087(34.66) = 86.81;$$

the second estimated LPSI value was obtained as

$$\widehat{I}_2 = 0.555(144.39) - 1.063(144.39) + 1.087(34.65) = 63.82, \text{etc.};$$

and the 20th estimated LPSI value was obtained as

$$\widehat{I}_{20} = 0.555(161.80) - 1.063(46.58) + 1.087(37.33) = 80.84.$$

This estimation procedure is valid for any number of genotypes. Table 2.3 presents the 20 genotypes ranked by the estimated LPSI values. Note that if we use 20% selection intensity for Table 2.2 data, we should select genotypes 12, 18, 1, 6, and 10, because their estimated LPSI values are higher than the remaining LPSI values for that set of genotypes. Using the idea described in Fig. 2.4, genotypes 12, 18, 1, 6, and 10 should be in the red zone, whereas the rest of the genotypes are in the white zone and should be culled. Here, the proportion selected is $q = 0.2$ and

Table 2.3 Number of genotypes, means of the trait (T1, T2 and T3) values and ranked values of the LPSI for part of a simulated data set

	Means of the trait values			Ranked
Number of genotypes	T1	T2	T3	LPSI values
12	171.38	41.17	35.13	89.52
18	167.50	41.69	36.73	88.55
1	164.46	39.63	34.66	86.81
6	166.26	40.44	29.93	81.81
10	163.63	45.43	35.73	81.35
20	161.80	46.58	37.33	80.84
16	160.70	42.51	32.99	79.85
3	157.48	48.04	37.9	77.52
8	160.00	42.91	31.23	77.12
9	158.51	46.32	34.52	76.25
11	156.16	46.75	35.58	75.62
4	167.30	47.98	30.49	74.97
19	159.17	50.6	36.25	73.93
5	164.11	49.89	32.03	72.85
15	159.63	49.01	31.72	70.96
17	157.07	45.49	28.4	69.68
13	153.17	54.18	36.23	66.79
2	144.39	50.77	34.65	63.82
7	154.59	52.22	30.31	63.22
14	149.89	52.33	31.13	61.39

$z = \frac{\exp\{-0.5u'^2\}}{\sqrt{2\pi}} = 0.31$, where $u' = \frac{81.35-75.64}{8.11} = 0.704$, 81.35 is the estimated LPSI value or the genotype number 10, 75.64 is the mean of the 20 LPSI values, and 8.11 is the standard deviation of the estimated LPSI values of the 20 genotypes presented in Tables 2.2 and 2.3.

Table 2.4 presents 25 genotypes and the means of the three traits obtained from the 500 simulated genotypes for cycle C1 and ranked by the estimated LPSI values. In this case, we used 5% selection intensity ($k_I = 2.063$). Also, the last four rows in Table 2.4 give:

1. The means of traits T_1, T_2, and T_3 (175.46, 39.26, and 38.83 respectively) of the selected individuals and the mean of the selected LPSI values (97.84).
2. The means of the three traits in the base population (161.88, 45.19, and 34.39) and the mean of the LPSI values in the base population (79.18)
3. The selection differentials for the three traits (13.58, −5.92, and 4.44) and the selection differential for the LPSI (18.66)
4. The LPSI expected genetic gain per trait (9.51, −5.48, and 4.22) and the LPSI selection response (19.21).

The variance of the estimated selection index for the 500 genotypes was $\widehat{V}(\widehat{I}) = \widehat{\mathbf{b}}'\mathbf{P}\widehat{\mathbf{b}} = 86.72$, from which the standard deviation of $\widehat{I}$ was 9.312. The

Table 2.4 Number of selected genotypes, selected means of the trait (T1, T2 and T3) values and ranked selected values of the LPSI from one simulated set of 500 genotypes with four repetitions

	Means of the trait values			Ranked
Number of genotypes	T1	T2	T3	LPSI values
353	189.68	38.16	36.13	103.97
370	178.27	34.38	37.79	103.45
480	174.84	42.72	45.12	100.66
300	177.38	39.15	40.34	100.65
273	181.18	35.94	35.14	100.52
275	167.94	36.82	42.2	99.92
148	173.37	37.07	39.62	99.86
137	185.48	46.48	42.55	99.77
351	173.79	38.38	40.52	99.68
236	182.85	37.88	34.96	99.2
217	175.13	38.48	39.16	98.84
356	171.09	39.6	41.98	98.47
167	175.39	38.73	37.73	97.17
230	169.73	37.1	38.69	96.8
243	171.9	41.53	41.45	96.29
55	170.02	36.92	37.76	96.15
68	172.56	37.18	36.7	96.13
36	175.8	38.86	36.34	95.75
164	173.61	38.37	36.42	95.14
140	170.53	42.52	41.97	95.05
146	177.4	39.64	35.5	94.89
432	174.01	40.73	38.26	94.84
378	176.62	42.69	38.47	94.44
288	172.14	39.31	37.26	94.23
386	175.77	42.89	38.81	94.13
Mean of selected individuals	175.46	39.26	38.83	97.84
Mean of all individuals	161.88	45.19	34.39	79.18
Selection differential	13.58	−5.92	4.44	18.66
Expected genetic gain for 5%	9.51	−5.48	4.22	19.21

The selection intensity was 5%

estimated standardized selection differentials for the LPSI can be obtained from Table A in Falconer and Mackay (1996), where, for 5% selection intensity, $k_I = 2.063$. This means that the estimated LPSI selection response was $\widehat{R} = 2.063(9.312) = 19.21$, whereas the expected genetic gain per trait, or multi-trait selection response, was $\widehat{\mathbf{E}}' = 2.063\left[\frac{\widehat{\mathbf{b}}'\widehat{\mathbf{G}}}{9.312}\right] = [9.51 \quad -5.48 \quad 4.22]$.

2.8.3 LPSI Efficiency Versus Base Index Efficiency

The estimated correlation between the LPSI and the net genetic merit was $\widehat{\rho}_{HI} = \frac{\widehat{\sigma}_I}{\widehat{\sigma}_H} = 0.894$, whereas the estimated correlation between the base index and the net genetic merit was $\widehat{\rho}_{HI_B} = 0.875$, thus $\widehat{\lambda} = \frac{\widehat{\rho}_{HI}}{\widehat{\rho}_{HI_B}} = 1.0217$ and, by Eq. (2.21), $\widehat{p} = 100(\widehat{\lambda} - 1) = 2.171$. This means that LPSI efficiency was only 2.2% higher than the base index efficiency for this data set.

Using the same data set described in Sect. 2.8.1 of this chapter, we conducted seven selection cycles (C1 to C7) for the four traits (T_1, T_2, T_3, and T_4) using the LPSI and the BLPSI. These results are presented in Table 2.5. To compare the LPSI efficiency versus BLPSI efficiency, we obtained the true selection response of the simulated data (second column in Table 2.5) and we estimated the LPSI and BLPSI selection response for each selection cycle (third column in Table 2.5); in addition, we estimated the LPSI and BLPSI expected genetic gain per trait for each selection cycle (columns 4 to 7 in Table 2.5). The first part of Table 2.5 shows the true selection response and the estimated values of the LPSI selection response and expected genetic gain per trait. In a similar manner, the second part of Table 2.5 shows the true selection response, the estimated values of the BLPSI selection

Table 2.5 The LPSI and BLPSI responses (true and estimated) and estimated expected genetic gain per trait for seven simulated selection cycles

	Selection response		Estimated expected genetic gain per trait			
Cycle	True	Estimated	T1	T2	T3	T4
LPSI						
1	17.84	17.81	7.90	−4.67	3.33	1.92
2	15.66	15.69	7.06	−3.59	3.17	1.86
3	14.44	14.22	6.67	−3.21	2.82	1.52
4	14.29	14.34	7.53	−3.45	2.07	1.29
5	13.86	13.64	7.14	−2.66	2.51	1.33
6	12.47	12.04	6.23	−2.62	1.98	1.21
7	12.44	11.61	5.38	−2.55	2.47	1.22
Average	14.43	14.19	6.85	−3.25	2.62	1.48
BLPSI						
1	17.84	22.15	8.38	−4.40	3.04	1.64
2	15.66	20.49	7.74	−3.33	2.82	1.53
3	14.44	19.33	7.29	−3.00	2.44	1.22
4	14.29	19.49	8.05	−3.17	1.89	1.05
5	13.86	18.93	7.64	−2.53	2.19	1.07
6	12.47	17.72	6.81	−2.40	1.72	0.93
7	12.44	17.28	5.89	−2.35	2.11	0.93
Average	14.43	19.34	7.40	−3.02	2.32	1.19

The selection intensity was 10% ($k_I = 1.755$)

response, and the expected genetic gain per trait. The average value of the true selection response was equal to 14.43, whereas the average values of the estimated LPSI and BLPSI selection response were 14.19 and 19.34 respectively. Note that 14.43–14.19 = 0.24, but 19.34–14.43 = 4.91. According to this result, the BLPSI over-estimated the true selection response of the simulated data by 34.7%. Thus, based on the Table 2.5 results and those presented in Fig. 2.6, we can conclude that the LPSI was more efficient than the BLPSI for this data set.

Finally, additional results can be seen in Chap. 10, where the LPSI was simulated for many selection cycles. Chapter 11 describes RIndSel: a program that uses R and the selection index theory to make selection.

2.9 The LPSI and Its Relationship with the Quadratic Phenotypic Selection Index

In the nonlinear selection index theory, the net genetic merit and the index are both nonlinear. There are many types of nonlinear indices; Goddard (1983) and Weller et al. (1996) have reviewed the general theory of nonlinear selection indices. In this chapter, we describe only the simplest of them: the quadratic index developed mainly by Wilton et al. (1968), Wilton (1968), and Wilton and Van Vleck (1969), which is related to the LPSI.

2.9.1 *The Quadratic Nonlinear Net Genetic Merit*

The most common form of writing the quadratic net genetic merit is

$$H_q = \alpha + \mathbf{w}'(\boldsymbol{\mu} + \mathbf{g}) + (\boldsymbol{\mu} + \mathbf{g})'\mathbf{A}(\boldsymbol{\mu} + \mathbf{g}), \tag{2.25}$$

where α is a constant, $\mathbf{g}$ is the vector of breeding values, which has normal distribution with zero mean and covariance matrix $\mathbf{G}$, $\boldsymbol{\mu}$ is the vector of population means, and $\mathbf{w}$ is a vector of economic weights. In addition, matrix $\mathbf{A}$ can be written as $\mathbf{A} = \begin{bmatrix} w_1 & 0.5w_{12} & \cdots & 0.5w_{1t} \\ 0.5w_{12} & w_2 & \cdots & 0.5w_{2t} \\ \vdots & \vdots & \ddots & \vdots \\ 0.5w_{1t} & 0.5w_{2t} & \cdots & w_t \end{bmatrix}$, where the diagonal ith values w_i (i = 1,2, . . ., t) is the relative economic weight of the genetic value of the squared trait i and w_{ij} (i,j = 1,2, . . ., t) is the economic weight of the cross products between the genetic values of traits i and j. The main difference between the linear net genetic merit (Eq. 2.2) and the net quadratic merit (Eq. 2.25) is that the latter depends on $\boldsymbol{\mu}$ and $(\boldsymbol{\mu} + \mathbf{g})'\mathbf{A}(\boldsymbol{\mu} + \mathbf{g})$.

2.9.2 *The Quadratic Index*

The quadratic phenotypic selection index is

$$I_q = \beta + \mathbf{b}'\mathbf{y} + \mathbf{y}'\mathbf{B}\mathbf{y} \tag{2.26}$$

where β is a constant, **y** is the vector of phenotypic values that has multivariate normal distribution with zero mean and covariance matrix **P**, $\mathbf{b}' = [b_1 \quad b_2 \quad \cdots \quad b_t]$ is a vector of coefficients, and $\mathbf{B} = \begin{bmatrix} b_1 & 0.5b_{12} & \cdots & 0.5b_{1t} \\ 0.5b_{12} & b_2 & \cdots & 0.5b_{2t} \\ \vdots & \vdots & \ddots & \vdots \\ 0.5b_{1t} & 0.5b_{2t} & \ldots & b_t \end{bmatrix}$. In matrix **B**, the diagonal *i*th values b_i (i = 1,2, . . ., t) is the index weight for the square of the phenotypic i and b_{ij} (i,j = 1,2, . . ., t) is the index weight for the cross products between the phenotype of the traits i and j.

2.9.3 *The Vector and the Matrix of Coefficients of the Quadratic Index*

As we saw in Sect. 2.3.2 of this chapter, to obtain the vector (**b**) and the matrix (**B**) of coefficients of the quadratic index that maximized the selection response, we can minimize the expectation of the square difference between the quadratic index (I_q) and the quadratic net genetic merit (H_q): $\Phi = E\{[I_q - E(I_q)] - [H_q - E(H_q)]\}^2$, or we can maximize the correlation between I_q and H_q, i.e., $\rho_{H_qI_q} = \frac{Cov(H_q, I_q)}{\sqrt{Var(I_q)}\sqrt{Var(H_q)}}$, where $Cov(H_q, I_q)$ is the covariance between I_q and H_q, $\sqrt{Var(I_q)}$ is the standard deviation of the variance of I_q, and $\sqrt{Var(H_q)}$ is the standard deviation of the variance of H_q. In this context, it is easier to maximize $\rho_{H_qI_q}$ than to minimize Φ. Vandepitte (1972) minimized Φ, but in this section we shall maximize $\rho_{H_qI_q}$.

Suppose that $\boldsymbol{\mu} = \mathbf{0}$, since α and β are constants that do not affect $\rho_{H_qI_q}$, we can write I_q and H_q as $I_q = \mathbf{b}'\mathbf{y} + \mathbf{y}'\mathbf{B}\mathbf{y}$ and $H_q = \mathbf{w}'\mathbf{g} + \mathbf{g}'\mathbf{A}\mathbf{g}$. Thus, under the assumption that **y** and **g** have multivariate normal distribution with mean **0** and covariance matrix **P** and **G**, respectively, $E(I_q) = tr(\mathbf{BP})$ and $E(H_q) = tr(\mathbf{AG})$ are the expectations of I_q and H_q, whereas $Var(I_q) = \mathbf{b}'\mathbf{P}\mathbf{b} + 2tr[(\mathbf{BP})^2]$ and $Var(H_q) = \mathbf{w}'\mathbf{G}\mathbf{w} + 2tr[(\mathbf{AG})^2]$ are the variances of I_q and H_q, respectively. The covariance between I_q and H_q is $Cov(H_q, I_q) = \mathbf{w}'\mathbf{G}\mathbf{b} + 2tr(\mathbf{BGAG})$ (Vandepitte 1972), where $tr(\circ)$ denotes the trace function of matrices.

According to the foregoing results, we can maximize the natural logarithm of $\rho_{H_qI_q}$ [$\ln\left(\rho_{H_qI_q}\right)$] with respect to vector **b** and matrix **B** assuming that **w**,**A**,**P**, and **G** are known. Hence, except for two proportional constants that do not affect the maximum value of $\rho_{H_qI_q}$ because this is invariant to the scale change, the results of the derivatives of $\ln\left(\rho_{H_qI_q}\right)$ with respect to **b** and **B** are

$$\mathbf{b} = \mathbf{P}^{-1}\mathbf{G}\mathbf{w} \text{ and } \mathbf{B} = \mathbf{P}^{-1}\mathbf{G}\mathbf{A}\mathbf{G}\mathbf{P}^{-1}, \tag{2.27}$$

respectively. In this case, $\mathbf{b} = \mathbf{P}^{-1}\mathbf{G}\mathbf{w}$ is the same as the LPSI vector of coefficients (see Eq. 2.8 for details); however, when $\boldsymbol{\mu} \neq \mathbf{0}$, $\mathbf{b} = \mathbf{P}^{-1}\mathbf{G}(\mathbf{w} + 2\mathbf{A}\boldsymbol{\mu}) = \mathbf{P}^{-1}\mathbf{G}\mathbf{w} + 2\mathbf{P}^{-1}\mathbf{G}\mathbf{A}\boldsymbol{\mu}$. In the latter case, **b** has the additional term $2\mathbf{P}^{-1}\mathbf{G}\mathbf{A}\boldsymbol{\mu}$, which is null when $\boldsymbol{\mu} = \mathbf{0}$ or $\mathbf{A} = \mathbf{0}$. Hence, when $\boldsymbol{\mu} \neq \mathbf{0}$ the quadratic index vector **b** shall have two components: $\mathbf{P}^{-1}\mathbf{G}\mathbf{w}$, which is the LPSI vector of coefficients, and $2\mathbf{P}^{-1}\mathbf{G}\mathbf{A}\boldsymbol{\mu}$, which is a function of the current population mean $\boldsymbol{\mu}$ multiplied by matrix **A**. Therefore, when $\boldsymbol{\mu} \neq \mathbf{0}$ and $\mathbf{A} \neq \mathbf{0}$, the quadratic index vector **b** will change when the $\boldsymbol{\mu}$ values change. However, $\boldsymbol{\mu}$ does not affect matrix **B**.

2.9.4 *The Accuracy and Maximized Selection Response of the Quadratic Index*

According to Eq. (2.27) results, $Var(I_q) = Cov(H_q, I_q) = \mathbf{b}'\mathbf{P}\mathbf{b} + 2tr[(\mathbf{B}\mathbf{P})^2]$, which means that the quadratic index accuracy and the maximized selection response can be written as:

$$\rho_{H_qI_q} = \frac{\sqrt{\mathbf{w}'\mathbf{G}\mathbf{P}^{-1}\mathbf{G}\mathbf{w} + 2tr\left[\left(\mathbf{P}^{-1}\mathbf{G}\mathbf{A}\mathbf{G}\right)^2\right]}}{\sqrt{\mathbf{w}'\mathbf{G}\mathbf{w} + 2tr\left[(\mathbf{A}\mathbf{G})^2\right]}} \tag{2.28}$$

and

$$R_q = k\sqrt{\mathbf{w}'\mathbf{G}\mathbf{P}^{-1}\mathbf{G}\mathbf{w} + 2tr\left[\left(\mathbf{P}^{-1}\mathbf{G}\mathbf{A}\mathbf{G}\right)^2\right]}, \tag{2.29}$$

respectively, where k is the selection intensity of the quadratic index. Equations (2.27) to (2.29) indicate that the LPSI and the quadratic index are related, and the only difference between them is the quadratic terms. Wilton et al. (1968) wrote Eq. (2.29) as: $R_q = k\sqrt{\mathbf{b}'\mathbf{P}\mathbf{b}} + k2tr\left[(\mathbf{B}\mathbf{P})^2\right]$.

References

Akbar MK, Lin CY, Gyles NR, Gavora JS, Brown CJ (1984) Some aspects of selection indices with constraints. Poult Sci 63:1899–1905

Anderson TW (2003) An introduction to multivariate statistical analysis, 3rd edn. Wiley, Hoboken, NJ

Baker RJ (1986) Selection indices in plant breeding. CRC Press, Boca Raton, FL

Blasco A (2001) The Bayesian controversy in animal breeding. J Anim Sci 79:2023–2046

Brascamp EW (1984) Selection indices with constraints. Anim Breed Abstr 52(9):645–654

Bulmer MG (1980) The mathematical theory of quantitative genetics. Lectures in biomathematics. University of Oxford, Clarendon Press, Oxford

Cartuche L, Pascual M, Gómez EA, Blasco A (2014) Economic weights in rabbit meat production. World Rabbit Sci 22:165–177

Cerón-Rojas JJ, Crossa J, Sahagún-Castellanos J, Castillo-González F, Santacruz-Varela A (2006) A selection index method based on eigen analysis. Crop Sci 46:1711–1721

Ceron-Rojas JJ, Crossa J, Arief VN, Basford K, Rutkoski J, Jarquín D, Alvarado G, Beyene Y, Semagn K, DeLacy I (2015) A genomic selection index applied to simulated and real data. Genes/Genomes/Genetics 5:2155–2164

Cochran WG (1951) Improvement by means of selection. In: Neyman J (ed) Proceedings of the second Berkeley symposium on mathematical statistics and probability. University of California Press, Berkeley, CA, pp 449–470

Costa MM, Di Mauro AO, Unêda-Trevisoli SH, Castro Arriel NH, Bárbaro IM, Dias da Silveira G, Silva Muniz FR (2008) Analysis of direct and indirect selection and indices in soybean segregating populations. Crop Breed Appl Biotechnol 8:47–55

Cotterill PP, Jackson N (1985) On index selection. I. Methods of determining economic weight. Silvae Genet 34:56–63

Cunningham EP, Tauebert H (2009) Measuring the effect of change in selection indices. J Dairy Sci 92:6192–6196

Falconer DS, Mackay TFC (1996) Introduction to quantitative genetics. Longman, New York

Fry JD (2004) Estimation of genetic variances and covariance by restricted maximum likelihood using PROC MIXED. In: Saxto AM (ed) Genetics analysis of complex traits using SAS. SAS Institute, Cary, NC, pp 11–34

Goddard ME (1983) Selection indices for non-linear profit functions. Theor Appl Genet 64:339–344

Haldane JBS (1919) The combination of linkage values and the calculation of distance between the loci of linked factors. J Genet 8:299–309

Hallauer AR, Carena MJ, Miranda Filho JB (2010) Quantitative genetics in Maize breeding. Springer, New York

Harris DL (1964) Expected and predicted progress from index selection involving estimates of population parameters. Biometrics 20(1):46–72

Hattaway JT (2010) Parameter estimation and hypothesis testing for the truncated normal distribution with applications to introductory statistics grades. All theses and dissertations. Paper 2053

Hayes JF, Hill WG (1980) A reparameterization of a genetic selection index to locate its sampling properties. Biometrics 36(2):237–248

Hayes JF, Hill WG (1981) Modification of estimates of parameters in the construction of genetic selection indices ('Bending'). Biometrics 37(3):483–493

Hazel LN (1943) The genetic basis for constructing selection indexes. Genetics 8:476–490

Hazel LN, Lush JL (1942) The efficiency of three methods of selection. J Hered 33:393–399

Henderson CR (1963) Selection index and expected genetic advance. In: Statistical genetics and plant breeding. National Academy of Science-National Research Council, Washington, DC, pp 141–163

Kabakoff RI (2011) R in action: data analysis and graphics with R. Manning Publications Co., Shelter Island, NY

Kempthorne O, Nordskog AW (1959) Restricted selection indices. Biometrics 15:10–19

Lin CY (1978) Index selection for genetic improvement of quantitative characters. Theor Appl Genet 52:49–56

Lin CY, Allaire FR (1977) Heritability of a linear combination of traits. Theor Appl Genet 51:1–3

Lynch M, Walsh B (1998) Genetics and analysis of quantitative traits. Sinauer, Sunderland, MA

MacNeil MD, Nugent RA, Snelling WM (1997) Breeding for profit: an Introduction to selection index concepts. Range Beef Cow Symposium. Paper 142
Magnussen S (1990) Selection index: economic weights for maximum simultaneous genetic gains. Theor Appl Genet 79:289–293
Mather K, Jinks JL (1971) Biometrical genetics: the study of continuous variation. Chapman and Hall, London
Morrissey MB, Kruuk LEB, Wilson AJ (2010) The danger of applying the breeder's equation in observational studies of natural populations. J Evol Biol 23:2277–2288
Mrode RA (2005) Linear models for the prediction of animal breeding values, 2nd edn. CABI Publishing, Cambridge, MA
Nordskog AW (1978) Some statistical properties of an index of multiple traits. Theor Appl Genet 52:91–94
Podlich DW, Cooper M (1998) QU-GENE: a simulation platform for quantitative analysis of genetic models. Bioinformatics 14:632–653
Rao CR (2002) Linear statistical inference and its applications, 2nd edn. Wiley, New York
Searle S, Casella G, McCulloch CE (2006) Variance components. Wiley, Hoboken, NJ
Smith HF (1936) A discriminant function for plant selection. In: Papers on quantitative genetics and related topics. Department of Genetics, North Carolina State College, Raleigh, NC, pp 466–476
Smith O, Hallauer AR, Russell WA (1981) Use of index selection in recurrent selection programs in maize. Euphytica 30:611–618
Tomar SS (1983) Restricted selection index in animal system: a review. Agric Rev 4(2):109–118
Vandepitte WM (1972) Factors affecting the accuracy of selection indexes for the genetic improvement of pigs. Retrospective Theses and Dissertations. Paper 6130
Wang J, van Ginkel M, Podlich DW, Ye G, Trethowan R, Pfeiffer W, DeLacy IH, Cooper M, Rajaram S (2003) Comparison of two breeding strategies by computer simulation. Crop Sci 43:1764–1773
Weller JI, Pasternak H, Groen AF (1996) Selection indices for non-linear breeding objectives, selection for optima. In: Proceedings international workshop on genetic improvement of functional traits in Cattle, Gembloux, January 1996. INTERBULL bulletin no. 12, pp 206–214
Williams JS (1962a) The evaluation of a selection index. Biometrics 18:375–393
Williams JS (1962b) Some statistical properties of a genetic selection. Biometrika 49(3):325–337
Wilton JW, Evans DA, Van Vleck LD (1968) Selection indices for quadratic models of total merit. Biometrics 24:937–949
Wilton JW (1968) Selection of dairy cows for economic merit. Faculty Papers and Publications in Animal Science, Paper 420
Wilton JW, Van Vleck LD (1969) Sire evaluation for economic merit. Faculty Papers and Publications in Animal Science, Paper 425
Xu S, Muir WM (1992) Selection index updating. Theor Appl Genet 83:451–458

3

RLPSI, PPG-LPSI and DG-LPSI: Theory and Practice

Abstract The linear phenotypic selection index (LPSI), the null restricted LPSI (RLPSI), and the predetermined proportional gains LPSI (PPG-LPSI) are the main phenotypic selection indices used to predict the net genetic merit and select parents for the next selection cycle. The LPSI is an unrestricted index, whereas the RLPSI and the PPG-LPSI allow restrictions equal to zero and predetermined proportional gain restrictions respectively to be imposed on the expected genetic gain values of the trait to make some traits change their mean values based on a predetermined level while the rest of the trait means remain without restrictions. One additional restricted index is the desired gains LPSI (DG-LPSI), which does not require economic weights and, in a similar manner to the PPG-LPSI, allows restrictions to be imposed on the expected genetic gain values of the trait to make some traits change their mean values based on a predetermined level. The aims of RLPSI and PPG-LPSI are to maximize the selection response, the expected genetic gains per trait, and provide the breeder with an objective rule for evaluating and selecting parents for the next selection cycle based on several traits. This chapter describes the theory and practice of the RLPSI, PPG-LPSI, and DG-LPSI. We show that the PPG-LPSI is the most general index and includes the LPSI and the RLPSI as particular cases. Finally, we describe the DG-LPSI as a modification of the PPG-LPSI. We illustrate the theoretical results of all the indices using real and simulated data.

3.1 The Null Restricted Linear Phenotypic Selection Index

Conditions to construct a valid null restricted linear phenotypic selection index (RLPSI) are the same as those described in Sect. 2.1 of Chap. 2. The main objective of the RLPSI is to optimize, under some null restrictions, the selection response, to predict the net genetic merit $H = \mathbf{w}'\mathbf{g}$ and select the individuals with the highest net genetic merit values as parents of the next generation. The RLPSI allows restrictions equal to zero to be imposed on the expected genetic gains of some traits, whereas other traits increase (or decrease) their expected genetic gains without imposing any restrictions. The RLPSI solves the LPSI equations subject to the condition that the covariance between the index and some linear functions of the genotypes involved be zero, thus preventing selection on the RLPSI from causing any genetic change in some expected genetic gains of the traits (Cunningham et al. 1970).

Vector $\mathbf{b} = \mathbf{P}^{-1}\mathbf{G}\mathbf{w}$ maximizes the LPSI selection response, expected genetic gains per trait, and the correlation between the LPSI and $H = \mathbf{w}'\mathbf{g}$. In this section, we show that the vector of the RLPSI coefficients, $\mathbf{b}_R = \mathbf{K}\mathbf{b}$:

1. Maximizes the RLPSI selection response.
2. Impose null restrictions on the RLPSI expected genetic gains per trait (or multi-trait selection response).
3. Maximizes the correlation with the true net genetic merit.
4. Minimizes the mean prediction error.

Vector $\mathbf{b}_R = \mathbf{K}\mathbf{b}$ is a linear transformation of the LPSI vector of coefficients ($\mathbf{b}$) made by the projector matrix $\mathbf{K}$. Matrix $\mathbf{K}$ is idempotent ($\mathbf{K} = \mathbf{K}^2$) and projects $\mathbf{b}$ into a space smaller than the original space of $\mathbf{b}$ because the restrictions imposed on the expected genetic gains per trait are equal to zero. The reduction of the space into which matrix $\mathbf{K}$ projects $\mathbf{b}$ is equal to the number of null restrictions imposed by the breeder on the expected genetic gain per trait, or multi-trait selection response (Cerón-Rojas et al. 2016).

The covariance between the breeding value vector ($\mathbf{g}$) and the LPSI ($I = \mathbf{b}'\mathbf{y}$) is $Cov(I, \mathbf{g}) = \mathbf{G}\mathbf{b}$. Suppose that the breeder is interested in improving only $(t - r)$ of t $(r < t)$ traits, leaving r of them fixed, that is, r expected genetic gains of the trait are equal to zero for a specific selection cycle. Thus, we want r covariances between the linear combinations of $\mathbf{g}$ ($\mathbf{U}'\mathbf{g}$) and the $I = \mathbf{b}'\mathbf{y}$ to be zero, i.e., $Cov(I, \mathbf{U}'\mathbf{g}) = \mathbf{U}'\mathbf{G}\mathbf{b} = \mathbf{0}$, where $\mathbf{U}'$ is a matrix with r 1's and $(t - r)$ 0's; 1 indicates that the trait is restricted and 0 that the trait is not restricted. That is, in the linear combinations of $\mathbf{g}$ ($\mathbf{U}'\mathbf{g}$), 1 is the coefficient of the genotypes that have covariance equal to zero with the LPSI, whereas the genotypes with coefficient 0 have no restriction on the expected genetic gains. We can solve this problem by maximizing the correlation between I and H (ρ_{HI}) or minimizing the mean squared difference between I and H($E[(H - I)^2]$) under the restriction $\mathbf{U}'\mathbf{G}\mathbf{b} = \mathbf{0}$.

3.1.1 *The Maximized RLPSI Parameters*

In the LPSI context, vector $\mathbf{b} = \mathbf{P}^{-1}\mathbf{G}\mathbf{w}$ minimizes the mean squared difference between I and H, $E[(H - I)^2] = \mathbf{w}'\mathbf{G}\mathbf{w} + \mathbf{b}'\mathbf{P}\mathbf{b} - 2\mathbf{w}'\mathbf{G}\mathbf{b}$. Let $\mathbf{C}' = \mathbf{U}'\mathbf{G}$ and $\mathbf{C}'\mathbf{b} = \mathbf{0}$; we need to minimize $E[(H - I)^2]$ with respect to $\mathbf{b}$ under the restriction $\mathbf{C}'\mathbf{b} = \mathbf{0}$. Thus, assuming that $\mathbf{P}$, $\mathbf{G}$, $\mathbf{U}'$ and $\mathbf{w}$ are known, we need to minimize the function

$$\Psi(\mathbf{b}, \mathbf{v}) = \mathbf{b}'\mathbf{P}\mathbf{b} + \mathbf{w}'\mathbf{G}\mathbf{w} - 2\mathbf{w}'\mathbf{G}\mathbf{b} + 2\mathbf{v}'\mathbf{C}'\mathbf{b} \tag{3.1}$$

with respect to vectors $\mathbf{b}$ and $\mathbf{v}' = [v_1 \ v_2 \ \cdots \ v_{r-1}]$, where $\mathbf{v}$ is a vector of Lagrange multipliers. The derivative results from $\mathbf{b}$ and $\mathbf{v}'$ are

$$\mathbf{Pb} + \mathbf{Cv} = \mathbf{Gw}$$

and

$$\mathbf{C'b} = \mathbf{0},$$

or, in matrix notation,

$$\begin{bmatrix} \mathbf{P} & \mathbf{C} \\ \mathbf{C'} & \mathbf{0} \end{bmatrix}\begin{bmatrix} \mathbf{b} \\ \mathbf{v} \end{bmatrix} = \begin{bmatrix} \mathbf{Gw} \\ \mathbf{0} \end{bmatrix} \quad \text{or} \quad \begin{bmatrix} \mathbf{0} & \mathbf{C'} \\ \mathbf{C} & \mathbf{P} \end{bmatrix}\begin{bmatrix} \mathbf{v} \\ \mathbf{b} \end{bmatrix} = \begin{bmatrix} \mathbf{0} \\ \mathbf{Gw} \end{bmatrix}. \tag{3.2}$$

In the latter case of Eq. (3.2), the solution is

$$\begin{bmatrix} \mathbf{v} \\ \mathbf{b}_R \end{bmatrix} = \begin{bmatrix} \mathbf{0} & \mathbf{C'} \\ \mathbf{C} & \mathbf{P} \end{bmatrix}^{-1}\begin{bmatrix} \mathbf{0} \\ \mathbf{Gw} \end{bmatrix}, \tag{3.3}$$

where $\begin{bmatrix} \mathbf{0} & \mathbf{C'} \\ \mathbf{C} & \mathbf{P} \end{bmatrix}^{-1}$ is the inverse of matrix $\begin{bmatrix} \mathbf{0} & \mathbf{C'} \\ \mathbf{C} & \mathbf{P} \end{bmatrix}$ and $\mathbf{b}_R$ is the RLPSI vector of coefficients. There is a mathematical algorithm (Searle 1966; Schott 2005) for finding matrix $\begin{bmatrix} \mathbf{0} & \mathbf{C'} \\ \mathbf{C} & \mathbf{P} \end{bmatrix}^{-1}$. It can be shown that

$$\begin{bmatrix} \mathbf{0} & \mathbf{C'} \\ \mathbf{C} & \mathbf{P} \end{bmatrix}^{-1} = \begin{bmatrix} (-\mathbf{C'P}^{-1}\mathbf{C})^{-1} & (\mathbf{C'P}^{-1}\mathbf{C})^{-1}\mathbf{C'P}^{-1} \\ \mathbf{P}^{-1}\mathbf{C}(\mathbf{C'P}^{-1}\mathbf{C})^{-1} & -\mathbf{P}^{-1}\mathbf{C}(\mathbf{C'P}^{-1}\mathbf{C})^{-1}\mathbf{C'P}^{-1} + \mathbf{P}^{-1} \end{bmatrix}, \tag{3.4}$$

whence the RLPSI vector of coefficients ($\mathbf{b}_R$) that minimizes $E[(H - I)^2]$ and maximizes ρ_{HI} under the restriction $\mathbf{C}'\mathbf{b} = \mathbf{0}$ can be written as

$$\mathbf{b}_R = \mathbf{Kb}, \tag{3.5}$$

where $\mathbf{K} = [\mathbf{I} - \mathbf{Q}]$, $\mathbf{Q} = \mathbf{P}^{-1}\mathbf{C}(\mathbf{C}'\mathbf{P}^{-1}\mathbf{C})^{-1}\mathbf{C}'$ and $\mathbf{b} = \mathbf{P}^{-1}\mathbf{Gw}$; $\mathbf{P}^{-1}$ is the inverse of matrix $\mathbf{P}$ and $\mathbf{I}$ is an identity matrix $t \times t$. When there are no restrictions on any traits, $\mathbf{U}'$ is a null matrix and $\mathbf{b}_R = \mathbf{b} = \mathbf{P}^{-1}\mathbf{Gw}$, the LPSI vector of coefficients. Thus, the RLPSI includes the LPSI as a particular case.

According to Eq. (3.5), the RLPSI can be written as

$$I_R = \mathbf{b}'_R\mathbf{y}, \tag{3.6}$$

whereas the maximized correlation between the RLPSI and the net genetic merit is

$$\rho_{HI_R} = \frac{\mathbf{w'Gb}_R}{\sqrt{\mathbf{w'Gw}}\sqrt{\mathbf{b}'_R\mathbf{Pb}_R}}. \tag{3.7}$$

According to conditions for constructing a valid RLPSI, the index $I_R = \mathbf{b}'_R\mathbf{y}$ should have normal distributions. Using 1 and 2 null restrictions, this assumption is illustrated in Fig. 3.1 for a real maize (*Zea mays*) F_2 population with 247 lines and

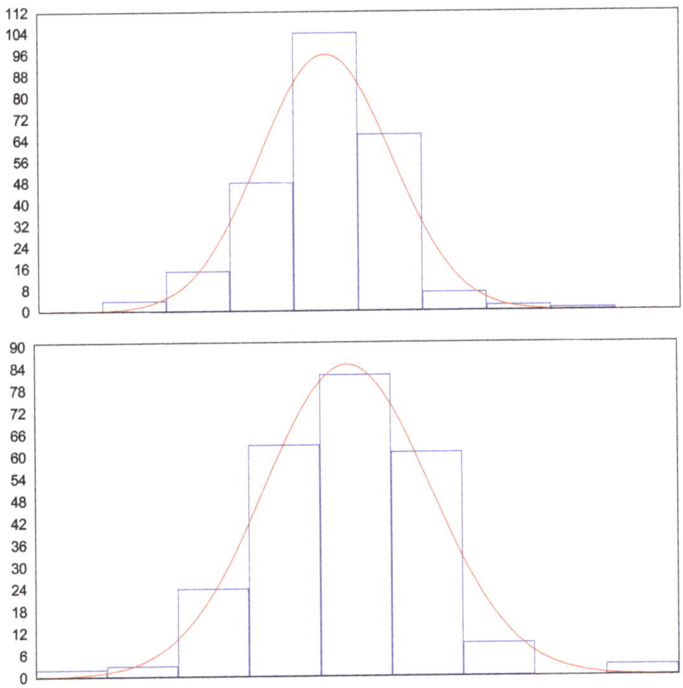

Fig. 3.1 (**a**) and (**b**) show the distributions of 247 values of the restricted linear phenotypic selection index (RLPSI), with one and two restrictions respectively, constructed with the phenotypic means of four maize (*Zea mays*) F_2 population traits: grain yield (ton ha^{-1}), plant height (cm), ear height (cm), and anthesis day (days), evaluated in one environment

four traits—grain yield (ton ha^{-1}); plant height (cm), ear height (cm), and anthesis day (days)—evaluated in one environment. Figure 3.1 indicates that, in effect, the RLPSI values approach normal distribution.

Under the null restrictions made by the breeder, $I_R = \mathbf{b}_R'\mathbf{y}$ should have maximum correlation with $H = \mathbf{w}'\mathbf{g}$ and should be useful for ranking and selecting among individuals with different net genetic merit; however, ρ_{HI_R} is lower than the correlation between LPSI and $H = \mathbf{w}'\mathbf{g}$ (ρ_{HI}) in each selection cycle because when the restriction $\mathbf{C}'\mathbf{b} = \mathbf{0}$ is imposed on the RLPSI vector of coefficients, the restricted traits do not affect the correlation ρ_{HI_R}. Using simulated data described in Sect. 2.8.1 of Chap. 2, we estimated ρ_{HI_R} and ρ_{HI} for seven selection cycles and compared the results in Fig. 3.2. Correlation ρ_{HI_R} values were estimated for one, two, and three null restrictions and in effect, they were lower than the estimated values of ρ_{HI} in all selection cycles (Fig. 3.2). Additional results can be seen in Chap. 10, where the RLPSI was simulated for many selection cycles. Chapter 11 describes RIndSel: a program that uses R (in this case R denotes a platform for data analysis, see Kabakoff 2011 for details) and the selection index theory to select individual candidates for selection.

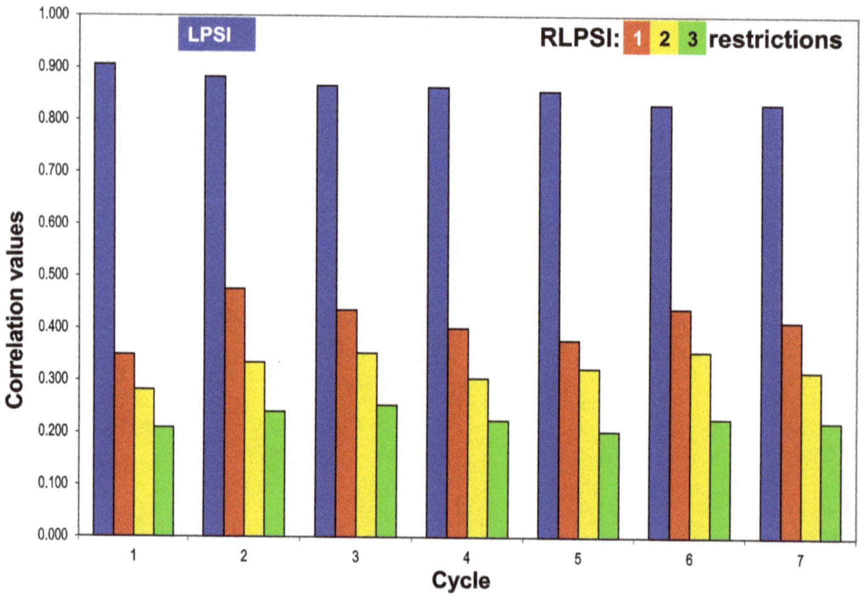

Fig. 3.2 Estimated correlation values between the linear phenotypic selection index (LPSI) and the net genetic merit ($H = \mathbf{w}'\mathbf{g}$); estimated correlation values between the RLPSI and H for one (*red*), two (*yellow*), and three (*green*) restrictions for four traits and 500 genotypes in one environment simulated for seven selection cycles

The maximized RLPSI selection response and the restricted expected genetic gain per trait can be written as

$$R_{\mathrm{R}} = k_I \sqrt{\mathbf{b}_R' \mathbf{P} \mathbf{b}_R} \tag{3.8}$$

and

$$\mathbf{E}_R = k_I \frac{\mathbf{G}\mathbf{b}_R}{\sqrt{\mathbf{b}_R' \mathbf{P} \mathbf{b}_R}}, \tag{3.9}$$

respectively, where k_I is the standardized selection differential or selection intensity associated with the RLPSI.

The maximized RLPSI selection response has the same form as the maximized LPSI selection response; thus, under r restrictions, Eq. (3.8) predicts the mean improvement in H owing to indirect selection on $I_R = \mathbf{b}_R' \mathbf{y}$ when $\mathbf{b}_R = \mathbf{K}\mathbf{b}$. The restriction effects are observed on the RLPSI expected genetic gains per trait (Eq. 3.9) where each restricted trait has an expected genetic gain equal to zero. In addition, because the RLPSI selection response and expected genetic gain per trait values are also affected by the restricted traits, they are lower than the LPSI selection response and expected genetic gain per trait values.

3.1.2 Statistical Properties of the RLPSI

Under the assumptions that $H = \mathbf{w}'\mathbf{g}$ and $I_R = \mathbf{b}_R'\mathbf{y}$ have a bivariate joint normal distribution, $\mathbf{b}_R = \mathbf{Kb}$, $\mathbf{b} = \mathbf{P}^{-1}\mathbf{Gw}$, and $\mathbf{P}$, $\mathbf{G}$, and $\mathbf{w}$ are known, the RLPSI has the following properties:

1. Matrices $\mathbf{Q} = \mathbf{P}^{-1}\mathbf{C}(\mathbf{C}'\mathbf{P}^{-1}\mathbf{C})^{-1}\mathbf{C}'$ and $\mathbf{K} = [\mathbf{I} - \mathbf{Q}]$ are projectors. That is, $\mathbf{Q}$ and $\mathbf{K}$ are idempotent ($\mathbf{Q} = \mathbf{Q}^2$ and $\mathbf{K} = \mathbf{K}^2$) and orthogonal ($\mathbf{KQ} = \mathbf{QK} = \mathbf{0}$). It can be shown that $\mathbf{Q} = \mathbf{Q}^2$, $\mathbf{K} = \mathbf{K}^2$, and $\mathbf{KQ} = \mathbf{QK} = \mathbf{0}$ noting that $\mathbf{Q}^2 = \mathbf{P}^{-1}\mathbf{C}(\mathbf{C}'\mathbf{P}^{-1}\mathbf{C})^{-1}\mathbf{C}'\mathbf{P}^{-1}\mathbf{C}(\mathbf{C}'\mathbf{P}^{-1}\mathbf{C})^{-1}\mathbf{C}' = \mathbf{P}^{-1}\mathbf{C}(\mathbf{C}'\mathbf{P}^{-1}\mathbf{C})^{-1}\mathbf{C}' = \mathbf{Q}$, $\mathbf{K}^2 = [\mathbf{I} - \mathbf{Q}][\mathbf{I} - \mathbf{Q}] = \mathbf{I} - 2\mathbf{Q} + \mathbf{Q}^2 = \mathbf{I} - \mathbf{Q} = \mathbf{K}$, and $\mathbf{KQ} = \mathbf{QK} = \mathbf{Q} - \mathbf{Q}^2 = \mathbf{0}$.
2. Matrix $\mathbf{Q}$ projects vector $\mathbf{b}$ into a space generated by the columns of matrix $\mathbf{C}$ owing to the restriction $\mathbf{C}'\mathbf{b} = \mathbf{0}$ used when $\Psi(\mathbf{b}, \mathbf{v})$ is maximized with respect to $\mathbf{b}$ and $\mathbf{v}$.
3. Matrix $\mathbf{K}$ projects $\mathbf{b}$ into a space perpendicular to the space generated by the $\mathbf{C}$ matrix columns (Rao 2002).
4. Because of the restriction $\mathbf{C}'\mathbf{b} = \mathbf{0}$, matrix $\mathbf{K}$ projects $\mathbf{b}$ into a space smaller than the original space of $\mathbf{b}$. The space reduction into which matrix $\mathbf{K}$ projects $\mathbf{b}$ is equal to the number of zeros that appears in Eq. (3.9).
5. Vector $\mathbf{b}_R = \mathbf{Kb}$ minimizes the mean square error under the restriction $\mathbf{C}'\mathbf{b} = \mathbf{0}$.
6. The variance of $I_R = \mathbf{b}_R'\mathbf{y}$ ($\sigma^2_{I_R} = \mathbf{b}_R'\mathbf{P}\mathbf{b}_R$) is equal to the covariance between $I_R = \mathbf{b}_R'\mathbf{y}$ and $H = \mathbf{w}'\mathbf{g}$ ($\sigma_{HI_R} = \mathbf{w}'\mathbf{G}\mathbf{b}_R$). First note that $\mathbf{K} = \mathbf{K}^2$, $\mathbf{K}'\mathbf{P} = \mathbf{PK}$, and $\mathbf{b}'\mathbf{P} = \mathbf{w}'\mathbf{G}$; then $\sigma^2_{I_R} = \mathbf{b}_R'\mathbf{P}\mathbf{b}_R = \mathbf{b}'\mathbf{K}'\mathbf{PKb} = \mathbf{b}'\mathbf{PK}^2\mathbf{b} = \mathbf{b}'\mathbf{PKb} = \mathbf{w}'\mathbf{G}\mathbf{b}_R = \sigma_{HI_R}$.
7. The maximized correlation between H and I_R is equal to $\rho_{HI_R} = \frac{\sigma_{I_R}}{\sigma_H}$. In point 6 of this subsection we showed that $\sigma_{HI_R} = \sigma^2_{I_R}$; then
$$\rho_{HI_R} = \frac{\mathbf{w}'\mathbf{G}\mathbf{b}_R}{\sqrt{\mathbf{w}'\mathbf{Gw}}\sqrt{\mathbf{b}_R'\mathbf{P}\mathbf{b}_R}} = \sqrt{\frac{\mathbf{b}_R'\mathbf{P}\mathbf{b}_R}{\mathbf{w}'\mathbf{Gw}}} = \frac{\sigma_{I_R}}{\sigma_H}.$$
8. The variance of the predicted error, $Var(H - I_R) = \left(1 - \rho^2_{HI_R}\right)\sigma^2_H$, is minimal. By point 6 $\sigma_{HI_R} = \sigma^2_{I_R}$, whence $Var(H - I_R) = \sigma^2_H - \sigma^2_{I_R} = \left(1 - \rho^2_{HI_R}\right)\sigma^2_H$.
9. RLPSI heritability is equal to $\mathrm{h}^2_{\mathrm{I_R}} = \dfrac{\mathbf{b}_R'\mathbf{G}\mathbf{b}_R}{\mathbf{b}_R'\mathbf{P}\mathbf{b}_R}$.

Points 1–4 show that in effect, the RLPSI projects the LPSI vector of coefficients into a space smaller than the original LPSI vector of coefficients. In addition, the RLPSI statistical properties denoted by points 5–9 are the same as the LPSI statistical properties. Thus, the RLPSI is a variant of the LPSI.

3.1.3 The RLPSI Matrix of Restrictions

The main difference between the RLPSI and the LPSI is the restriction $\mathbf{U}'\mathbf{Gb} = \mathbf{0}$ used to obtain the RLPSI vector of coefficients. This restriction is introduced through matrix $\mathbf{U}'$ $(t - 1) \times t$, which is called matrix of null restrictions and is very important in an RLPSI context. The form and size of matrix $\mathbf{U}'$ depends on the number of restricted traits. For example, suppose that we restrict only one of t traits; then we can restrict the first of them as $\mathbf{U}' = [1 \quad 0 \quad 0 \quad \cdots \quad 0]$, the second as $\mathbf{U}' = [0 \quad 1 \quad 0 \quad \cdots \quad 0]$, the third as $\mathbf{U}' = [0 \quad 0 \quad 1 \quad \cdots \quad 0]$, etc. When we restrict two of t traits, matrix $\mathbf{U}'$ could be constructed as follows. We can restrict the first and second traits as $\mathbf{U}' = \begin{bmatrix} 1 & 0 & 0 & \cdots & 0 \\ 0 & 1 & 0 & \cdots & 0 \end{bmatrix}$, the first and third traits as $\mathbf{U}' = \begin{bmatrix} 1 & 0 & 0 & \cdots & 0 \\ 0 & 0 & 1 & \cdots & 0 \end{bmatrix}$, the second and third traits as $\mathbf{U}' = \begin{bmatrix} 0 & 1 & 0 & \cdots & 0 \\ 0 & 0 & 1 & \cdots & 0 \end{bmatrix}$, etc. If we restrict three of t traits, matrix $\mathbf{U}'$ will have the following form when the first, second, and third traits are restricted, $\mathbf{U}' = \begin{bmatrix} 1 & 0 & 0 & 0 & \cdots & 0 \\ 0 & 1 & 0 & 0 & \cdots & 0 \\ 0 & 0 & 1 & 0 & \cdots & 0 \end{bmatrix}$; if the first, second, and fourth traits are restricted, $\mathbf{U}' = \begin{bmatrix} 1 & 0 & 0 & 0 & \cdots & 0 \\ 0 & 1 & 0 & 0 & \cdots & 0 \\ 0 & 0 & 0 & 1 & \cdots & 0 \end{bmatrix}$, if the second, the third and the fourth traits are restricted, $\mathbf{U}' = \begin{bmatrix} 0 & 1 & 0 & 0 & \cdots & 0 \\ 0 & 0 & 1 & 0 & \cdots & 0 \\ 0 & 0 & 0 & 1 & \cdots & 0 \end{bmatrix}$, etc. The procedure to construct matrix $\mathbf{U}'$ is valid for any number of restricted traits.

There are $\sum_{r=0}^{t} \binom{t}{r} = 2^t$ (Leon-Garcia 2008) possible forms for constructing matrix $\mathbf{U}'$, where $\binom{t}{r} = \frac{t!}{r!(t-r)!}$ and $t! = t(t-1)(t-2)(t-3)\cdots(t-(t-1))$. Note, however, that when $r = 0$, $\mathbf{U}'$ is a null matrix, and when $r = t$, all traits are restricted and then the RLPSI values are null. Thus, the breeder should be interested only in $2^t - 2$ possible ways of constructing matrix $\mathbf{U}'$.

3.1.4 Numerical Examples

To illustrate the RLPSI theoretical results, we use the data set described in Sect. 2.8.1 of Chap. 2. We used that data set for seven phenotypic selection cycles (C1 to C7), each with four traits (T_1, T_2, T_3 and T_4), 500 genotypes and four replicates for each genotype. The economic weights for T_1, T_2, T_3, and T_4 were 1, -1, 1, and 1 respectively. The estimated phenotypic ($\widehat{\mathbf{P}}$) and genetic ($\widehat{\mathbf{G}}$) covariance matrices for traits T_1, T_2, T_3, and T_4 obtained for the first selection cycle (C1) of the simulated data were

$$\widehat{\mathbf{P}} = \begin{bmatrix} 62.50 & -12.74 & 8.53 & 2.73 \\ -12.74 & 17.52 & -3.38 & -2.28 \\ 8.53 & -3.38 & 12.31 & 0.16 \\ 2.73 & -2.28 & 0.16 & 7.27 \end{bmatrix} \text{ and}$$

$$\widehat{\mathbf{G}} = \begin{bmatrix} 36.21 & -12.93 & 8.35 & 2.74 \\ -12.93 & 13.04 & -3.40 & -2.24 \\ 8.35 & -3.40 & 9.96 & 0.16 \\ 2.74 & -2.24 & 0.16 & 6.64 \end{bmatrix},$$

respectively. We can restrict T_1 with matrix $\mathbf{U}_1' = [1 \quad 0 \quad 0 \quad 0]$; T_1 and T_2 with matrix $\mathbf{U}_2' = \begin{bmatrix} 1 & 0 & 0 & 0 \\ 0 & 1 & 0 & 0 \end{bmatrix}$, and T_1, T_2 and T_3 with matrix $\mathbf{U}_3' = \begin{bmatrix} 1 & 0 & 0 & 0 \\ 0 & 1 & 0 & 0 \\ 0 & 0 & 1 & 0 \end{bmatrix}$.

Matrix $\mathbf{C}' = \mathbf{U}'\widehat{\mathbf{G}}$ associated with $\mathbf{U}_1'$, $\mathbf{U}_2'$, and $\mathbf{U}_3'$ can be obtained as $\mathbf{C}_1' = \mathbf{U}_1'\mathbf{G} = [36.21 \quad -12.93 \quad 8.35 \quad 2.74]$,

$$\mathbf{C}_2' = \mathbf{U}_2'\widehat{\mathbf{G}} = \begin{bmatrix} 36.21 & -12.93 & 8.35 & 2.74 \\ -12.93 & 13.04 & -3.40 & -2.24 \end{bmatrix}, \text{ and}$$

$$\mathbf{C}_3' = \mathbf{U}_3'\widehat{\mathbf{G}} = \begin{bmatrix} 36.21 & -12.93 & 8.35 & 2.74 \\ -12.93 & 13.04 & -3.04 & -2.24 \\ 8.35 & -3.40 & 9.96 & 0.16 \end{bmatrix}.$$

The estimated LPSI vector of coefficients was $\widehat{\mathbf{b}}' = \mathbf{w}'\widehat{\mathbf{G}}\widehat{\mathbf{P}}^{-1} = [0.55 \quad -1.05 \quad 1.09 \quad 1.06]$.

The estimated matrices $\widehat{\mathbf{Q}} = \widehat{\mathbf{P}}^{-1}\mathbf{C}(\mathbf{C}'\widehat{\mathbf{P}}^{-1}\mathbf{C})^{-1}\mathbf{C}'$ and $\widehat{\mathbf{K}} = [\mathbf{I}_4 - \widehat{\mathbf{Q}}]$ (where $\mathbf{I}_4$ is an identity matrix 4 × 4) for 1 null restriction, were

$$\widehat{\mathbf{Q}}_1 = \widehat{\mathbf{P}}^{-1}\mathbf{C}_1(\mathbf{C}_1'\widehat{\mathbf{P}}^{-1}\mathbf{C}_1)^{-1}\mathbf{C}_1' = \begin{bmatrix} 0.72 & -0.26 & 0.17 & 0.05 \\ -0.51 & 0.18 & -0.12 & -0.04 \\ 0.39 & -0.14 & 0.09 & 0.03 \\ 0.14 & -0.05 & 0.03 & 0.01 \end{bmatrix} \text{ and}$$

$$\widehat{\mathbf{K}}_1 = [\mathbf{I}_4 - \widehat{\mathbf{Q}}_1] = \begin{bmatrix} 0.28 & 0.26 & -0.17 & -0.05 \\ 0.51 & 0.82 & 0.12 & 0.04 \\ -0.39 & 0.14 & 0.91 & -0.03 \\ -0.14 & 0.05 & -0.03 & 0.99 \end{bmatrix}.$$

Thus, the estimated RLPSI vector of coefficients was $\widehat{\mathbf{b}}_{R_1}' = (\widehat{\mathbf{K}}_1\widehat{\mathbf{b}})' = [-0.35 \quad -0.41 \quad 0.59 \quad 0.89]$, whence the estimated RLPSI for 1 null restriction can be written as $\widehat{I}_{R_1} = -0.35T_1 - 0.41T_2 + 0.59T_3 + 0.89T_4$. The average values of T_1, T_2, T_3, and T_4 were 164.46, 39.63, 34.66, and 23.11 (Table 3.1) respectively; then,

Table 3.1 Ten genotypes, mean values of four traits, and unranked and ranked values of the restricted linear phenotypic selection index (RLPSI) obtained from 500 simulated genotypes (each with four repetitions) and four traits (T1, T2, T3, and T4) in one environment for one selection cycle

	Means of the trait values				
Number of genotypes	T1	T2	T3	T4	RLPSI values
1	164.46	39.63	34.66	23.11	−33.24 (unranked)
2	144.39	50.77	34.65	19.56	−33.94 (unranked)
3	157.48	48.04	37.9	19.03	−35.96 (unranked)
4	167.3	47.98	30.49	24.75	−38.73 (unranked)
5	164.11	49.89	32.03	25.32	−36.98 (unranked)
6	166.26	40.44	29.93	20.55	−39.29 (unranked)
7	154.59	52.22	30.31	18.86	−41.33 (unranked)
8	160	42.91	31.23	20.95	−36.98 (unranked)
9	158.51	46.32	34.52	18.36	−38.2 (unranked)
10	163.63	45.43	35.73	19.57	−37.85 (unranked)
1	164.46	39.63	34.66	23.11	−33.24 (ranked)
2	144.39	50.77	34.65	19.56	−33.94 (ranked)
3	157.48	48.04	37.9	19.03	−35.96 (ranked)
5	164.11	49.89	32.03	25.32	−36.98 (ranked)
8	160	42.91	31.23	20.95	−36.98 (ranked)
10	163.63	45.43	35.73	19.57	−37.85 (ranked)
9	158.51	46.32	34.52	18.36	−38.2 (ranked)
4	167.3	47.98	30.49	24.75	−38.73 (ranked)
6	166.26	40.44	29.93	20.55	−39.29 (ranked)
7	154.59	52.22	30.31	18.86	−41.33 (ranked)

$$\widehat{I}_{R_1} = -0.35(164.46) - 0.41(39.63) + 0.59(34.66) + 0.89(23.11) = -33.24.$$

In Table 3.1 we present ten genotypes, the mean values of four traits, and the unranked and ranked values of the RLPSI from 500 genotypes in one environment simulated for one selection cycle. The first part of Table 3.1 presents the ten unranked genotypes, whereas the second part presents the ten genotypes ranked by the estimated RLPSI values.

Assuming a selection intensity of 10% ($k_I = 1.755$), the estimated selection response and the estimated expected genetic gain per trait for 1 null restriction were $\widehat{R}_{R_1} = 1.755\sqrt{\widehat{\mathbf{b}}'_{R_1}\widehat{\mathbf{P}}\widehat{\mathbf{b}}_{R_1}} = 6.87$ and $\widehat{\mathbf{E}}'_{R_1} = 1.755\frac{\widehat{\mathbf{b}}'_{R_1}\widehat{\mathbf{G}}}{\sqrt{\widehat{\mathbf{b}}'_{R_1}\widehat{\mathbf{P}}\widehat{\mathbf{b}}_{R_1}}} = [0 \quad -2.2 \quad 2.03 \quad 2.66]$, respectively, and the estimated correlation between the RLPSI and the net genetic merit was $\widehat{\rho}_{HI_{R_1}} = \sqrt{\frac{\widehat{\mathbf{b}}'_{R_1}\widehat{\mathbf{P}}\widehat{\mathbf{b}}_{R_1}}{\mathbf{w}'\widehat{\mathbf{G}}\mathbf{w}}} = 0.35.$

In a similar manner to that for 1 null restriction, it is possible to obtain the estimated matrices $\widehat{\mathbf{Q}} = \widehat{\mathbf{P}}^{-1}\mathbf{C}\left(\mathbf{C}'\widehat{\mathbf{P}}^{-1}\mathbf{C}\right)^{-1}\mathbf{C}'$ and $\widehat{\mathbf{K}} = \left[\mathbf{I}_4 - \widehat{\mathbf{Q}}\right]$, and the estimated

RLPSI vector of coefficients for 2 and 3 null restrictions. Thus, for 2 and 3 null restrictions, the estimated selection responses were $\widehat{R}_{R_2} = 1.755\sqrt{\widehat{\mathbf{b}}'_{R_2}\widehat{\mathbf{P}}\widehat{\mathbf{b}}_{R_2}} = 5.54$ and $\widehat{R}_{R_3} = 1.755\sqrt{\widehat{\mathbf{b}}'_{R_3}\widehat{\mathbf{P}}\widehat{\mathbf{b}}_{R_3}} = 4.12$ respectively, whereas the estimated expected genetic gains per trait were $\widehat{\mathbf{E}}'_{R_2} = 1.755\frac{\widehat{\mathbf{b}}'_{R_2}\mathbf{G}}{\sqrt{\widehat{\mathbf{b}'}_{R_2}\widehat{\mathbf{P}}\widehat{\mathbf{b}}_{R_2}}} = [0 \quad 0 \quad 2.773 \quad 2.768]$ and $\widehat{\mathbf{E}}'_{R_3} = 1.755\frac{\widehat{\mathbf{b}}'_{R_3}\mathbf{G}}{\sqrt{\widehat{\mathbf{b}'}_{R_3}\widehat{\mathbf{P}}\widehat{\mathbf{b}}_{R_3}}} = [0 \quad 0 \quad 0 \quad 4.12]$.

Note that the estimated RLPS selection response decreased when the number of restrictions increased. Also, the number of zeros in the expected genetic gain per trait increased from 1 to 3 depending on the number of null restrictions. The same is true for the estimated correlation between the RLPSI and the net genetic merit (Fig. 3.2).

Table 3.2 presents the estimated LPSI selection response and its heritabilities, and the estimated RLPSI selection response and its heritabilities for 1, 2, and 3 null restrictions for seven simulated selection cycles using a selection intensity of 10% ($k_I = 1.755$). Note that the averages of the estimated RLPSI selection response for the seven selection cycles were 6.76, 5.30, and 3.70 for 1, 2, and 3 null restrictions respectively, and that 3.70, the average value for 3 null restrictions, is only 54.73% of the average value for 1 null restriction (6.76). However, the estimated RLPSI heritabilities for 1, 2, and 3 null restrictions tend to increase. This is because the simulated true heritabilities of traits T_1, T_2, T_3, and T_4 were 0.4, 0.6, 0.6, and 0.8 respectively, whereas the averages of the estimated heritabilities of traits T_1, T_2, T_3, and T_4 were 0.70, 0.78, and 0.87 for 1, 2, and 3 null restrictions respectively.

Table 3.3 presents the estimated LPSI expected genetic gain per trait and the estimated RLPSI expected genetic gain per trait for 1, 2, and 3 null restrictions for

Table 3.2 Estimated linear phenotypic selection index (LPSI) selection response and its heritability, and estimated restricted LPSI (RLPSI) selection response and its heritability for one, two, and three null restrictions for seven simulated selection cycles

			RLPSI					
	LPSI		selection response for one, two, and three restrictions			Heritability for one, two, and three restrictions		
Cycle	selection response	Heritability	1	2	3	1	2	3
1	17.81	0.84	6.87	5.54	4.13	0.65	0.77	0.89
2	15.69	0.80	8.45	5.94	4.27	0.76	0.80	0.90
3	14.22	0.77	7.17	5.79	4.16	0.71	0.80	0.88
4	14.34	0.76	6.68	5.06	3.72	0.71	0.79	0.89
5	13.64	0.75	6.02	5.16	3.24	0.67	0.76	0.86
6	12.04	0.71	6.37	5.17	3.31	0.70	0.79	0.86
7	11.61	0.72	5.77	4.44	3.09	0.68	0.74	0.84
Average	14.19	0.76	6.76	5.30	3.70	0.70	0.78	0.87

The selection intensity was 10% ($k_I = 1.755$)

Table 3.3 Estimated LPSI expected genetic gain per trait, and estimated RLPSI expected genetic gain per trait for one, two, and three null restrictions for seven simulated selection cycles

	LPSI expected gain per trait				RLPSI expected gain per trait for one restriction			
Cycle	T1	T2	T3	T4	T1	T2	T3	T4
1	7.90	−4.67	3.33	1.92	0	−2.18	2.03	2.66
2	7.06	−3.59	3.17	1.86	0	−3.41	2.33	2.71
3	6.67	−3.21	2.82	1.52	0	−2.30	3.12	1.74
4	7.53	−3.45	2.07	1.29	0	−2.88	1.42	2.38
5	7.14	−2.66	2.51	1.33	0	−1.83	2.38	1.81
6	6.23	−2.62	1.98	1.21	0	−2.41	2.09	1.87
7	5.38	−2.55	2.47	1.22	0	−2.24	1.34	2.19
Average	6.85	−3.25	2.62	1.48	0	−2.46	2.10	2.19
	RLPSI expected gain per traits for two restrictions				RLPSI expected gain per traits for three restrictions			
Cycle	T1	T2	T3	T4	T1	T2	T3	T4
1	0	0	2.77	2.77	0	0	0	4.13
2	0	0	2.87	3.07	0	0	0	4.27
3	0	0	3.11	2.68	0	0	0	4.16
4	0	0	2.35	2.70	0	0	0	3.72
5	0	0	3.12	2.04	0	0	0	3.24
6	0	0	2.84	2.33	0	0	0	3.31
7	0	0	2.07	2.37	0	0	0	3.09
Average	0	0	2.73	2.57	0	0	0	3.70

The selection intensity was 10% ($k_I = 1.755$)

seven simulated selection cycles using a selection intensity of 10% ($k_I = 1.755$). In effect, due to the restriction $\mathbf{C}'\mathbf{b} = \mathbf{0}$, matrix $\mathbf{K}$ projects $\mathbf{b}$ into a space smaller than the original space of $\mathbf{b}$ and the space reduction into which matrix $\mathbf{K}$ projects $\mathbf{b}$ is equal to the number of zeros that appear in the RLPSI expected genetic gain per trait.

It can be shown that in the three restrictions case (Table 3.3) the estimated RLPSI expected genetic gain pert traits (or multi-trait selection response) is equal to the one trait selection response (Eqs. 2.4 and 2.5) when only trait T4 is selected. This means that in effect, when we imposed three restriction over the RLPSI expected genetic gains pert trait, we reduced one space of four dimensions to one space of only one dimension.

3.2 The Predetermined Proportional Gains Linear Phenotypic Selection Index

This index is called the *predetermined proportional gains phenotypic selection index* (PPG-LPSI) because the breeder pre-sets optimal levels for certain traits before the selection is carried out. The conditions for constructing a valid PPG-LPSI are the same as those described for the LPSI in Sect. 2.1 of Chap. 2. Some of the main objectives of the PPG-LPSI are to optimize the expected genetic gain per trait, predict the net genetic merit $H = \mathbf{w}'\mathbf{g}$, and select the individuals with the highest net genetic merit values as parents of the next generation. The PPG-LPSI allows restrictions different from zero to be imposed on the expected genetic gains of some traits, whereas other traits increase (or decrease) their expected genetic gains without imposing any restrictions. The PPG-LPSI solves the LPSI equations subject to the condition that the covariance between the LPSI and some linear functions of the genotypes involved be equal to a vector of predetermined constants or genetic gains defined by the breeder (Cunningham et al. 1970).

Let $\mathbf{d}' = [d_1 \quad d_2 \quad \cdots \quad d_r]$ be a vector $r \times 1$ of the predetermined proportional gains and assume that μ_q is the population mean of the qth trait before selection. One objective could be to change μ_q to $\mu_q + d_q$, where d_q is a predetermined change in μ_q (in the RLPSI, $d_q = 0$, $q = 1, 2, \cdots, r$, where r is the number of predetermined proportional gains). We can solve this problem in a similar manner to that used with the RLPSI. That is, minimizing the mean squared difference between I and H(E[(H − I)^2]) under the restriction $\mathbf{D}'\mathbf{U}'\mathbf{G}\mathbf{b} = \mathbf{0}$, where $\mathbf{D}' = \begin{bmatrix} d_r & 0 & \cdots & 0 & -d_1 \\ 0 & d_r & \cdots & 0 & -d_2 \\ \vdots & \vdots & \ddots & \vdots & \vdots \\ 0 & 0 & \cdots & d_r & -d_{r-1} \end{bmatrix}$ is a Mallard (1972) matrix $(r - 1) \times r$ of predetermined proportional gains, d_q ($q = 1, 2..., r$) is the qth element of vector $\mathbf{d}'$, $\mathbf{U}'$ is the RLPSI matrix of restrictions of 1's and 0's described earlier in this chapter, $\mathbf{G}$ is the covariance matrix of genotypic values, and $\mathbf{b}$ is the LPSI vector of coefficients. Also, it is possible to minimize $E[(H - I)^2]$ under the restriction $\mathbf{U}'\mathbf{G}\mathbf{b} = \theta\mathbf{d}$ (Tallis 1985), where θ is a proportionality constant, which is a scalar to be determined a posteriori (Lin 2005), that is, θ is indeterminate a priori (Itoh and Yamada 1987). Both approaches are very similar but the equations obtained when introducing the $\mathbf{D}'\mathbf{U}'\mathbf{G}\mathbf{b} = \mathbf{0}$ restriction are simpler than when introducing $\mathbf{U}'\mathbf{G}\mathbf{b} = \theta\mathbf{d}$ restrictions into the process of minimizing $E[(H - I)^2]$. The $\mathbf{D}'\mathbf{U}'\mathbf{G}\mathbf{b} = \mathbf{0}$ restriction leads to a set of equations similar to Eq. (3.5) whereas the $\mathbf{U}'\mathbf{G}\mathbf{b} = \theta\mathbf{d}$ restriction leads to a set of equations that are difficult to solve.

3.2.1 *The Maximized PPG-LPSI Parameters*

Let $\mathbf{M}' = \mathbf{D}'\mathbf{C}'$ be the Mallard (1972) matrix of predetermined restrictions, where $\mathbf{C}' = \mathbf{U}'\mathbf{G}$. Under the restriction $\mathbf{M}'\mathbf{b} = \mathbf{0}$, we can minimize $E[(I - H)^2]$, assuming that $\mathbf{P}$, $\mathbf{G}$, $\mathbf{U}'$, $\mathbf{D}'$, and $\mathbf{w}$ are known; that is, we need to minimize the function

$$\Phi(\mathbf{b}, \mathbf{v}) = \mathbf{b}'\mathbf{P}\mathbf{b} + \mathbf{w}'\mathbf{G}\mathbf{w} - 2\mathbf{w}'\mathbf{G}\mathbf{b} + 2\mathbf{v}'\mathbf{M}'\mathbf{b} \tag{3.10}$$

with respect to vectors $\mathbf{b}$ and $\mathbf{v}' = [v_1 \quad v_2 \quad \cdots \quad v_{r-1}]$, where $\mathbf{v}$ is a vector of Lagrange multipliers. Note that the only difference between Eqs. (3.1) and (3.10) is matrix $\mathbf{D}'$ and that matrix $\mathbf{M}' = \mathbf{D}'\mathbf{C}'$ has the same function in Eq. (3.10) that matrix $\mathbf{C}' = \mathbf{U}'\mathbf{G}$ had in Eq. (3.1). Then, the derivative results of Eq. (3.10) from $\mathbf{b}$ and $\mathbf{v}$ should be similar to those of Eq. (3.1), i.e.,

$$\begin{bmatrix} \mathbf{P} & \mathbf{M} \\ \mathbf{M}' & \mathbf{0} \end{bmatrix} \begin{bmatrix} \mathbf{b} \\ \mathbf{v} \end{bmatrix} = \begin{bmatrix} \mathbf{Gw} \\ \mathbf{0} \end{bmatrix}$$

whence the vector that minimizes $E[(H - I)^2]$ under the restriction $\mathbf{M}'\mathbf{b} = \mathbf{0}$ is

$$\mathbf{b}_M = \mathbf{K}_M\mathbf{b}, \tag{3.11}$$

where $\mathbf{K}_M = [\mathbf{I}_t - \mathbf{Q}_M]$, $\mathbf{Q}_M = \mathbf{P}^{-1}\mathbf{M}(\mathbf{M}'\mathbf{P}^{-1}\mathbf{M})^{-1}\mathbf{M}' = \mathbf{P}^{-1}\mathbf{C}\mathbf{D}(\mathbf{D}'\mathbf{C}'\mathbf{P}^{-1}\mathbf{C}\mathbf{D})^{-1}\mathbf{D}'\mathbf{C}'$, and $\mathbf{I}_t$ is an identity matrix of size $t \times t$. When $\mathbf{D} = \mathbf{U}$, $\mathbf{b}_M = \mathbf{b}_R$ (the RLPSI vector of coefficients), and when $\mathbf{D} = \mathbf{U}$ and $\mathbf{U}'$ is a null matrix, $\mathbf{b}_M = \mathbf{b}$ (the LPSI vector of coefficients). Thus, the Mallard (1972) index is more general than the RLPSI and is an optimal PPG-LPSI. In addition, it includes the LPSI and the RLPSI as particular cases.

Instead of using restriction $\mathbf{M}'\mathbf{b} = \mathbf{0}$ to minimize $E[(I - H)^2]$, we can use restriction $\mathbf{C}'\mathbf{b} = \theta\mathbf{d}$ and minimize

$$\Phi_T(\mathbf{b}, \mathbf{v}) = \mathbf{b}'\mathbf{P}\mathbf{b} + \mathbf{w}'\mathbf{G}\mathbf{w} - 2\mathbf{w}'\mathbf{G}\mathbf{b} + 2\mathbf{v}'(\mathbf{C}'\mathbf{b} - \theta\mathbf{d}) \tag{3.12}$$

with respect to $\mathbf{b}$, $\mathbf{v}'$, and θ (Tallis 1985; Lin 2005) assuming that $\mathbf{P}$, $\mathbf{G}$, $\mathbf{U}'$, $\mathbf{d}$, and $\mathbf{w}$ are known. The derivative results in matrix notation are

$$\begin{bmatrix} \mathbf{b}_T \\ \mathbf{v} \\ \theta \end{bmatrix} = \begin{bmatrix} \mathbf{P} & \mathbf{C} & \mathbf{0}_{t\times 1} \\ \mathbf{C}' & \mathbf{0}_{r\times t} & -\mathbf{d} \\ \mathbf{0}'_{1\times t} & -\mathbf{d}' & 0 \end{bmatrix}^{-1} \begin{bmatrix} \mathbf{Gw} \\ \mathbf{0} \\ 0 \end{bmatrix}, \tag{3.13}$$

where $\mathbf{0}_{t\times 1}$ is a null vector $t \times 1$, $\mathbf{0}_{r\times t}$ is a null matrix $r \times t$, and $\mathbf{0}$ is a null column vector $(r - 1) \times 1$; 0 is the standard zero value. The inverse matrix of coefficients $\begin{bmatrix} \mathbf{P} & \mathbf{C} & \mathbf{0}_{t\times 1} \\ \mathbf{C}' & \mathbf{0}_{r\times t} & -\mathbf{d} \\ \mathbf{0}'_{1\times t} & -\mathbf{d}' & 0 \end{bmatrix}^{-1}$ in Eq. (3.13) is not easy to obtain; for this reason, Tallis (1985) obtained his results in two steps. That is, Tallis (1985) first derived Eq. (3.12) with respect to $\mathbf{b}$ and $\mathbf{v}'$, whence he obtained

$$\mathbf{b}_T = \mathbf{b}_R + \theta\boldsymbol{\delta}, \tag{3.14}$$

where $\mathbf{b}_R = \mathbf{Kb}$ (Eq. 3.5), $\boldsymbol{\delta} = \mathbf{P}^{-1}\mathbf{C}(\mathbf{C}'\mathbf{P}^{-1}\mathbf{C})^{-1}\mathbf{d}$, and $\mathbf{d}' = [d_1 \quad d_2 \quad \cdots \quad d_r]$. Next, he derived $E\left[(\mathbf{b}'_T\mathbf{y} - H)^2\right]$ only with respect to θ, and his result was

$$\theta = \frac{\mathbf{b}'\mathbf{C}(\mathbf{C}'\mathbf{P}^{-1}\mathbf{C})^{-1}\mathbf{d}}{\mathbf{d}'(\mathbf{C}'\mathbf{P}^{-1}\mathbf{C})^{-1}\mathbf{d}}, \tag{3.15}$$

where $\mathbf{b} = \mathbf{P}^{-1}\mathbf{Gw}$ is the LPSI vector of coefficients, $\mathbf{C}' = \mathbf{U}'\mathbf{G}$, $\mathbf{d}$ is the vector of the predetermined proportional gains imposed by the breeder and $\mathbf{P}^{-1}$ is the inverse of matrix $\mathbf{P}$. When $\theta = 0$, $\mathbf{b}_T = \mathbf{b}_R$, and if $\theta = 0$ and $\mathbf{U}'$ is the null matrix, $\mathbf{b}_T = \mathbf{b}$. That is, the PPG-LPSI obtained by Tallis (1985) is more general than the RLPSI and the LPSI. The foregoing results indicate that Eq. (3.14) consists of three parts:

1. Vector $\mathbf{b}_R = \mathbf{Kb}$, which represents the weights of the RLPSI with the restriction that the expected genetic gain per trait be equal to zero.
2. Vector $\boldsymbol{\delta} = \mathbf{P}^{-1}\mathbf{C}(\mathbf{C}'\mathbf{P}^{-1}\mathbf{C})^{-1}\mathbf{d}$, which should represent the weights of the PPG-LPSI leading to the greatest improvement in the desired direction independently of economic weights.
3. θ represents the regression coefficient of $H = \mathbf{w}'\mathbf{g}$ on $\boldsymbol{\delta} = \mathbf{P}^{-1}\mathbf{C}(\mathbf{C}'\mathbf{P}^{-1}\mathbf{C})^{-1}\mathbf{d}$ (Itoh and Yamada 1987).

When $\theta = 1$, Eq. (3.14) is equal to

$$\mathbf{b}_{T_0} = \mathbf{b}_R + \boldsymbol{\delta}. \tag{3.16}$$

The latter equation was the original result obtained by Tallis (1962). Tallis (1962) derived Eq. (3.12) with respect to vectors $\mathbf{b}$ and $\mathbf{v}$ under the restriction $\mathbf{U}'\mathbf{G}\mathbf{b} = \mathbf{d}$, i.e., without θ or $\theta = 1$. Later, James (1968) maximized the correlation between I and $H(\rho_{HI})$ under the Tallis (1962) restriction and once more obtained Eq. (3.16). Mallard (1972) showed that Eq. (3.16) is not optimal, i.e., it does not minimize $E[(I - H)^2]$ and does not maximize ρ_{HI}, and gave the optimal solution, which we have presented here in Eq. (3.11). Later, using restriction $\mathbf{U}'\mathbf{G}\mathbf{b} = \theta\mathbf{d}$, Tallis (1985) obtained Eq. (3.14), which also is optimal.

Figure 3.3 presents the estimated correlation values between PPG-LPSI and the net genetic merit ($H = \mathbf{w}'\mathbf{g}$) for the optimal PPG-LPSI (Eq. 3.14) and non-optimal PPG-LPSI (Eq. 3.16) using one ($d_1 = 7$), two ($\mathbf{d}' = [7 \quad -3]$), and three ($\mathbf{d}' = [7 \quad -3 \quad 5]$) predetermined restrictions, four traits and 500 simulated genotypes in

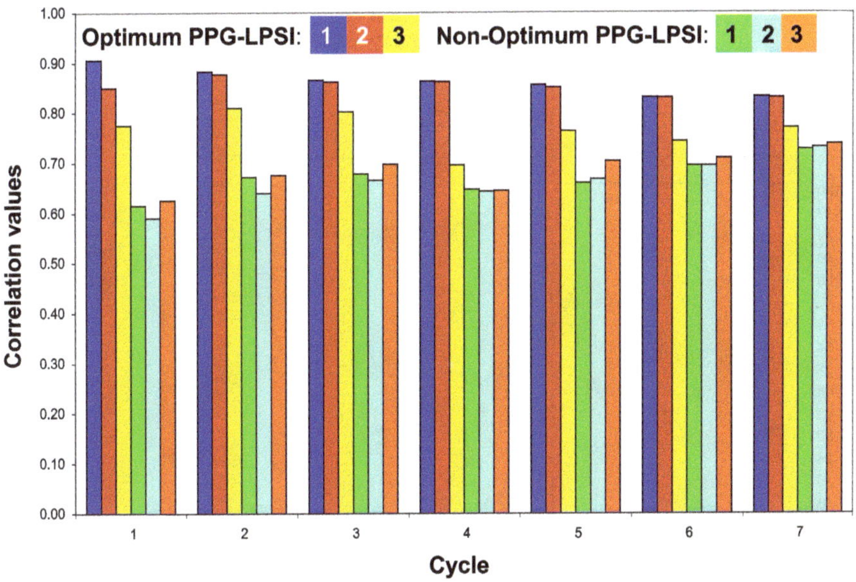

Fig. 3.3 Estimated correlation values between the predetermined proportional gain linear phenotypic selection index (PPG-LPSI) and the net genetic merit ($H = \mathbf{w}'\mathbf{g}$) for the optimal and non-optimal PPG-LPSI using 1 ($d_1 = 7$), 2 ($\mathbf{d}' = [7 \quad -3]$) and 3 ($\mathbf{d}' = [7 \quad -3 \quad 5]$) predetermined restrictions, 4 traits and 500 simulated genotypes in 1 environment for 7 selection cycles

one environment for seven selection cycles (see Sect. 2.8.1 of Chap. 2). Note that in effect, the non-optimal PPG-LPSI has lower correlations than the optimal PPG-LPSI for the seven simulated selection cycles.

Let $\mathbf{b}_P = \mathbf{b}_M = \mathbf{b}_T$ be the PPG-LPSI vector of coefficients. Then, the optimal PPG-LPSI can be written as

$$I_P = \mathbf{b}_P'\mathbf{y}, \tag{3.17}$$

whereas the maximized correlation between the PPG-LPSI and the net genetic merit is

$$\rho_{HI_P} = \frac{\mathbf{w}'\mathbf{G}\mathbf{b}_P}{\sqrt{\mathbf{w}'\mathbf{G}\mathbf{w}}\sqrt{\mathbf{b}_P'\mathbf{P}\mathbf{b}_P}}. \tag{3.18}$$

According to the conditions for constructing a valid PPG-LPSI described in Sect. 2.1 of Chap. 2, the index $I_P = \mathbf{b}_P'\mathbf{y}$ should have normal distributions. Figure 3.4 presents the distribution of 500 estimated PPG-LPSI values with two ($\mathbf{d}' = [7 \quad -3]$) and three ($\mathbf{d}' = [7 \quad -3 \quad 5]$) predetermined restrictions respectively, obtained from one selection cycle, with four traits and 500 genotypes simulated in one environment

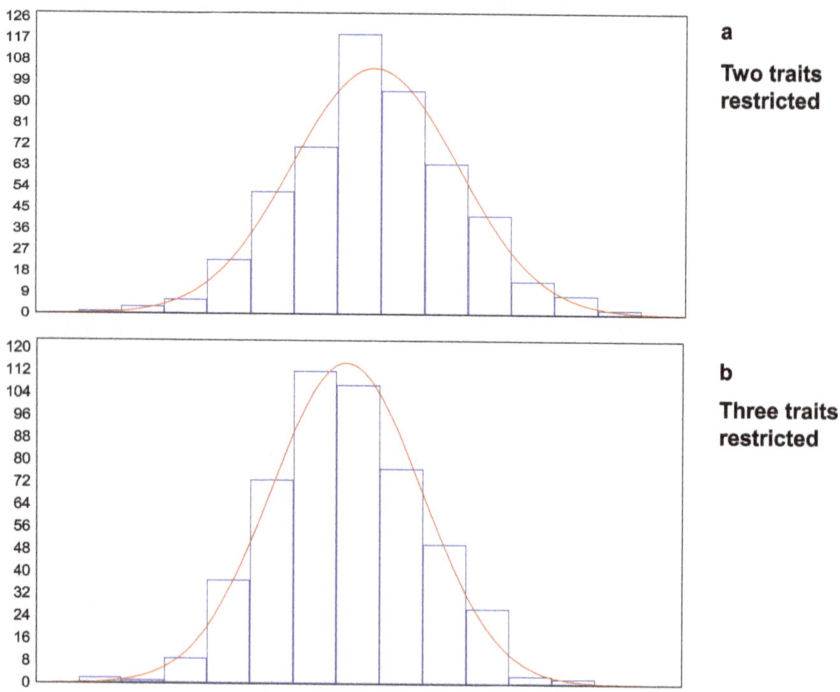

Fig. 3.4 (**a**) and (**b**) show the distribution of 500 estimated predetermined proportional gain linear phenotypic selection index values with two ($\mathbf{d}' = [7 \quad -3]$) and three ($\mathbf{d}' = [7 \quad -3 \quad 5]$) predetermined restrictions respectively, obtained from one selection cycle for 500 genotypes and four traits simulated in one environment

(see Chap. 2, Sect. 2.8.1 for details). Figure 3.4 indicates that, in effect, the PPG-LPSI values approach normal distribution.

Under the predetermined restrictions imposed by the breeder, $I_P = \mathbf{b}'_P\mathbf{y}$ should have maximal correlation with $H = \mathbf{w}'\mathbf{g}$ and it should be useful for ranking and selecting among individuals with different net genetic merits. However, for more than two restrictions the proportionality constant (θ) could be lower than 1; in that case, ρ_{HI_P} is lower than the correlation between LPSI and $H = \mathbf{w}'\mathbf{g}$ (ρ_{HI}). In addition, when the restriction $\mathbf{M}'\mathbf{b} = \mathbf{0}$ or $\mathbf{U}'\mathbf{G}\mathbf{b} = \theta\mathbf{d}$ is imposed on the PPG-LPSI vector of coefficients, the restricted traits decrease their effect on the correlation between PPG-LPSI and $H = \mathbf{w}'\mathbf{g}$. Using the simulated data set described in Sect. 2.8.1 of Chap. 2, we estimated ρ_{HI_P} and ρ_{HI} for seven selection cycles and compared the results in Fig. 3.5. Correlation ρ_{HI_P} values were estimated using one ($d_1 = 7$), two ($\mathbf{d}' = [7 \quad -3]$), and three ($\mathbf{d}' = [7 \quad -3 \quad 5]$) predetermined restrictions. Figure 3.5 indicates that when the number of predetermined restrictions is equal to or higher than two, the estimated values of ρ_{HI_P} decrease more than when only one predetermined restriction is imposed on the PPG-LPSI.

The maximized PPG-LPSI selection response and expected genetic gains per trait can be written as

$$R_P = k_I\sqrt{\mathbf{b}'_M\mathbf{P}\mathbf{b}_M} = k_I\sqrt{\mathbf{b}'_T\mathbf{P}\mathbf{b}_T} \tag{3.19}$$

and

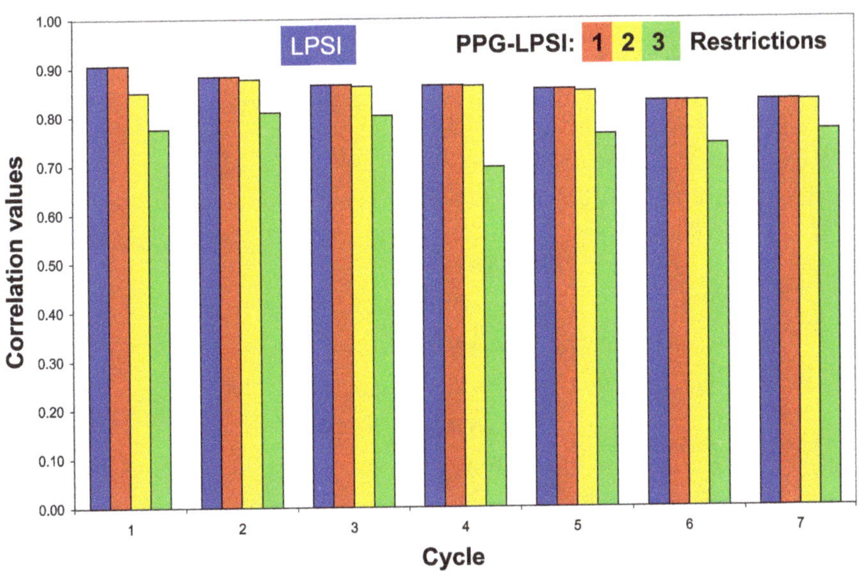

Fig. 3.5 Estimated correlation values between the LPSI and the net genetic merit ($H = \mathbf{w}'\mathbf{g}$); and estimated correlation values between the PPG-LPSI and H with one ($d_1 = 7$), two ($\mathbf{d}' = [7 \quad -3]$), and three ($\mathbf{d}' = [7 \quad -3 \quad 5]$) predetermined restrictions obtained from seven selection cycles for four traits and 500 simulated genotypes in one environment

$$\mathbf{E}_\mathrm{P} = k_I \frac{\mathbf{G}\mathbf{b}_M}{\sqrt{\mathbf{b}'_M \mathbf{P}\mathbf{b}_M}} = k_I \frac{\mathbf{G}\mathbf{b}_T}{\sqrt{\mathbf{b}'_T \mathbf{P}\mathbf{b}_T}}, \tag{3.20}$$

respectively, where k_I is the standardized selection differential or selection intensity associated with the PPG-LPSI.

The maximized PPG-LPS selection response (Eq. 3.19) has the same form as the maximized LPSI selection response. Thus, under r predetermined restrictions, Eq. (3.19) predicts the mean improvement in H due to indirect selection on $I_P = \mathbf{b}'_P\mathbf{y}$. Predetermined restriction effects are observed on the PPG-LPSI expected genetic gain per trait (Eq. 3.20). The main difference between the RLPSI and the PPG-LPSI is the vector of predetermined proportional gains.

3.2.2 Statistical Properties of the PPG-LPSI

Assuming that $H = \mathbf{w}'\mathbf{g}$ and $I_P = \mathbf{b}'_P\mathbf{y}$ have a bivariate joint normal distribution, $\mathbf{b}_P = \mathbf{K}_M\mathbf{b}$, $\mathbf{b} = \mathbf{P}^{-1}\mathbf{G}\mathbf{w}$, and $\mathbf{P}$, $\mathbf{G}$ and $\mathbf{w}$ are known, the PPG-LPSI has the same properties as the RLPSI. Some of the main PPG-LPSI properties are:

1. Matrices $\mathbf{Q}_M = \mathbf{P}^{-1}\mathbf{M}(\mathbf{M}'\mathbf{P}^{-1}\mathbf{M})^{-1}\mathbf{M}'$ and $\mathbf{K}_M = [\mathbf{I} - \mathbf{Q}_M]$ have the same function as matrices $\mathbf{Q} = \mathbf{P}^{-1}\mathbf{C}(\mathbf{C}'\mathbf{P}^{-1}\mathbf{C})^{-1}\mathbf{C}'$ and $\mathbf{K} = [\mathbf{I} - \mathbf{Q}]$ in the RLPSI.
2. Matrices $\mathbf{Q}_M$ and $\mathbf{K}_M$ are both projectors, i.e., they are idempotent ($\mathbf{K}_M = \mathbf{K}_M^2$ and $\mathbf{Q}_M = \mathbf{Q}_M^2$), unique and orthogonal, i.e., $\mathbf{K}_M\mathbf{Q}_M = \mathbf{Q}_M\mathbf{K}_M = \mathbf{0}$.
3. Matrix $\mathbf{Q}_M$ projects $\mathbf{b}$ into a space generated by the columns of matrix $\mathbf{M}$ due to the restriction $\mathbf{M}'\mathbf{b} = \mathbf{0}$ that is introduced when $\Phi(\mathbf{b}, \mathbf{v})$ is maximized with respect to $\mathbf{b}$, whereas matrix $\mathbf{K}_M$ projects $\mathbf{b}$ into a space that is perpendicular to the space generated by the columns of matrix $\mathbf{M}$ (Rao 2002). Thus, the function of matrix $\mathbf{K}_M$ is to transform vector $\mathbf{b} = \mathbf{P}^{-1}\mathbf{G}\mathbf{w}$ into vector $\mathbf{b}_P = \mathbf{K}_M\mathbf{b}$.
4. The variance of $I_P = \mathbf{b}'_P\mathbf{y}$ ($\sigma^2_{I_P} = \mathbf{b}'_P\mathbf{P}\mathbf{b}_P$) is equal to the covariance between $I_P = \mathbf{b}'_P\mathbf{y}$ and $H = \mathbf{w}'\mathbf{a}$ ($\sigma_{HI_P} = \mathbf{w}'\mathbf{G}\mathbf{b}_P$). As $\mathbf{K}_M = \mathbf{K}_M^2, \mathbf{K}'_M\mathbf{P} = \mathbf{P}\mathbf{K}_M$ and $\mathbf{b}'\mathbf{P} = \mathbf{w}'\mathbf{G}$, then
$$\sigma^2_{I_P} = \mathbf{b}'_P\mathbf{P}\mathbf{b}_P = \mathbf{b}'\mathbf{K}'_M\mathbf{P}\mathbf{K}_M\mathbf{b} = \mathbf{b}'\mathbf{P}\mathbf{K}_M^2\mathbf{b} = \mathbf{b}'\mathbf{P}\mathbf{K}_M\mathbf{b} = \mathbf{w}'\mathbf{G}\mathbf{b}_P = \sigma_{HI_P}.$$
5. The maximized correlation between H and $I_P = \mathbf{b}'_P\mathbf{y}$ is equal to $\rho_{HI_P} = \frac{\sigma_{I_P}}{\sigma_H}$. In point 4 of this subsection, we showed that $\sigma_{HI_P} = \sigma^2_{I_P}$, then
$$\rho_{HI_P} = \frac{\mathbf{w}'\mathbf{G}\mathbf{b}_P}{\sqrt{\mathbf{w}'\mathbf{G}\mathbf{w}}\sqrt{\mathbf{b}'_P\mathbf{P}\mathbf{b}_P}} = \sqrt{\frac{\mathbf{b}'_P\mathbf{P}\mathbf{b}_P}{\mathbf{w}'\mathbf{G}\mathbf{w}}} = \frac{\sigma_{I_P}}{\sigma_H}.$$
6. The variance of the predicted error, $Var(H - I_P) = \left(1 - \rho^2_{HI_P}\right)\sigma^2_H$, is minimal. By point 4 of this subsection, $\sigma_{HI_P} = \sigma^2_{I_P}$, then $Var(H - I_R) = \sigma^2_H - \sigma^2_{I_P} = \left(1 - \rho^2_{HI_P}\right)\sigma^2_H$.
7. The heritability of the PPG-LPSI is equal to $\mathrm{h}^2_{\mathrm{I_P}} = \frac{\mathbf{b}'_P\mathbf{G}\mathbf{b}_P}{\mathbf{b}'_P\mathbf{P}\mathbf{b}_P}$.

Points 1–3 show that in effect, the PPG-LPSI projects the LPSI vector of coefficients into a different space than the original LPSI vector of coefficients. In addition, the PPG-LPSI statistical properties denoted by points 4–7 are the same as the LPSI statistical properties. Thus, the PPG-LPSI is a variant of the LPSI.

3.2.3 There Is Only One Optimal PPG-LGSI

Let $\mathbf{S} = \mathbf{C}'\mathbf{P}^{-1}\mathbf{C}$, under the restriction $\mathbf{D}'\mathbf{d} = \mathbf{0}$, Itoh and Yamada (1987) showed that $\mathbf{D}(\mathbf{D}'\mathbf{S}\mathbf{D})^{-1}\mathbf{D}' = \mathbf{S}^{-1} - \mathbf{S}^{-1}\mathbf{d}(\mathbf{d}'\mathbf{S}^{-1}\mathbf{d})^{-1}\mathbf{d}'\mathbf{S}^{-1}$, whence substituting $\mathbf{S}^{-1} - \mathbf{S}^{-1}\mathbf{d}(\mathbf{d}'$-$\mathbf{S}^{-1}\mathbf{d})^{-1}\mathbf{d}'\mathbf{S}^{-1}$ for $\mathbf{D}(\mathbf{D}'\mathbf{S}\mathbf{D})^{-1}\mathbf{D}'$ in matrix $\mathbf{Q}_M$, Eq. (3.11) can be written as Eq. (3.14), i.e., $\mathbf{b}_\mathrm{M} = \mathbf{b}_\mathrm{T}$. Therefore, the Mallard (1972) and Tallis (1985) vectors of coefficients are the same. In addition, Itoh and Yamada (1987) showed that the Harville (1975) vector of coefficients can written as $\frac{\mathbf{b}_\mathrm{T}}{\sigma_{I_T}}$ (Eq. 2.21d), where σ_{I_T} is the standard deviation of the variance of the Tallis (1985) PPG-LPSI. Thus, in reality, there is only one optimal PPG-LPSI.

Itoh and Yamada (1987) also pointed out that matrix $\mathbf{D}' = \begin{bmatrix} d_r & 0 & \cdots & 0 & -d_1 \\ 0 & d_r & \cdots & 0 & -d_2 \\ \vdots & \vdots & \ddots & \vdots & \vdots \\ 0 & 0 & \cdots & d_r & -d_{r-1} \end{bmatrix}$ is only one example of several possible Mallard (1972) $\mathbf{D}'$ matrices. They showed that any matrix $\mathbf{D}'$ that satisfies condition $\mathbf{D}'\mathbf{d} = \mathbf{0}$ is another Mallard (1972) matrix of predetermined proportional gains. According to Itoh and Yamada (1987), matrices

$$\mathbf{D}' = \begin{bmatrix} d_2 & -d_1 & 0 & \cdots & 0 & 0 \\ 0 & d_3 & -d_2 & \cdots & 0 & 0 \\ \vdots & \vdots & \vdots & \vdots & \vdots & 0 \\ 0 & 0 & 0 & 0 & d_r & d_{r-1} \end{bmatrix} \quad \text{and}$$

$$\mathbf{D}' = \begin{bmatrix} d_2 & -d_1 & 0 & \cdots & 0 \\ d_3 & 0 & -d_1 & \cdots & 0 \\ \vdots & \vdots & \vdots & \vdots & \vdots \\ d_r & 0 & 0 & 0 & -d_1 \end{bmatrix}$$

are also Mallard (1972) matrices of predetermined proportional gains because they satisfy condition $\mathbf{D}'\mathbf{d} = \mathbf{0}$. However, matrix $\mathbf{D}' = \begin{bmatrix} d_r & 0 & \cdots & 0 & -d_1 \\ 0 & d_r & \cdots & 0 & -d_2 \\ \vdots & \vdots & \ddots & \vdots & \vdots \\ 0 & 0 & \cdots & d_r & -d_{r-1} \end{bmatrix}$ is "easier" to construct.

Harville (1975) maximized the correlation between I and H (ρ_{IH}) under the restriction $\mathbf{C}'\mathbf{b} = \theta\mathbf{d}$ and was the first to point out the importance of the proportionality constant (θ) in the PPG-LPSI. Mallard (1972) showed that the restriction $\mathbf{U}'\mathbf{Gb} = \mathbf{d}$ does not maximize the correlation with the net genetic merit ($H = \mathbf{w}'\mathbf{g}$) and Harville (1975) indicated that the restriction $\mathbf{U}'\mathbf{Gb} = \mathbf{d}$ only changes the sign of the genetic expected gain (or multi-trait selection response) but does not maximize the correlation between $I = \mathbf{b}'\mathbf{y}$ and $H = \mathbf{w}'\mathbf{g}$. According to Mallard (1972), Harville (1975), and Tallis (1985), the PPG-LPSI is optimal only under the restriction $\mathbf{U}'\mathbf{Gb} = \theta\mathbf{d}$.

Itoh and Yamada (1987) pointed out several problems associated with the Tallis (1985) PPG-PSI:

1. When the number of restrictions imposed on the PPG-PSI expected genetic gains increases, θ tends to zero and then the accuracy of the PPG-PSI decreases.
2. The θ values could be negative, in which case PPG-PSI results have no meaning in practice.
3. The PPG-PSI may cause the population means to shift in the opposite direction to the predetermined desired direction; this may happen because of the opposite directions between the economic values and the predetermined desired direction.

Itoh and Yamada (1987) thought that one possible solution to those problems could be to use the linear phenotypic selection index with desired gains.

3.2.4 Numerical Examples

The estimated phenotypic ($\widehat{\mathbf{P}}$) and genetic ($\widehat{\mathbf{G}}$) covariance matrices described in Sect. 3.1.4 of this chapter for RLPSI are used as the first example. First, Eq. (3.11) is described to obtain the PPG-LPSI vector of coefficients. Let $\mathbf{d}_2' = [7 \quad -3]$ be the vector for 2 predetermined restrictions, then, the Mallard (1972) matrix is $\mathbf{D}' = [-3 \quad -7]$, while matrix $\mathbf{U}'$ is $\mathbf{U}_2' = \begin{bmatrix} 1 & 0 & 0 & 0 \\ 0 & 1 & 0 & 0 \end{bmatrix}$. Matrix $\mathbf{M}' = \mathbf{D}'\mathbf{U}'\widehat{\mathbf{G}}$ for 2 predetermined restrictions will be $\mathbf{M}' = \mathbf{D}'\mathbf{U}'_2\widehat{\mathbf{G}} = [-18.12 \quad -52.49 \quad -1.25 \quad 7.46]$, whence

$$\widehat{\mathbf{Q}}_M = \widehat{\mathbf{P}}^{-1}\mathbf{M}\left(\mathbf{M}'\widehat{\mathbf{P}}^{-1}\mathbf{M}\right)^{-1}\mathbf{M}' = \begin{bmatrix} 0.084 & 0.242 & 0.006 & -0.034 \\ 0.313 & 0.906 & 0.022 & -0.129 \\ 0.037 & 0.106 & 0.003 & -0.015 \\ -0.019 & -0.055 & -0.001 & 0.008 \end{bmatrix} \text{ and}$$

$$\widehat{\mathbf{K}}_M = \left[\mathbf{I}_4 - \widehat{\mathbf{Q}}_M\right] = \begin{bmatrix} 0.916 & -0.242 & -0.006 & 0.034 \\ -0.313 & 0.094 & -0.022 & 0.129 \\ -0.037 & 0.106 & 0.997 & 0.015 \\ 0.019 & 0.055 & 0.001 & 0.992 \end{bmatrix};$$

$\mathbf{I}_4$ is an identity matrix of size 4×4.

The estimated LPSI and PPG-LPSI vectors of coefficients were $\widehat{\mathbf{b}}' = [0.554 \quad -1.053 \quad 1.090 \quad 1.058]$ and $\widehat{\mathbf{b}}'_M = \left(\widehat{\mathbf{K}}_M\widehat{\mathbf{b}}\right)' = [0.793 \quad -0.159 \quad 1.1941.004]$ respectively, and the estimated PPG-LPSI was $\widehat{I}_M = 0.793T_1 - 0.159T_2 + 1.194T_3 + 1.004T_4$. The standard deviation of the estimated variance of $\widehat{I}_M$ was $\widehat{\sigma}_{I_M} = \sqrt{\widehat{\mathbf{b}}'_M\widehat{\mathbf{P}}\widehat{\mathbf{b}}_M} = 9.526$, whereas the estimated correlation value between the PPG-LPSI and the net genetic merit was $\widehat{\rho}_{HI_P} = \frac{\widehat{\sigma}_{I_M}}{\widehat{\sigma}_H} = 0.85$, where $\widehat{\sigma}_H = \sqrt{\mathbf{w}\widehat{\mathbf{G}}\mathbf{w}} = 11.202$ is the estimated standard deviation of the variance of the net genetic merit.

Suppose that the selection intensity was 10% ($k_I = 1.755$); then, the estimated PPG-LPSI expected genetic gain per trait and the estimated selection response are $\widehat{\mathbf{E}}'_M = 1.755\frac{\widehat{\mathbf{b}}'_M\mathbf{G}}{\sqrt{\widehat{\mathbf{b}}'_M\widehat{\mathbf{P}}\widehat{\mathbf{b}}_M}} = [8.013 \quad -3.434 \quad 3.541 \quad 1.730]$ and $\widehat{R}_M = (1.755)\sqrt{\widehat{\mathbf{b}}'_M\widehat{\mathbf{P}}\widehat{\mathbf{b}}_M} = (1.755)(9.526) = 16.717$ respectively.

Now, let $\mathbf{d}'_3 = [7 \quad -3 \quad 5]$ be the vector for three predetermined restrictions, then there are three possible predetermined Mallard matrices, i.e., $\mathbf{D}'_1 = \begin{bmatrix} 5 & 0 & -7 \\ 0 & 5 & 3 \end{bmatrix}$, $\mathbf{D}'_2 = \begin{bmatrix} -3 & -7 & 0 \\ 0 & 5 & 3 \end{bmatrix}$, and $\mathbf{D}'_3 = \begin{bmatrix} -3 & -7 & 0 \\ 5 & 0 & -7 \end{bmatrix}$, and matrix $\mathbf{U}'$ for three restrictions is $\mathbf{U}'_3 = \begin{bmatrix} 1 & 0 & 0 & 0 \\ 0 & 1 & 0 & 0 \\ 0 & 0 & 1 & 0 \end{bmatrix}$. Thus, for three predetermined restrictions matrix $\mathbf{M}' = \mathbf{D}'\mathbf{U}'\widehat{\mathbf{G}}$ shall have three possible forms, i.e., $\mathbf{M}'_1 = \mathbf{D}'_1\mathbf{U}'_3\widehat{\mathbf{G}} = \begin{bmatrix} 122.60 & -40.85 & -27.97 & 12.58 \\ -39.60 & 55.00 & 12.88 & -10.72 \end{bmatrix}$, but $\mathbf{M}'_2 = \mathbf{D}'_2\mathbf{U}'_3\widehat{\mathbf{G}} = \mathbf{M}'_3 = \begin{bmatrix} -18.12 & -52.49 & -1.25 & 7.46 \\ 122.60 & -40.85 & -27.97 & 12.58 \end{bmatrix}$. Note that matrix $\mathbf{M}'_1$ is different from matrices $\mathbf{M}'_2$ and $\mathbf{M}'_3$, and that the two latter are the same; however, both matrices should lead to the same estimated PPG-LPSI vector of coefficients and to the same estimated PPG-LPSI expected genetic gain per trait and selection response. It can be shown that for matrices $\mathbf{M}'_1$, $\mathbf{M}'_2$, and $\mathbf{M}'_3$, matrices $\widehat{\mathbf{Q}}_M$ and $\widehat{\mathbf{K}}_M = \left[\mathbf{I}_4 - \widehat{\mathbf{Q}}_M\right]$ are the same and can be written as

$$\widehat{\mathbf{Q}}_M = \begin{bmatrix} 0.771 & 0.080 & -0.145 & 0.026 \\ 0.123 & 0.951 & 0.063 & -0.145 \\ -1.131 & 0.382 & 0.258 & -0.117 \\ 0.118 & -0.087 & -0.031 & 0.020 \end{bmatrix} \quad \text{and}$$

$$\widehat{\mathbf{K}}_M = \begin{bmatrix} 0.229 & -0.080 & 0.145 & -0.026 \\ -0.123 & 0.049 & -0.063 & 0.145 \\ 1.131 & -0.382 & 0.742 & 0.117 \\ -0.118 & 0.087 & 0.031 & 0.980 \end{bmatrix}.$$

The estimated LPSI vector of coefficients was equal to $\widehat{\mathbf{b}}' = [0.554 \quad -1.053 \quad 1.090 \quad 1.058]$, whereas the estimated PPG-LPSI vector of coefficients was $\widehat{\mathbf{b}}'_M = \left(\widehat{\mathbf{K}}_M\widehat{\mathbf{b}}\right)' = [0.342 \quad -0.035 \quad 1.960 \quad 0.914]$. The estimated PPG-LPSI was $\widehat{I}_M = 0312T_1 - 0.035T_2 + 1.960T_3 + 0.914T_4$ and the standard deviation of the estimated variance of $\widehat{I}_M$ was $\widehat{\sigma}_{I_M} = \sqrt{\widehat{\mathbf{b}}'_M\widehat{\mathbf{P}}\widehat{\mathbf{b}}_M} = 8.68$. The estimated correlation value between the PPG-LPSI and the net genetic merit was $\widehat{\rho}_{HI_P} = \frac{\widehat{\sigma}_{I_M}}{\widehat{\sigma}_H} = 0.775$, where $\widehat{\sigma}_H = \sqrt{\mathbf{w}\widehat{\mathbf{G}}\mathbf{w}} = 11.202$ is the estimated standard deviation of the variance of the net genetic merit.

Using a selection intensity of 10% ($k_I = 1.755$), the estimated PPG-LPSI expected genetic gain per trait and the estimated selection response were $\widehat{\mathbf{E}}'_M = 1.755\frac{\widehat{\mathbf{b}}'_M\mathbf{G}}{\sqrt{\widehat{\mathbf{b}}'_M\widehat{\mathbf{P}}\widehat{\mathbf{b}}_M}} = [6.410 \quad -2.747 \quad 4.579 \quad 1.496]$ and $\widehat{R}_M = (1.755)\sqrt{\widehat{\mathbf{b}}'_M\widehat{\mathbf{P}}\widehat{\mathbf{b}}_M}$ $= (1.755)(8.68) = 15.32$, respectively.

According to Eq. (3.14), the estimated Tallis (1985) vector of coefficients can be obtained as $\widehat{\mathbf{b}}_T = \widehat{\mathbf{b}}_R + \widehat{\theta}\widehat{\boldsymbol{\delta}}$, where $\widehat{\mathbf{b}}_R = \widehat{\mathbf{K}}\widehat{\mathbf{b}}$ is the estimated RLPSI, $\widehat{\boldsymbol{\delta}} = \widehat{\mathbf{P}}^{-1}\mathbf{C}\left(\mathbf{C}'\widehat{\mathbf{P}}^{-1}\mathbf{C}\right)^{-1}\mathbf{d}$, $\widehat{\theta} = \frac{\widehat{\mathbf{b}}'\mathbf{C}\left(\mathbf{C}'\widehat{\mathbf{P}}^{-1}\mathbf{C}\right)^{-1}\mathbf{d}}{\mathbf{d}'\left(\mathbf{C}'\widehat{\mathbf{P}}^{-1}\mathbf{C}\right)^{-1}\mathbf{d}}$ is the estimated constant of proportionality, $\widehat{\mathbf{b}} = \widehat{\mathbf{P}}^{-1}\widehat{\mathbf{G}}\mathbf{w}$ is the estimated LPSI vector of coefficients, and $\mathbf{d}' = [d_1 \quad d_2 \quad \cdots \quad d_r]$ is the vector of predetermined restrictions.

In Sect. 3.1.4 of this chapter we described how to obtain $\widehat{\mathbf{b}}_R = \widehat{\mathbf{K}}\widehat{\mathbf{b}}$, and we also obtained matrix $\mathbf{C}' = \mathbf{U}'\widehat{\mathbf{G}}$ for two and three null restrictions as $\mathbf{C}'_2 = \mathbf{U}'_2\widehat{\mathbf{G}} = \begin{bmatrix} 36.21 & -12.93 & 8.35 & 2.74 \\ -12.93 & 13.04 & -3.40 & -2.24 \end{bmatrix}$ and $\mathbf{C}'_3 = \mathbf{U}'_3\widehat{\mathbf{G}} =$ $\begin{bmatrix} 36.21 & -12.93 & 8.35 & 2.74 \\ -12.93 & 13.04 & -3.04 & -2.24 \\ 8.35 & -3.40 & 9.96 & 0.16 \end{bmatrix}$, whence the $\widehat{\mathbf{b}}_R = \widehat{\mathbf{K}}\widehat{\mathbf{b}}$ values for two and three null restrictions were $\widehat{\mathbf{b}}'_{R_2} = [-0.164 \quad 0.162 \quad 0.680 \quad 0.856]$ and $\widehat{\mathbf{b}}'_{R_3} = [-0.032 \quad 0.136 \quad 0.059 \quad 0.890]$ respectively.

The $\widehat{\theta}$ and $\widehat{\boldsymbol{\delta}}$ values for two and three predetermined restrictions were $\widehat{\theta}_2 = \frac{\widehat{\mathbf{b}}'\mathbf{C}_2\left(\mathbf{C}'_2\widehat{\mathbf{P}}^{-1}\mathbf{C}_2\right)^{-1}\mathbf{d}_2}{\mathbf{d}'_2\left(\mathbf{C}'_2\widehat{\mathbf{P}}^{-1}\mathbf{C}_2\right)^{-1}\mathbf{d}_2} = 6.213$, $\widehat{\theta}_3 = \frac{\widehat{\mathbf{b}}'\mathbf{C}_3\left(\mathbf{C}'_3\widehat{\mathbf{P}}^{-1}\mathbf{C}_3\right)^{-1}\mathbf{d}_3}{\mathbf{d}'_3\left(\mathbf{C}'_3\widehat{\mathbf{P}}^{-1}\mathbf{C}_3\right)^{-1}\mathbf{d}_3} = 4.529$, $\widehat{\boldsymbol{\delta}}'_2 = \left(\widehat{\mathbf{P}}^{-1}\mathbf{C}_2\left(\mathbf{C}'_2\widehat{\mathbf{P}}^{-1}\mathbf{C}_2\right)^{-1}\mathbf{d}_2\right)' = [0.153 \quad -0.052 \quad 0.083 \quad 0.024]$, and $\widehat{\boldsymbol{\delta}}'_3 = \left(\widehat{\mathbf{P}}^{-1}\mathbf{C}_3\left(\mathbf{C}'_3\widehat{\mathbf{P}}^{-1}\mathbf{C}_3\right)^{-1}\mathbf{d}_3\right)' = [0.083 \quad -0.038 \quad 0.420 \quad 0.005]$. With these results, the estimated Tallis (1985) vectors of coefficients for two and three predetermined restrictions were $\widehat{\mathbf{b}}'_{T_2} = [0.793 \quad -0.159 \quad 1.194 \quad 1.004]$ and $\widehat{\mathbf{b}}'_{T_3}$ $= [0.342 \quad -0.035 \quad 1.960 \quad 0.914]$ respectively. These latter two vectors of coefficients are the same as the vectors of coefficients obtained using the Mallard (1972) method for two and three predetermined restrictions. These results corroborate that, in effect, the Mallard (1972) and Tallis (1985) PPG-LPSIs are the same.

With the data set described in Sect. 2.8.1 of Chap. 2 we constructed Table 3.4, which presents the estimated LPSI selection response and heritability, and the estimated PPG-LPSI selection response and heritability for one, two, and three predetermined restrictions for seven simulated selection cycles using a selection intensity of 10% ($k_I = 1.755$). The averages of the estimated PPG-LPSI selection responses were 14.19, 14.00, and 12.58 for one, two, and three restrictions respectively. Note that 14.19 is also the average value for the estimated LPSI selection response. This means that the PPG-LPSI and the LPSI selection responses are the same for only one predetermined restriction. However, the estimated PPG-LPSI selection responses for two and three restrictions tend to decrease (Table 3.4). The same is true for the estimated PPG-LPSI heritability. That is, the estimated PPG-LPSI heritability for one predetermined restriction is equal to the estimated LPSI heritability. The estimated PPG-LPSI heritability for two predetermined restrictions decreased, but increased for three predetermined restrictions (Table 3.4). This is because the simulated true heritabilities of traits T_1, T_2, T_3, and T_4 were 0.4, 0.6, 0.6, and 0.8 respectively.

Table 3.5 presents the estimated LPSI expected genetic gain per trait without restrictions, and the estimated PPG-LPSI expected genetic gain per trait for one, two, and three predetermined restrictions for seven simulated selection cycles using a selection intensity of 10% ($k_I = 1.755$). Once again, note that for one predetermined restriction, the estimated PPG-LPSI expected genetic gains were equal to the estimated LPSI expected genetic gains, and for two predetermined restrictions, the estimated PPG-LPSI expected genetic gains were similar to the estimated LPSI expected genetic gains; however, for three predetermined restrictions, the estimated PPG-LPSI expected genetic gains tended to decrease.

Table 3.4 Estimated LPSI selection response and heritability, and estimated predetermined proportional gain LPSI (PPG-LPSI) selection response and heritability for one, two, and three predetermined restrictions for seven simulated selection cycles

			PPG-LPSI					
	LPSI		Selection response for one, two, and three restrictions			Heritability for one, two, and three restrictions		
Cycle	Selection response	Heritability	1	2	3	1	2	3
1	17.81	0.84	17.81	16.72	15.23	0.84	0.77	0.83
2	15.69	0.80	15.69	15.59	14.39	0.80	0.78	0.83
3	14.22	0.77	14.22	14.16	13.18	0.77	0.76	0.80
4	14.34	0.76	14.34	14.33	11.56	0.76	0.75	0.78
5	13.64	0.75	13.64	13.56	12.16	0.75	0.75	0.79
6	12.04	0.71	12.04	12.04	10.77	0.71	0.71	0.76
7	11.61	0.72	11.61	11.59	10.75	0.72	0.71	0.76
Average	14.19	0.76	14.19	14.00	12.58	0.76	0.75	0.79

The selection intensity was 10% ($k_I = 1.755$) and the vectors of predetermined proportional gains for one, two, and three predetermined restrictions were $\mathbf{d}_1' = 7$, $\mathbf{d}_2' = [7 \quad -3]$ and $\mathbf{d}_3' = [7 \quad -3 \quad 5]$ respectively

Table 3.5 Estimated LPSI expected genetic gain per trait, and estimated PPG-LPSI expected genetic gain per trait for one, two, and three predetermined restrictions for seven simulated selection cycles

	LPSI expected gain per trait				PPG-LPSI expected gain per trait for one restriction			
Cycle	T1	T2	T3	T4	T1	T2	T3	T4
1	7.90	−4.67	3.33	1.92	7.90	−4.67	3.33	1.92
2	7.06	−3.59	3.17	1.86	7.06	−3.59	3.17	1.86
3	6.67	−3.21	2.82	1.52	6.67	−3.21	2.82	1.52
4	7.53	−3.45	2.07	1.29	7.53	−3.45	2.07	1.29
5	7.14	−2.66	2.51	1.33	7.14	−2.66	2.51	1.33
6	6.23	−2.62	1.98	1.21	6.23	−2.62	1.98	1.21
7	5.38	−2.55	2.47	1.22	5.38	−2.55	2.47	1.22
Average	6.85	−3.25	2.62	1.48	6.85	−3.25	2.62	1.48
	PPG-LPSI expected gain per trait for two restrictions				PPG-LPSI expected gain per trait for three restrictions			
Cycle	T1	T2	T3	T4	T1	T2	T3	T4
1	8.01	−3.43	3.54	1.73	6.41	−2.75	4.58	1.50
2	7.39	−3.17	3.22	1.81	5.89	−2.52	4.21	1.77
3	6.86	−2.94	2.77	1.60	5.48	−2.35	3.91	1.45
4	7.65	−3.28	2.12	1.27	4.76	−2.04	3.40	1.35
5	6.88	−2.95	2.41	1.33	5.08	−2.18	3.63	1.28
6	6.20	−2.66	1.98	1.21	4.39	−1.88	3.14	1.36
7	5.50	−2.36	2.53	1.19	4.41	−1.89	3.15	1.30
Average	6.93	−2.97	2.65	1.45	5.20	−2.23	3.72	1.43

The selection intensity was 10% ($k_I = 1.755$) and the vectors of predetermined proportional gains for one, two, and three restrictions were $\mathbf{d}' = 7$, $\mathbf{d}' = [7 \quad -3]$ and $\mathbf{d}' = [7 \quad -3 \quad 5]$ respectively

The first part of Table 3.6 presents the estimated correlation of the net genetic merit ($H = \mathbf{w}'\mathbf{g}$) with the estimated LPSI and RLPSI values for one, two, and three null restrictions. In addition, this first part presents the estimated LPSI versus RLPSI efficiency $p = 100(\lambda_R - 1)$ (Eq. 2.21, Chap. 2). The second part of Table 3.6 presents the estimated correlation of $H = \mathbf{w}'\mathbf{g}$ with the estimated LPSI and PPG-LPSI values for one, two, and three predetermined restrictions, and the estimated LPSI versus RLPSI efficiency $p = 100(\lambda_P - 1)$. Finally, the third part of Table 3.6 presents the estimated variance of the predicted error (VPE) of the LPSI ($\left(1 - \rho_{HI}^2\right)\sigma_H^2$), the RLPSI ($\left(1 - \rho_{HI_R}^2\right)\sigma_H^2$), and the PPG-LPSI ($\left(1 - \rho_{HI_P}^2\right)\sigma_H^2$) for one, two, and three restrictions for seven simulated selection cycles.

The estimated VPE of the RLPSI is higher than that of the LPSI and PPG-LPSI for one, two, and three restrictions for the seven simulated selection cycles; however, the estimated VPE of PPG-LPSI is only greater than that of the LPSI for two and three predetermined restrictions.

Table 3.6 Correlation of the net genetic merit with the LPSI, the RLPSI, and the PPG-LPSI for one, two, and three null and predetermined restrictions; LPSI versus RLPSI efficiency and LPSI versus PPG-LPSI efficiency, and estimated variance of the predicted error (VPE) of the LPSI, the RLPSI, and the PPG-LPSI for one, two, and three restrictions for seven simulated selection cycles

		RLPSI correlation for one, two, and three null restrictions			LPSI versus RLPSI efficiency in percentage terms for one, two, and three null restrictions		
Cycle	LPSI Correlation	1	2	3	1	2	3
1	0.91	0.35	0.28	0.21	159.16	221.34	331.65
2	0.88	0.48	0.33	0.24	85.69	164.19	267.25
3	0.87	0.44	0.35	0.25	98.42	145.51	241.61
4	0.86	0.40	0.30	0.22	114.77	183.56	285.28
5	0.86	0.38	0.32	0.20	126.47	164.15	321.00
6	0.83	0.44	0.36	0.23	89.09	132.96	264.22
7	0.83	0.41	0.32	0.22	101.23	161.60	275.26
Average	0.86	0.41	0.32	0.23	110.69	167.62	283.75
		PPG-LPSI correlation for one, two, and three predetermined restrictions			LPSI vs. PPG-LPSI efficiency in percentage terms for one, two, and three predetermined restrictions		
Cycle	LPSI Correlation	1	2	3	1	2	3
1	0.91	0.91	0.85	0.77	0	17.13	22.74
2	0.88	0.88	0.88	0.81	0	3.44	10.42
3	0.87	0.87	0.86	0.80	0	3.35	10.21
4	0.86	0.86	0.86	0.70	0	2.32	22.96
5	0.86	0.86	0.85	0.76	0	0.30	10.09
6	0.83	0.83	0.83	0.74	0	0.83	11.13
7	0.83	0.83	0.83	0.77	0	2.35	7.74
Average	0.86	0.86	0.85	0.77	0	4.25	13.61
		RLPSI VPE for one, two, and three null restrictions			PPG-LPSI VPE for one, two, and three predetermined restrictions		
Cycle	LPSI VPE	1	2	3	1	2	3
1	22.53	110.16	115.52	119.96	22.53	50.44	57.14
2	22.66	79.40	91.13	96.65	22.66	27.88	37.03
3	21.95	70.92	76.70	81.97	21.95	26.14	33.55
4	22.84	75.16	81.33	85.14	22.84	25.84	45.46
5	22.13	70.75	73.86	79.11	22.13	22.49	32.69
6	21.18	55.07	59.56	64.68	21.18	21.95	30.13
7	19.47	52.44	56.85	60.14	19.47	21.45	25.53
Average	21.82	73.41	79.28	83.95	21.82	28.03	37.36

Thus, according to the results obtained for the LPSI, the RLPSI, and the PPG-LPSI, the best predictor of the net genetic merit was the LPSI followed by the PPG-LPSI and the RLPSI.

3.3 The Desired Gains Linear Phenotypic Selection Index

The most important aspect of the desired gains linear phenotypic selection index (DG-LPSI) is that it does not require economic weights. Note that the LPSI expected genetic gain per trait $\mathbf{E} = k_I \frac{\mathbf{Gb}}{\sigma_I}$ is maximized when $\mathbf{b} = \mathbf{P}^{-1}\mathbf{Gw}$ and is proportional to k_I and σ_I. Now let $\mathbf{Gb}$ be written as

$$\mathbf{Gb} = \mathbf{d}, \tag{3.21}$$

where $\mathbf{d}$ is the vector of desired gains. From Eq. (3.21), $\mathbf{E}$ can be written as

$$\mathbf{E} = k_I \frac{\mathbf{d}}{\sigma_I}. \tag{3.22}$$

Equation (3.22) indicates that $\mathbf{E}$ is inversely proportional to σ_I; then we can minimize σ_I with respect to $\mathbf{b}$ subject to the constraints $\mathbf{Gb} = \mathbf{d}$ and then $\mathbf{E}$ is maximized (Brascamp 1984; Itoh and Yamada 1986). That is, we need to take the derivative of the function

$$\Phi_{DG}(\mathbf{b}, \mathbf{v}) = \mathbf{0.5}(\mathbf{b}'\mathbf{Pb}) + \mathbf{v}'(\mathbf{Gb} - \mathbf{d}) \tag{3.23}$$

with respect to $\mathbf{b}$ and $\mathbf{v}$, where $\mathbf{v}$ is a vector of Lagrange multipliers, assuming that $\mathbf{P}$, $\mathbf{G}$, and $\mathbf{d}$ are known. The restriction $\mathbf{Gb} = \mathbf{d}$ in Eq. (3.23) is similar to the Tallis (1985) restriction $\mathbf{U}'\mathbf{Gb} = \theta\mathbf{d}$, but with $\mathbf{U}' = \mathbf{I}$ and $\theta = 1$, or $\theta = \frac{k_I}{\sigma_I}$ (Tallis 1962).

It can be shown that the vector that minimizes σ_I and maximizes $\mathbf{E}$ can be written as

$$\mathbf{b}_{DG} = \mathbf{P}^{-1}\mathbf{G}\left(\mathbf{GP}^{-1}\mathbf{G}\right)^{-1}\mathbf{d}. \tag{3.24}$$

thus, in effect, as $\mathbf{Gb} = \mathbf{d}$, $\mathbf{b}_{DG} = \mathbf{P}^{-1}\mathbf{G}(\mathbf{GP}^{-1}\mathbf{G})^{-1}\mathbf{d} = \mathbf{P}^{-1}\mathbf{G}(\mathbf{GP}^{-1}\mathbf{G})^{-1}\mathbf{Gb} = \mathbf{b}$. In Eq. (3.24) we are assuming that the traits in the index are the same as those in the net genetic merit. However, this may not be the case, that is, the number of traits could be different from the number of genotypes. In the latter case, Eq. (3.21) should be written as $\mathbf{G}'\mathbf{b} = \mathbf{d}$ and Eq. (3.24) as $\mathbf{b}_{DG} = \mathbf{P}^{-1}\mathbf{G}(\mathbf{G}'\mathbf{P}^{-1}\mathbf{G})^{-1}\mathbf{d}$ (Itoh and Yamada 1986).

According to Itoh and Yamada (1986, 1988), Eq. (3.24) does not maximize the correlation between I and H (ρ_{IH}) nor the selection response because the covariance between I and H is not defined, given that $Cov(H, I) = \mathbf{w}'\mathbf{Gb}$ requires the economic weight vector $\mathbf{w}'$ and DG-LPSI does not use economic weights. However, note that because $\mathbf{Gb} = \mathbf{d}$, the variance of the DG-LPSI is $Var(I_{DG}) = \mathbf{d}'(\mathbf{GP}^{-1}\mathbf{G})^{-1}\mathbf{d} = \mathbf{b}'\mathbf{Pb}$.

In practice, $\mathbf{d}$ is chosen arbitrarily and then we are in the same situation as when economic weights need to be selected. Pesek and Baker (1969), Yamada et al. (1975), and Itoh and Yamada (1986, 1988) argued that this should not be a problem for experienced breeders because they must know the relative merits and demerits of their strains. However, this may be true only for some breeders and the selection of

d is always subjective. Another problem with this index is that, as it is not associated with $H = \mathbf{w}'\mathbf{g}$, it is not a predictor of $H = \mathbf{w}'\mathbf{g}$.

3.4 Applicability of the LPSI, RLPSI, and PPG-LPSI

In the context of animal breeding, Hazel (1943) pointed out that because any index is constructed from data on a herd in one locality, it may not be widely applicable. The reasons for this are:

1. Relative economic values for a trait may vary according to the particular locality or nature of the enterprise.
2. The genetic constitution of herds may differ, especially when they are under distinctly nonrandom mating systems such as intense inbreeding.
3. Different managerial practices may cause standard deviations for the traits to vary in different herds. The standard deviations for subjective traits such as market conformation measured by judging or by scores may vary because different judges vary the range over which they spread their scores.
4. Few herds are large enough to provide enough data to make the sampling errors of the genetic constants small. These limitations are applicable to the LPSI, RLPSI, and PPG-LPSI, and to all selection indices described in this book.

References

Brascamp EW (1984) Selection indices with constraints. Anim Breed Abstr 52(9):645–654
Cerón-Rojas JJ, Crossa J, Sahagún-Castellanos J (2016) Statistical sampling properties of the coefficients of three phenotypic selection indices. Crop Sci 56:51–58
Cunningham EP, Moen RA, Gjedrem T (1970) Restriction of selection indexes. Biometrics 26 (1):67–74
Harville DA (1975) Index selection with proportionality constraints. Biometrics 31(1):223–225
Hazel LN (1943) The genetic basis for constructing selection indexes. Genetics 8:476–490
Itoh Y, Yamada Y (1986) Re-examination of selection index for desired gains. Genet Sel Evol 18 (4):499–504
Itoh Y, Yamada Y (1987) Comparisons of selection indices achieving predetermined proportional gains. Genet Sel Evol 19(1):69–82
Itoh Y, Yamada Y (1988) Selection indices for desired relative genetic gains with inequality constraints. Theor Appl Genet 75:731–735
James JW (1968) Index selection with restriction. Biometrics 24:1015–1018
Kabakoff RI (2011) R in action: data analysis and graphics with R. Manning Publications Co., Shelter Island, NY
Leon-Garcia A (2008) Probability, statistics, and random processes for electrical engineering, 3rd edn. Pearson Education, Upper Saddle River, NJ
Lin CY (2005) A simultaneous procedure for deriving selection indexes with multiple restrictions. J Anim Sci 83:531–536
Mallard J (1972) The theory and computation of selection indices with constraints: a critical synthesis. Biometrics 28:713–735

Pesek J, Baker RJ (1969) Desired improvement in relation to selection indices. Can J Plant Sci 49:803–804
Rao CR (2002) Linear statistical inference and its applications, 2nd edn. Wiley, New York
Schott JR (2005) Matrix analysis for statistics, 2nd edn. Wiley, Hoboken, NJ
Searle SR (1966) Matrix algebra for the biological sciences. Wiley, New York
Tallis GM (1962) A selection index for optimum genotype. Biometrics 18:120–122
Tallis GM (1985) Constrained selection. Jpn J Genet 60(2):151–155
Yamada Y, Yokouchi K, Nishida A (1975) Selection index when genetic gains of individual traits are of primary concern. Jpn J Genet 50(1):33–41

4

LMSI and GW-LMSI

Abstract There are two main linear marker selection indices employed in marker-assisted selection (MAS) to predict the net genetic merit and to select individual candidates as parents for the next generation: the linear marker selection index (LMSI) and the genome-wide LMSI (GW-LMSI). Both indices maximize the selection response, the expected genetic gain per trait, and the correlation with the net genetic merit; however, applying the LMSI in plant or animal breeding requires genotyping the candidates for selection; performing a linear regression of phenotypic values on the coded values of the markers such that the selected markers are statistically linked to quantitative trait loci that explain most of the variability in the regression model; constructing the marker score, and combining the marker score with phenotypic information to predict and rank the net genetic merit of the candidates for selection. On the other hand, the GW-LMSI is a single-stage procedure that treats information at each individual marker as a separate trait. Thus, all marker information can be entered together with phenotypic information into the GW-LMSI, which is then used to predict the net genetic merit and select candidates. We describe the LMSI and GW-LMSI theory and show that both indices are direct applications of the linear phenotypic selection index theory to MAS. Using real and simulated data we validated the theory of both indices.

4.1 The Linear Marker Selection Index

4.1.1 Basic Conditions for Constructing the LMSI

In Chap. 2, Sect. 2.1, we indicated ten basic conditions for constructing a valid linear phenotypic selection index (LPSI). These ten conditions are also necessary for the linear marker selection index (LMSI); however, in addition to those conditions, the LMSI also requires the following conditions:

1. The markers and the quantitative trait loci (QTL) should be in linkage disequilibrium in the population under selection.
2. The QTL effects should be combined additively both within and between loci.

3. The QTL should be in coupling mode, that is, one of the initial lines should have all the alleles that have a positive effect on the chromosome, and the other lines should have all the negative effects.
4. The traits of interest should be affected by a few QTL with large effects (and possibly a number of very small QTL effects) rather than many small QTL effects.
5. The heritability of the traits should be low.
6. Markers correlated with the traits of interest should be identified.

Under these conditions, the LMSI should be more efficient than the LPSI, at least in the first selection cycles (Whittaker 2003; Moreau et al. 2007).

4.1.2 *The LMSI Parameters*

Let $y_i = g_i + e_i$ be the *i*th trait ($i = 1, 2, \ldots, t$, $t =$ number of traits), where $e_i \sim N(0, \sigma^2_{e_i})$ is the residual with expectation equal to zero and variance value $\sigma^2_{e_i}$, and N stands for normal distribution. Assuming that the QTL effects combine additively both within and between loci, the *i*th unobservable genetic value g_i can be written as

$$g_i = \sum_{k=1}^{N_Q} \alpha_k q_k, \tag{4.1}$$

where α_k is the effect of the *k*th QTL, q_k is the number of favorable alleles at the *k*th QTL (2, 1 or 0), and N_Q is the number of QTL affecting the *i*th trait of interest.

If the QTL effect values are not observable, the g_i values in Eq. (4.1) are also not observable; however, we can use a linear combination of the markers linked to the QTL (s_i) that affect the *i*th trait to predict the g_i value as

$$s_i = \sum_{j=1}^{M} \theta_j x_j, \tag{4.2}$$

where s_i is a predictor of g_i, θ_j is the regression coefficient of the linear regression model, x_j is the coded value of the *j*th markers (e.g., 1, 0, and -1 for marker genotypes *AA*, *Aa* and *aa* respectively), and M is the number of selected markers linked to the QTL that affect the *i*th trait. Equation (4.2) is called the *marker score* (Lande and Thompson 1990; Whittaker 2003) and this is the main reason why the LMSI is not equal to the LPSI described in Chap. 2. The number of selected markers is only a subset of potential markers linked to QTL in the population under selection; thus, the s_i values should be lower than or equal to the g_i values. One way of estimating the s_i values is to perform a linear regression of phenotypic values on the coded values of the markers, select markers that are statistically linked to quantitative trait loci that explain most of the variability in the regression model, and then obtain the estimated value of s_i ($\widehat{s}_i$) as the sum of the products of the QTL effects linked to markers and multiplied by the marker coded values associated with the ith trait. Some authors (e.g., Moreau et al. 2007) call $\widehat{s}_i$ the molecular score; in this book, we call s_i the marker score and $\widehat{s}_i$ the estimated marker score.

The objective of the LMSI is to predict the net genetic merit of each individual and select the individuals with the highest net genetic merit for further breeding. In the LMSI context, the net genetic merit can be written as

$$H = \mathbf{w}'\mathbf{g} + \mathbf{w}_2'\mathbf{s} = \begin{bmatrix} \mathbf{w}' & \mathbf{w}_2' \end{bmatrix} \begin{bmatrix} \mathbf{g} \\ \mathbf{s} \end{bmatrix} = \mathbf{a}'\mathbf{z}, \tag{4.3}$$

where $\mathbf{g}' = \begin{bmatrix} g_1 & \cdots & g_q \end{bmatrix}$ is the vector of breeding values; $\mathbf{w}' = \begin{bmatrix} w_1 & \cdots & w_t \end{bmatrix}$ is the vector of economic weights associated with $\mathbf{g}$; $\mathbf{w}_2' = \begin{bmatrix} 0_1 & \cdots & 0_t \end{bmatrix}$ is a null vector associated with the vector of marker scores $\mathbf{s}' = \begin{bmatrix} s_1 & \cdots & s_t \end{bmatrix}$; s_i is the ith marker score; $\mathbf{a}' = \begin{bmatrix} \mathbf{w}' & \mathbf{w}_2' \end{bmatrix}$ and $\mathbf{z} = \begin{bmatrix} \mathbf{g}' & \mathbf{s}' \end{bmatrix}$.

The information provided by the marker score can be used in breeding programs to increase the accuracy of predicting the net genetic merit of the individuals under selection. The LMSI combines the phenotypic and marker scores to predict H in each selection cycle and can be written as

$$I_M = \boldsymbol{\beta}_y'\mathbf{y} + \boldsymbol{\beta}_s'\mathbf{s} = \begin{bmatrix} \boldsymbol{\beta}_y' & \boldsymbol{\beta}_s' \end{bmatrix} \begin{bmatrix} \mathbf{y} \\ \mathbf{s} \end{bmatrix} = \boldsymbol{\beta}'\mathbf{t}, \tag{4.4}$$

where $\boldsymbol{\beta}_y'$ and $\boldsymbol{\beta}_s$ are vectors of phenotypic and marker score weights respectively; $\mathbf{y}' = \begin{bmatrix} y_1 & \cdots & y_t \end{bmatrix}$ is the vector of trait phenotypic values and $\mathbf{s}$ was defined in Eq. (4.3); $\boldsymbol{\beta}' = \begin{bmatrix} \boldsymbol{\beta}_y' & \boldsymbol{\beta}_s' \end{bmatrix}$ and $\mathbf{t}' = \begin{bmatrix} \mathbf{y}' & \mathbf{s}' \end{bmatrix}$.

The LMSI selection response can be written as

$$R_M = k_I \sigma_H \rho_{I_M H} = k_I \sigma_H \frac{\mathbf{a}'\mathbf{Z}_M\boldsymbol{\beta}}{\sqrt{\mathbf{a}'\mathbf{Z}_M\mathbf{a}}\sqrt{\boldsymbol{\beta}'\mathbf{T}_M\boldsymbol{\beta}}}, \tag{4.5}$$

where k_I is the standardized selection differential of the LMSI, $\sigma_H = \sqrt{\mathbf{a}'\mathbf{Z}_M\mathbf{a}}$ and $\sqrt{\boldsymbol{\beta}'\mathbf{T}_M\boldsymbol{\beta}}$ are the standard deviations of the variances of H and I_M, whereas $\rho_{I_M H}$ and $\mathbf{a}'\mathbf{Z}_M\boldsymbol{\beta}$ are the correlation and the covariance between H and I_M respectively; $\mathbf{T}_M = Var\begin{bmatrix} \mathbf{y} \\ \mathbf{s} \end{bmatrix} = \begin{bmatrix} \mathbf{P} & \mathbf{S} \\ \mathbf{S} & \mathbf{S} \end{bmatrix}$ and $\mathbf{Z}_M = Var\begin{bmatrix} \mathbf{g} \\ \mathbf{s} \end{bmatrix} = \begin{bmatrix} \mathbf{C} & \mathbf{S} \\ \mathbf{S} & \mathbf{S} \end{bmatrix}$ are block matrices of covariance where $\mathbf{P} = Var(\mathbf{y})$, $\mathbf{S} = Var(\mathbf{s})$, and $\mathbf{C} = Var(\mathbf{g})$ are the covariance matrices of phenotypic values ($\mathbf{y}$), the marker score ($\mathbf{s}$), and the genetic value ($\mathbf{g}$) respectively in the population. Vectors $\mathbf{a}$ and $\boldsymbol{\beta}$ were defined in Eqs. (4.3) and (4.4) respectively.

The LMSI expected genetic gain per trait can be written as

$$\mathbf{E}_M = k_I \frac{\mathbf{Z}_M \boldsymbol{\beta}}{\sqrt{\boldsymbol{\beta}' \mathbf{T}_M \boldsymbol{\beta}}}. \tag{4.6}$$

All the parameters in Eq. (4.6) were previously defined.

4.1.3 The Maximized LMSI Parameters

Suppose that **P**, **S** and **C** are known matrices; then, matrices $\mathbf{T}_M$ and $\mathbf{Z}_M$ are known and, according to the LPSI theory (Chap. 2 for details), the LMSI vector of coefficients ($\boldsymbol{\beta}_M$) that maximizes $\rho_{I_M H}$, R_M, and $\mathbf{E}_M$ can be written as

$$\boldsymbol{\beta} = \mathbf{T}_M^{-1} \mathbf{Z}_M \mathbf{a}, \tag{4.7}$$

whence the maximized selection response and the maximized correlation (or LMSI accuracy) between H and I_M can be written as

$$R_M = k_I \sqrt{\boldsymbol{\beta}' \mathbf{T}_M \boldsymbol{\beta}}, \tag{4.8a}$$

and

$$\rho_{I_M H} = \frac{\sigma_{I_M}}{\sigma_H}, \tag{4.8b}$$

respectively, where $\sigma_{I_M} = \sqrt{\boldsymbol{\beta}' \mathbf{T}_M \boldsymbol{\beta}}$ is the standard deviation of the variance of I_M and $\sigma_H = \sqrt{\mathbf{a}' \mathbf{Z}_M \mathbf{a}}$ is the deviation of the variance of H. Equations (4.8a) and (4.8b) show that the LMSI is a direct application of the LPSI theory in the marker-assisted selection (MAS) context.

Let $\mathbf{Q} = \mathbf{T}_M^{-1} \mathbf{Z}_M$; then, matrix **Q** can be written as

$$\mathbf{Q} = \begin{bmatrix} (\mathbf{P} - \mathbf{S})^{-1}(\mathbf{C} - \mathbf{S}) & \mathbf{0} \\ \mathbf{I} - (\mathbf{P} - \mathbf{S})^{-1}(\mathbf{C} - \mathbf{S}) & \mathbf{I} \end{bmatrix}, \tag{4.9}$$

whence $\boldsymbol{\beta} = \mathbf{Q}\mathbf{a}$, and as $\mathbf{w}_2' = [0_1 \quad \cdots \quad 0_t]$, we can write the two vectors of $\boldsymbol{\beta}' = [\boldsymbol{\beta}_y' \quad \boldsymbol{\beta}_s']$ as

$$\boldsymbol{\beta}_y = (\mathbf{P} - \mathbf{S})^{-1}(\mathbf{C} - \mathbf{S})\mathbf{w} \quad \text{and} \quad \boldsymbol{\beta}_s = \left[\mathbf{I} - (\mathbf{P} - \mathbf{S})^{-1}(\mathbf{C} - \mathbf{S})\right]\mathbf{w}. \tag{4.10a}$$

Another way of writing the marker score vector weights is

$$\boldsymbol{\beta}_s = \mathbf{w} - \boldsymbol{\beta}_y, \tag{4.10b}$$

where $\boldsymbol{\beta}_y = (\mathbf{P} - \mathbf{S})^{-1}(\mathbf{C} - \mathbf{S})\mathbf{w}$. By Eq. (4.10b), the optimal LMSI can be written as

$$I_M = \mathbf{w}'\mathbf{s} + \boldsymbol{\beta}_y'(\mathbf{y} - \mathbf{s}). \tag{4.11}$$

Equation (4.11) indicates that, in practice, to estimate the optimal LMSI, we only need to estimate the vector of coefficients $\boldsymbol{\beta}_y$. By Eq. (4.10a), Eq. (4.8a) can be written as

$$R_M = k_I\sqrt{\mathbf{w}'\mathbf{C}(\mathbf{P}-\mathbf{S})^{-1}(\mathbf{C}-\mathbf{S})\mathbf{w} + \mathbf{w}'\mathbf{S}\left[\mathbf{I}-(\mathbf{P}-\mathbf{S})^{-1}(\mathbf{C}-\mathbf{S})\right]\mathbf{w}}. \tag{4.12}$$

Thus, by Eqs. (4.10a) and (4.12), when $\mathbf{S}$ is a null matrix, vector $\boldsymbol{\beta}_y$ is equal to $\boldsymbol{\beta}_y = \mathbf{P}^{-1}\mathbf{C}\mathbf{w} = \mathbf{b}$ and $R_M = k_I\sqrt{\mathbf{b}'\mathbf{P}\mathbf{b}} = R_I$, which are the LPSI vector of coefficients and its selection response respectively.

Assume that when the number of markers and genotypes tend to infinity, $\mathbf{S}$ tends to $\mathbf{C}$; then, at the limit, we can suppose that $\mathbf{S} = \mathbf{C}$, and by this latter result, R_M is equal to

$$k_I\sqrt{\mathbf{w}'\mathbf{C}\mathbf{w}}. \tag{4.13}$$

That is, Eq. (4.13) is the maximum value of the LMSI selection response when the numbers of markers and genotypes tend to infinity. Thus, the possible LMSI selection response values of Eq. (4.12) should be between $k_I\sqrt{\mathbf{b}'\mathbf{P}\mathbf{b}}$ and $k_I\sqrt{\mathbf{w}'\mathbf{C}\mathbf{w}}$, i.e.,

$$k_I\sqrt{\mathbf{b}'\mathbf{P}\mathbf{b}} \le R_M \le k_I\sqrt{\mathbf{w}'\mathbf{C}\mathbf{w}}, \tag{4.14}$$

or between 1 and $\frac{\sqrt{\mathbf{w}'\mathbf{C}\mathbf{w}}}{\sqrt{\mathbf{b}'\mathbf{P}\mathbf{b}}} = \frac{\sigma_H}{\sigma_I}$, that is,

$$1 \le R_M \le \frac{\sigma_H}{\sigma_I}. \tag{4.15}$$

Note that $\frac{\sigma_H}{\sigma_I} = \frac{1}{\rho_{HI}}$, where ρ_{HI} is the maximized correlation between the net genetic merit (H) and the LPSI (I) described in Chap. 2. Equation (4.15) indicates that LMSI efficiency tends to infinity when the ρ_{HI} value tends to zero and is an additional way of denoting the paradox of LMSI efficiency described by Knapp (1998), which implies that LMSI efficiency tends to infinity when the ρ_{HI} value tends to zero.

4.1.4 The LMSI for One Trait

For the one-trait case, matrices $\mathbf{T}_M$, $\mathbf{Z}_M$, and $\mathbf{Q}$ can be written as

$$\mathbf{T}_M = \begin{bmatrix} \sigma_y^2 & \sigma_s^2 \\ \sigma_s^2 & \sigma_s^2 \end{bmatrix}, \quad \mathbf{Z}_M = \begin{bmatrix} \sigma_g^2 & \sigma_s^2 \\ \sigma_s^2 & \sigma_s^2 \end{bmatrix} \quad \text{and} \quad \mathbf{Q} = \begin{bmatrix} \frac{\sigma_g^2 - \sigma_s^2}{\sigma_y^2 - \sigma_s^2} & 0 \\ \frac{\sigma_y^2 - \sigma_g^2}{\sigma_y^2 - \sigma_s^2} & 1 \end{bmatrix}, \tag{4.16}$$

where σ_y^2, σ_g^2, and σ_s^2 are the phenotypic, genetic, and marker score variances respectively. By Eqs. (4.10a) and (4.10b), when $\mathbf{a}' = [1 \quad 0]$, the elements of vector $\boldsymbol{\beta} = \mathbf{Qa}$ are

$$\beta_y = \frac{\sigma_g^2 - \sigma_s^2}{\sigma_y^2 - \sigma_s^2} \quad \text{and} \quad \beta_s = 1 - \beta_y, \tag{4.17a}$$

whence the optimal LMSI can be written as

$$I_M = s + \beta_y(y - s); \tag{4.17b}$$

whereas by Eq. (4.12), the maximized LMSI selection response can be written as

$$R_M = k_I \sqrt{\frac{\sigma_g^2\left(\sigma_g^2 - \sigma_s^2\right) + \sigma_s^2\left(\sigma_y^2 - \sigma_g^2\right)}{\sigma_y^2 - \sigma_s^2}}. \tag{4.18}$$

When $\sigma_s^2 = 0, \beta_y = \frac{\sigma_g^2}{\sigma_y^2} = h^2, I_M = h^2 y$, and $R_M = k\frac{\sigma_g^2}{\sigma_y} = k\sigma_y h^2 = R$, the selection response for the one-trait case without markers.

4.1.5 *Efficiency of LMSI Versus LPSI Efficiency for One Trait*

Suppose that the intensity of selection is the same in both indices; then, to compare LMSI versus LPSI efficiency for predicting the net genetic merit, we can use the ratio $\lambda_M = \frac{\rho_{I_M H}}{\rho_{HI}} = \frac{R_M}{R_I}$ (Bulmer 1980; Moreau et al. 1998), where R_I is the maximized LPSI selection response. In percentage terms, the LMSI versus LPSI efficiency can be written as

$$p_M = 100(\lambda_M - 1). \tag{4.19}$$

When $p_M = 0$, the efficiency of both indices is the same; when $p_M > 0$, the efficiency of the LMSI is higher than that of the LPSI, and when $p_M < 0$, LPSI efficiency is higher than LMSI efficiency for predicting the net genetic merit.

In the case of one trait, Lande and Thompson (1990) showed that LMSI efficiency (not in percentage terms) with respect to phenotypic efficiency can be written as

$$\lambda_M = \frac{R_M}{R} = \sqrt{\frac{q}{h^2} + \frac{(1-q)^2}{1-qh^2}}, \tag{4.20}$$

where R_M was defined in Eq. (4.18), $R = k\sigma_y h^2$, h^2 is the trait heritability, and $q = \frac{\sigma_s^2}{\sigma_g^2}$ is the proportion of additive genetic variance explained by the markers. According to Eq. (4.20), the advantage of the LMSI over phenotypic selection increases as the population size increases and heritability decreases, because in such cases, $q = \frac{\sigma_s^2}{\sigma_g^2}$ tends to 1 and Eq. (4.20) approaches $\frac{1}{h}$. Therefore, the LMSI is most efficient for traits with low heritability and when the marker score explains a large proportion of the genetic variance. Thus, note that when h^2 tends to zero, $\frac{1}{h}$ tends to infinity; this means that in the asymptotic context, LMSI efficiency with respect to phenotypic efficiency for one trait (Eq. 4.20) tends to infinity and this is the LMSI paradox pointed out by Knapp (1998). There are other problems associated with the LMSI: it increases the selection response only in the short term and can result in lower cumulative responses in the longer term than phenotypic selection, as the LMSI fixes the QTL at a faster rate than phenotypic selection. In addition, it requires the weights (Eq. 4.17a) to be updated, because in each generation the frequency of the QTL changes (Dekkers and Settar 2004).

4.1.6 Statistical LMSI Properties

Assume that H and I_M have bivariate joint normal distribution, $\boldsymbol{\beta} = \mathbf{T}_M^{-1}\mathbf{Z}_M\mathbf{a}$, and that $\mathbf{P}$, $\mathbf{C}$, $\mathbf{S}$, and $\mathbf{w}$ are known; then, the statistical LMSI properties are the same as the LPSI properties described in Chap. 2. That is,

1. $\sigma_{I_M}^2 = \sigma_{HI_M}$: the variance of I_M ($\sigma_{I_M}^2$) and the covariance between H and I_M (σ_{HI_M}) are the same.
2. The maximized correlation between H and I_M (or I_M accuracy) is $\rho_{HI_M} = \frac{\sigma_{I_M}}{\sigma_H}$.
3. The variance of the predicted error, $Var(H - I_M) = \left(1 - \rho_{HI_M}^2\right)\sigma_H^2$, is minimal.
4. The total variance of H explained by I_M is $\sigma_{I_M}^2 = \rho_{HI_M}^2\sigma_H^2$.
5. The heritability of I_M is $\mathrm{h}_{\mathrm{M}}^2 = \frac{\boldsymbol{\beta}_M'\mathbf{Z}_M\boldsymbol{\beta}_M}{\boldsymbol{\beta}_M'\mathbf{T}_M\boldsymbol{\beta}_M}$.

Properties 1 to 4 are the same as LPSI properties 1 to 4, but, because the LMSI jointly incorporates the phenotypic and marker information to predict the net genetic merit, LMSI accuracy should be higher than LPSI accuracy. The same is true of the LMSI selection response and expected genetic gain per trait when compared with the LPSI selection response and expected genetic gain per trait.

4.2 The Genome-Wide Linear Selection Index

The genome-wide linear marker selection index (GW-LMSI) is a single-stage procedure that treats information at each individual marker as a separate trait. Thus, all marker information can be entered together with phenotypic information into the GW-LMSI, which is then used to predict the net genetic merit. In a similar manner to the LMSI, the GW-LMSI exploits the linkage disequilibrium between markers and the QTL produced when inbred lines are crossed.

4.2.1 The GW-LMSI Parameters

In a similar manner to the LPSI, the main objective of the GW-LMSI is to predict the net genetic merit values of each individual and select the best individuals for further breeding. In the GW-LMSI context, the net genetic merit can be written as

$$H = \mathbf{w}'\mathbf{g} + \mathbf{w}_2'\mathbf{m} = \begin{bmatrix} \mathbf{w}' & \mathbf{w}_2' \end{bmatrix} \begin{bmatrix} \mathbf{g} \\ \mathbf{m} \end{bmatrix} = \mathbf{a}_W'\mathbf{z}_W, \tag{4.21}$$

where $\mathbf{g}' = [g_1 \quad \cdots \quad g_t]$ ($j = 1, 2, \ldots, t$ = number of traits) is the vector of breeding values, $\mathbf{w}' = [w_1 \quad \cdots \quad w_t]$ is the vector of economic weights associated with the breeding values, and $\mathbf{w}_2' = [0_1 \quad \cdots \quad 0_m]$ is a null vector associated with the coded values of the markers $\mathbf{m}' = [m_1 \quad \cdots \quad m_m]$, where m_j ($j = 1, 2, \ldots, m$ = number of markers) is the jth marker in the training population; $\mathbf{a}_W' = [\mathbf{w}' \quad \mathbf{w}_2']$ and $\mathbf{z}_W = [\mathbf{g}' \quad \mathbf{m}']$.

The GW-LMSI (I_W) combines the phenotypic value and the molecular information linked to the individual traits to predict H values in each selection cycle. It can be written as

$$I_W = \boldsymbol{\beta}_y'\mathbf{y} + \boldsymbol{\beta}_m'\mathbf{m} = \begin{bmatrix} \boldsymbol{\beta}_y' & \boldsymbol{\beta}_m' \end{bmatrix} \begin{bmatrix} \mathbf{y} \\ \mathbf{m} \end{bmatrix} = \boldsymbol{\beta}_W'\mathbf{t}_W, \tag{4.22}$$

where $\boldsymbol{\beta}_y'$ and $\boldsymbol{\beta}_m$ are vectors of phenotypic and marker weights respectively; $\mathbf{y}' = [y_1 \quad \cdots \quad y_t]$ is the vector of phenotypic values and $\mathbf{m}$ was defined in Eq. (4.21); $\boldsymbol{\beta}_W' = \begin{bmatrix} \boldsymbol{\beta}_y' & \boldsymbol{\beta}_m' \end{bmatrix}$ and $\mathbf{t}_W' = [\mathbf{y}' \quad \mathbf{m}']$.

The GW-LSI selection response can be written as

$$R_W = k_I \sigma_H \rho_{I_W H} = k_I \sigma_H \frac{\mathbf{a}_W'\boldsymbol{\Psi}\boldsymbol{\beta}_W}{\sqrt{\mathbf{a}_W'\boldsymbol{\Psi}\mathbf{a}_W}\sqrt{\boldsymbol{\beta}_W'\boldsymbol{\Phi}\boldsymbol{\beta}_W}}, \tag{4.23a}$$

where k_I is the standardized selection differential of the GW-LMSI, $\sigma_H^2 = \mathbf{a}_W'\boldsymbol{\Psi}\mathbf{a}_W$ and $Var(I_W) = \boldsymbol{\beta}_W'\boldsymbol{\Phi}\boldsymbol{\beta}_W$ are the variance of H and I_W, whereas $\rho_{I_W H} = \frac{\mathbf{a}_W'\boldsymbol{\Psi}\boldsymbol{\beta}_W}{\sqrt{\mathbf{a}_W'\boldsymbol{\Psi}\mathbf{a}_W}\sqrt{\boldsymbol{\beta}_W'\boldsymbol{\Phi}\boldsymbol{\beta}_W}}$ and $\mathbf{a}_W'\boldsymbol{\Psi}\boldsymbol{\beta}_W$ are the correlation and the covariance between

H and I_W respectively; $\mathbf{\Phi} = Var\begin{bmatrix} \mathbf{y} \\ \mathbf{m} \end{bmatrix} = \begin{bmatrix} \mathbf{P} & \mathbf{W}' \\ \mathbf{W} & \mathbf{M} \end{bmatrix}$ and $\mathbf{\Psi} = Var\begin{bmatrix} \mathbf{g} \\ \mathbf{m} \end{bmatrix} = \begin{bmatrix} \mathbf{C} & \mathbf{W}' \\ \mathbf{W} & \mathbf{M} \end{bmatrix}$ are block covariance matrices where $\mathbf{P} = Var(\mathbf{y})$, $\mathbf{M} = Var(\mathbf{m})$, $\mathbf{C} = Var(\mathbf{g})$, and $\mathbf{W} = Cov(\mathbf{y}, \mathbf{m}) = Cov(\mathbf{g}, \mathbf{m})$ are the covariance matrices of phenotypic values (**y**), the molecular marker (**m**) coded values, and the genetic (**g**) values, whereas **W** is the covariance matrix between **y** and **m**, and between **g** and **m**. The size of matrices **P** and **C** is $t \times t$, but the sizes of matrices **M** and **W** are $m \times m$ and $m \times t$ respectively.

From a theoretical point of view, Crossa and Cerón-Rojas (2011) showed that matrix **M** can be written as

$$\mathbf{M} = \begin{bmatrix} 1 & (1-2\delta_{11}) & \cdots & (1-2\delta_{1N}) \\ (1-2\delta_{21}) & 1 & \cdots & (1-2\delta_{2N}) \\ \vdots & \vdots & \ddots & \vdots \\ (1-2\delta_{N1}) & (1-2\delta_{N2}) & \cdots & 1 \end{bmatrix}, \tag{4.23b}$$

where $(1 - 2\delta_{ij})$ is the covariance (or correlation) and δ_{ij} the recombination frequency between the ith and jth marker ($i, j = 1, 2, \ldots, m$ = number of markers). According to Crossa and Cerón-Rojas (2011), matrix **W** can be written as

$$\mathbf{W} = \begin{bmatrix} (1-2r_{11})\alpha_{11} & (1-2r_{11})\alpha_{12} & \cdots & (1-2r_{1N})\alpha_{1N_Q} \\ (1-2r_{21})\alpha_{21} & (1-2r_{22})\alpha_{22} & \cdots & (1-2r_{2N})\alpha_{2N_Q} \\ \vdots & \vdots & \ddots & \vdots \\ (1-2r_{t1})\alpha_{t1} & (1-2r_{N2})\alpha_{t2} & \cdots & (1-2r_{NN})\alpha_{tN_Q} \end{bmatrix}, \tag{4.23c}$$

where $(1 - 2r_{ik})\alpha_{qk}$ ($i = 1, 2, \ldots, m$, $k = 1, 2, \ldots, N_Q$ = number of QTL, $q = 1, 2, \ldots, t$) is the covariance between the qth trait and the ith marker; r_{ik} is the recombination frequency between the ith marker and the kth QTL; and α_{qk} is the effect of the kth QTL over the qth trait.

The GW-LMSI expected genetic gain per trait can be written as

$$\mathbf{E}_{LW} = k_I \frac{\mathbf{\Psi\beta}}{\sqrt{\mathbf{\beta}'\mathbf{\Phi\beta}}}. \tag{4.24}$$

All parameters in Eq. (4.24) were previously defined.

Matrix $\mathbf{\Phi}$ could be singular, i.e., its inverse ($\mathbf{\Phi}^{-1}$) could not exist because matrix **W** is singular. Suppose that matrices $\mathbf{\Phi}$ and $\mathbf{\Psi}$ are known; then, according to the LPSI theory, the GW-LMSI vector of coefficients ($\mathbf{\beta}_W$) that maximizes $\rho_{I_W H}$ can be written as

$$\boldsymbol{\beta}_W = \boldsymbol{\Phi}^{-}\boldsymbol{\Psi}\mathbf{a}_W, \tag{4.25a}$$

where matrix $\boldsymbol{\Phi}^{-}$ denotes a generalized inverse of $\boldsymbol{\Phi}$. By Eq. (4.25a), the maximized GW-LMSI selection response is

$$R_W = k_I\sqrt{\boldsymbol{\beta}'_W\boldsymbol{\Phi}\boldsymbol{\beta}_W}. \tag{4.25b}$$

Equations (4.25a) and (4.25b) show that the GW-LMSI is a direct application of the LPSI to MAS. By Eq. (4.25a), the maximized correlation between H and I_W is

$$\rho_{I_W H} = \frac{\sigma_{I_W}}{\sigma_H}, \tag{4.25c}$$

where $\sigma_{I_W} = \sqrt{\boldsymbol{\beta}'_W\boldsymbol{\Phi}\boldsymbol{\beta}_W}$ is the standard deviation of the variance of I_W and $\sigma_H = \sqrt{\mathbf{a}'_W\boldsymbol{\Psi}\mathbf{a}_W}$ is the standard deviation of the variance of H.

4.2.2 Relationship Between the GW-LMSI and the LPSI

Matrix $\boldsymbol{\Phi}^{-}$ can be written as

$$\boldsymbol{\Phi}^{-} = \begin{bmatrix} \mathbf{L}^{-} & -\mathbf{L}^{-}\mathbf{W}'\mathbf{M}^{-} \\ -\mathbf{M}^{-}\mathbf{W}\mathbf{L}^{-} & \mathbf{M}^{-} + \mathbf{M}^{-}\mathbf{W}\mathbf{L}^{-}\mathbf{W}'\mathbf{M}^{-} \end{bmatrix}, \tag{4.26}$$

where $\mathbf{L}^{-}$ is a generalized inverse of matrix $\mathbf{L} = \mathbf{P} - \mathbf{W}'\mathbf{M}^{-}\mathbf{W}$, and $\mathbf{M}^{-}$ is a generalized inverse of matrix $\mathbf{M}$. In matrix $\boldsymbol{\Phi}^{-}$, the inverse of matrix $\mathbf{W}$ is not required and the standard inverse of matrix $\mathbf{M}$ ($\mathbf{M}^{-1}$) may exist. In the latter case, the standard inverse of matrix $\mathbf{L}$ ($\mathbf{L}^{-1}$) exists and can be written as $\mathbf{L}^{-1} = (\mathbf{P} - \mathbf{W}'\mathbf{M}^{-1}\mathbf{W})^{-1} = \mathbf{P}^{-1} + \mathbf{P}^{-1}\mathbf{W}'[\mathbf{M} - \mathbf{W}\mathbf{P}^{-1}\mathbf{W}']^{-1}\mathbf{W}\mathbf{P}^{-1}$ (Searle et al. 2006).

By Eq. (4.26) and because $\mathbf{w}'_2 = [0_1 \ \cdots \ 0_N]$, the vector components of $\boldsymbol{\beta}'_W = [\boldsymbol{\beta}'_y \ \ \boldsymbol{\beta}'_m]$, or $\boldsymbol{\beta}_W = \boldsymbol{\Phi}^{-}\boldsymbol{\Psi}\mathbf{a}_W$, can be written as

$$\boldsymbol{\beta}_y = [\mathbf{L}^{-}\mathbf{C} - \mathbf{L}^{-}\mathbf{W}'\mathbf{M}^{-}\mathbf{W}]\mathbf{w} \tag{4.27}$$

and

$$\boldsymbol{\beta}_m = [(\mathbf{M}^{-} + \mathbf{M}^{-}\mathbf{W}\mathbf{L}^{-}\mathbf{W}'\mathbf{M}^{-})\mathbf{W} - \mathbf{M}^{-}\mathbf{W}\mathbf{L}^{-}\mathbf{C}]\mathbf{w}, \tag{4.28}$$

where $\mathbf{w}$ is the vector of economic weights. Suppose that there is no marker information; then, matrices $\mathbf{M}$ and $\mathbf{W}$ are null and Eq. (4.27) is equal to $\boldsymbol{\beta}_y = \mathbf{P}^{-1}\mathbf{C}\mathbf{w} = \mathbf{b}$ (the LPSI vector of coefficients), whereas $\boldsymbol{\beta}_m = \mathbf{0}$ and $R_W = k_I\sqrt{\boldsymbol{\beta}'_W\boldsymbol{\Phi}\boldsymbol{\beta}_W} = k_I\sqrt{\mathbf{b}'\mathbf{P}\mathbf{b}} = R_I$, the LPSI selection response. Now suppose that the markers explain all the genetic variability; in this case, $\boldsymbol{\beta}_y = \mathbf{0}$ and $\boldsymbol{\beta}_m = (\mathbf{X}'\mathbf{X})^{-}\mathbf{X}'\mathbf{Y}$, the matrix of linear regression coefficients in the multivariate context, where $(\mathbf{X}'\mathbf{X})^{-}$ is a generalized inverse matrix of $\mathbf{X}'\mathbf{X}$ and $\mathbf{Y}$ is a matrix of phenotypic observations.

4.2.3 *Statistical Properties of GW-LMSI*

Assume that H and I_W have bivariate joint normal distribution, $\boldsymbol{\beta}_W = \boldsymbol{\Phi}^{-}\boldsymbol{\Psi}\mathbf{a}_W$, and $\mathbf{P}$, $\mathbf{C}$, $\mathbf{M}$, $\mathbf{W}$, and $\mathbf{w}$ are known; then, the statistical GW-LMSI properties are the same as the LMSI properties. That is,

1. $\sigma^2_{I_W} = \sigma_{HI_W}$, i.e., the variance of I_W ($\sigma^2_{I_W}$) and the covariance between H and I_W (σ_{HI_W}) are the same.
2. The maximized correlation between H and I_W, or I_W accuracy, is $\rho_{HI_W} = \frac{\sigma_{I_W}}{\sigma_H}$.
3. The variance of the predicted error, $Var(H - I_W) = \left(1 - \rho^2_{HI_W}\right)\sigma^2_H$, is minimal.
4. The total variance of H explained by I_W is $\sigma^2_{I_W} = \rho^2_{HI_W}\sigma^2_H$.

According to Lange and Whittaker (2001), GW-LMSI efficiency should be greater than LMSI efficiency. However, this would be true only if matrices $\mathbf{P}$, $\mathbf{C}$, $\mathbf{M}$, and $\mathbf{W}$ are known and trait heritability is very low.

4.3 Estimating the LMSI Parameters

When covariance matrices $\mathbf{P}$, $\mathbf{C}$, and $\mathbf{S}$, and the vector of economic weights ($\mathbf{w}$) are known, there is no error in the estimation of the LMSI parameters (selection response, expected genetic gain, etc.); the same is true for the GW-LMSI when, in addition to $\mathbf{P}$, $\mathbf{C}$, and $\mathbf{w}$, the covariance matrices $\mathbf{M}$ and $\mathbf{W}$ are known. In such cases, the relative efficiency of the LMSI (GW-LMSI) depends only on the heritability of the traits and on the portion of phenotypic variation associated with markers. Using simulated data, Lange and Whittaker (2001) found that GW-LMSI efficiency was higher than LMSI efficiency when trait heritability was 0.2 and matrices $\mathbf{P}$, $\mathbf{C}$, $\mathbf{M}$, and $\mathbf{W}$ were known. When $\mathbf{P}$, $\mathbf{C}$, $\mathbf{S}$, $\mathbf{M}$, and $\mathbf{W}$ are unknown, it is necessary to estimate them; then, the LMSI and GW-LMSI vector of coefficients and the effects associated with markers are estimated with some error. This error leads to lower LMSI and GW-LMSI efficiency than expected under the assumption that the parameters are known; however, in the latter case, Lange and Whittaker (2001) also found that GW-LMSI efficiency was greater than that of the LMSI when trait heritability was 0.05. Moreover, in the LMSI there is additional bias in the estimation of the parameters because only markers with significant effects are included in the index (Moreau et al. 1998).

In Chap. 2, we described the restricted maximum likelihood (REML) method for estimating matrices $\mathbf{P}$ and $\mathbf{C}$. Some authors (Lande and Thompson 1990; Charcosset and Gallais 1996; Hospital et al. 1997; Moreau et al. 1998, 2007) have described methods for estimating marker scores, the variance of the marker scores, the LMSI vector of coefficients, etc., in the context of one trait; however, up to now there have been no reports on the estimation of matrix **S** in the multi-trait case. Lange and Whittaker (2001) only indicated that matrix **S** can be estimated as $\widehat{\mathbf{S}} = Var(\widehat{\mathbf{s}})$, where $\widehat{\mathbf{s}}$ is a vector of estimated marker scores associated with several individual traits.

The main problems associated with the estimated LMSI parameters are:

1. The estimated values of the covariance matrix **S** ($\widehat{\mathbf{S}}$) tend to overestimate the genetic covariance matrix (**C**).
2. The estimated variances of the marker scores can be negative.

When the first point is true, the estimated LMSI selection response and efficiency could be negative because the estimated matrix $\widehat{\mathbf{T}}_M = \begin{bmatrix} \widehat{\mathbf{P}} & \widehat{\mathbf{S}} \\ \widehat{\mathbf{S}} & \widehat{\mathbf{S}} \end{bmatrix}$ is not positive definite (all eigenvalues positive) and the estimated matrix $\widehat{\mathbf{Z}}_M = \begin{bmatrix} \widehat{\mathbf{G}} & \widehat{\mathbf{S}} \\ \widehat{\mathbf{S}} & \widehat{\mathbf{S}} \end{bmatrix}$ is not positive semi-definite (no negative eigenvalues). In addition, the results can lead to all weights being placed on the molecular score and the weights on the phenotype values can be negative (Moreau et al. 2007). When the second point is true, the variance of the marker scores is not useful. The two problems indicated above could be caused by using the same data set to select markers and to estimate marker effects, and there is no simple way of solving them. Lande and Thompson (1990) proposed that the markers used to obtain $\widehat{\mathbf{S}}$ be selected a priori as those with the most highly significant partial regression coefficients from among all the markers in the linkage group analyzed in the previous generation. Zhang and Smith (1992, 1993) proposed using two independent sets of markers: one to estimate marker effects and the other to select markers. Additional solutions to these problems were described by Moreau et al. (2007).

In this subsection, we describe methods (in the univariate and multivariate context) for estimating molecular marker effects, marker scores, and their variance and covariance, and for estimating the LMSI and GW-LMSI vector of coefficients, selection response, expected genetic gain, and accuracy. This subsection is only for illustration; we use the same data set to select markers, and to estimate marker effects and the variance of marker scores.

4.3.1 Estimating the Marker Score

According to Eqs. (4.11) and (4.17b), when the vector of economic weights is equal to $\mathbf{a}' = [1 \quad 0]$, the LMSI for the ith trait y_i ($i = 1, 2, \cdots, t$; t = number of traits) value can be written as $I_{M_{li}} = s_i + \beta_{y_i}(y_i - s_i)$ ($l = 1, 2, \cdots, n$; n = number of individuals or genotypes), where $\beta_{yi} = \frac{\sigma^2_{g_i} - \sigma^2_{s_i}}{\sigma^2_{y_i} - \sigma^2_{s_i}} = \frac{h^2_i(1 - q_i)}{1 - q_i h^2_i}$ is the LMSI coefficient, $h^2_i = \frac{\sigma^2_{g_i}}{\sigma^2_{y_i}}$ is the heritability of the ith trait, and $q_i = \frac{\sigma^2_{s_i}}{\sigma^2_{g_i}}$ is the proportion of genetic variance explained by the QTL or markers associated with the ith trait; $s_i = \sum_{j=1}^{M} \theta_j x_j$ ($j = 1, 2, \cdots, M$; M = number of selected markers) is the ith individual trait marker score; and $\sigma^2_{y_i}$, $\sigma^2_{g_i}$, and $\sigma^2_{s_i}$ are the ith variances of the phenotypic, genetic, and marker score values respectively.

The simplest way of estimating the ith marker score s_i is to perform a multiple linear regression of phenotypic values (y_i) on the coded values of the markers (x_j) and then select the markers statistically linked to the ith QTL that explain most of the variability in the regression model and use them to construct $s_i = \sum_{j \in M} \theta_j x_j$.

We can fit the model $y^*_i = \sum_{j \in M} \theta_j x_j + e$, where $y^*_i = y_i - \bar{y}_i$ and $\bar{y}_i$ are the average values of the ith trait, by maximum likelihood or least squares. When estimating θ_j, the main problem is to choose the set of markers M based on criteria for declaring markers as significant and then use the estimated values of θ_j ($\widehat{\theta}_j$) to estimate the ith marker score s_i as $\widehat{s}_i = \sum_{j \in M} \widehat{\theta}_j x_j$. The values of $\widehat{s}_i$ may increase or decrease according to the number of markers (x_j) included in the model, and $\widehat{s}_i$ affects LMSI selection response and efficiency by means of the estimated variance of $\widehat{s}_i$ ($\widehat{\sigma}^2_{s_i}$) (Figs. 4.1 and 4.2).

According to the least squares method of estimation, $\widehat{\boldsymbol{\theta}} = (\mathbf{X}'\mathbf{X})^{-1}\mathbf{X}'\mathbf{y}^*$ is an estimator of the vector of regression coefficients $\boldsymbol{\theta}' = [\theta_1 \quad \theta_2 \quad \cdots \quad \theta_m]$, where m ($m < n$) is the number of markers, $\mathbf{X}$ is a matrix $n \times m$ of coded marker values (e.g., 1, 0 and -1 for marker genotypes AA, Aa, and aa respectively) and $\mathbf{y}^*$ is a vector $n \times 1$ of phenotypic values centered based on its average values. Only a subset $M(M < m)$ of the m markers is statistically linked to the QTL and then only a subset M of the estimated vector $\widehat{\boldsymbol{\theta}}$ values is selected to estimate s_i as $\widehat{s}_i = \sum_{j=1}^{M} \widehat{\theta}_j x_j$.

To illustrate how to obtain $\widehat{s}_i = \sum_{j \in M} \widehat{\theta}_j x_j$, we use a real maize (*Zea mays*) F_2 population with 247 genotypes (each one with two repetitions), 195 molecular markers, and four traits – grain yield (GY, ton ha^{-1}); plant height (PHT, cm), ear height (EHT, cm), and anthesis day (AD, days) – evaluated in one environment. In an F_2 population, the marker homozygous loci for the allele from the first parental line can be coded by 1, whereas the marker homozygous loci for the allele from the second parental line can be coded by -1, and the marker heterozygous loci by 0.

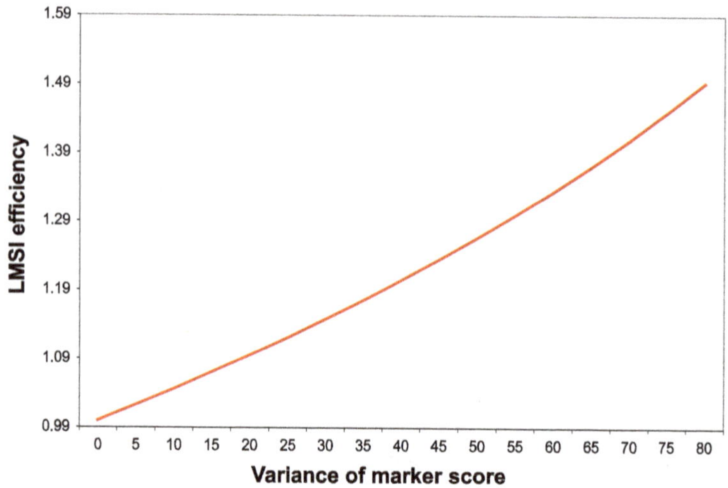

Fig. 4.1 Efficiency of the linear molecular selection index with respect to phenotypic selection for the one-trait case for different values of the variance of the marker score when the phenotypic and genetic variances are fixed

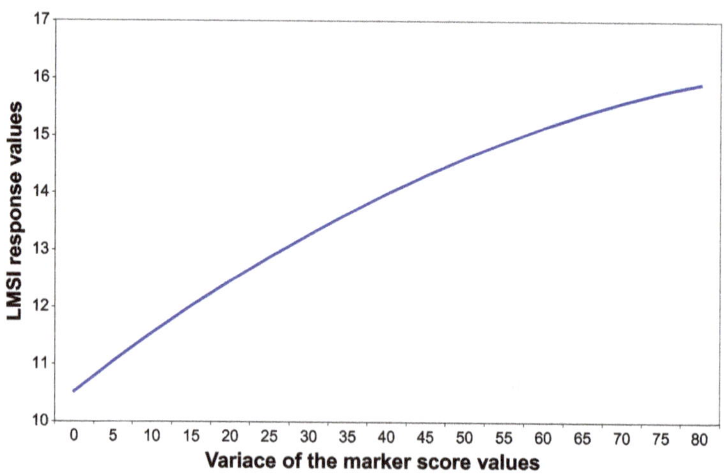

Fig. 4.2 Selection response values of the linear molecular selection index for the one-trait case for different values of the variance of the marker score when the phenotypic and genetic variances are fixed

For this example, we used trait PHT. Only seven markers were statistically linked to the PHT. The estimated vector of regression coefficients for these seven markers was $\widehat{\boldsymbol{\theta}}' = [5.46 \quad -4.54 \quad 0.98 \quad 7.39 \quad -7.75 \quad -1.91 \quad -3.53]$. Table 4.1 presents the first 20 genotypes, the coded values of the seven selected markers, and the first 20 estimated $\widehat{s}_{PHT}$ values of the 247 genotypes in the maize (*Zea mays*) F_2

Table 4.1 Number of selected genotypes, coded values of seven selected markers, and estimated marker score values obtained from a maize (*Zea mays*) F_2 population with 247 genotypes and 195 molecular markers

Number of genotypes	Coded values of the selected markers							Marker score
	M1	M2	M3	M4	M5	M6	M7	
1	0	0	0	0	0	1	−1	1.62
2	−1	−1	0	0	0	−1	0	0.99
3	0	0	0	0	0	0	1	−3.53
4	1	1	0	0	0	−1	−1	6.37
5	1	1	0	−1	−1	−1	−1	6.72
6	0	0	1	0	0	0	0	0.98
7	1	1	0	1	1	0	0	0.57
8	0	0	0	0	0	0	0	0
9	0	0	1	0	0	1	0	−0.93
10	0	0	1	1	0	0	1	4.84
11	0	0	0	0	0	0	0	0
12	−1	−1	0	0	0	0	0	−0.92
13	0	0	0	0	0	0	0	0
14	1	1	0	−1	−1	0	−1	4.81
15	0	0	1	−1	−1	0	0	1.34
16	0	0	0	0	0	0	0	0
17	−1	−1	0	0	0	0	1	−4.46
18	−1	−1	0	0	0	0	1	−4.46
19	−1	−1	1	0	0	−1	1	−1.56
20	0	0	0	0	0	0	−1	3.53

population. According to $\widehat{\boldsymbol{\theta}}'$ and the coded values of the seven markers, the first estimated $\widehat{s}_{PHT}$ value was obtained as $\widehat{s}_{PHT1} = -1.91(1) + -3.53(-1) = 1.62$; the second estimated $\widehat{s}_{PHT}$ value was obtained as $\widehat{s}_{PHT2} = 5.46(-1)+ -4.54(-1) - 1.91(-1) = 0.99$, etc. The 20th estimated $\widehat{s}_{PHT}$ value was obtained as $\widehat{s}_{PHT20} = -3.53(-1) = 3.53$. This estimation procedure is valid for any number of genotypes and markers.

Figure 4.3 shows the distribution of the 247 estimated marker scores associated with traits PHT and EHT of the maize F_2 population. Note that the estimated marker score values approach normal distribution.

4.3.2 Estimating the Variance of the Marker Score

There are many methods of estimating the variance of the marker score associated with the *i*th trait ($\sigma^2_{s_i}$); the first one was proposed by Lande and Thompson (1990). According to these authors, $\sigma^2_{s_i}$ can be estimated as

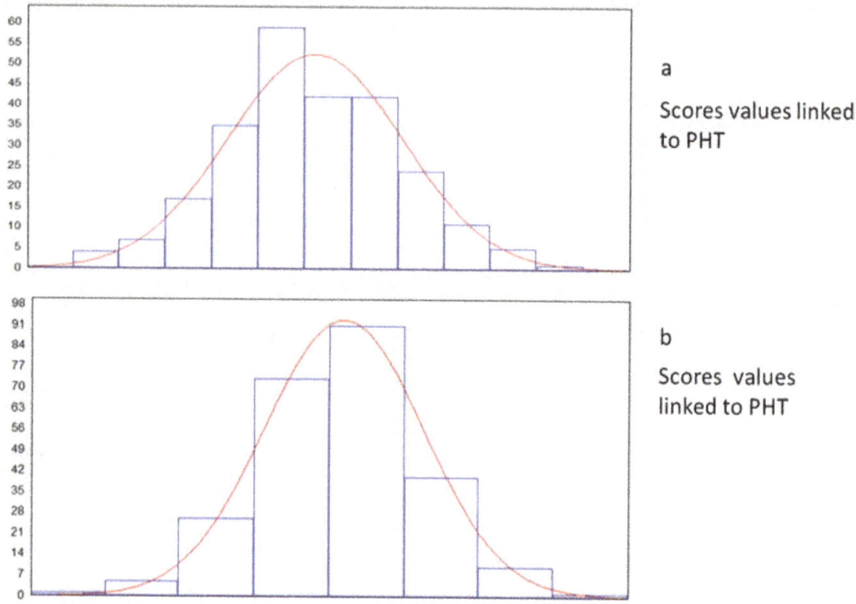

Fig. 4.3 Distribution of the marker scores associated with traits (**a**) plant height and (**b**) ear height of a maize (*Zea mays*) F_2 population. Note that the distribution of frequencies of the marker score values approaches normal distribution

$$\widehat{\sigma}^2_{\widehat{s}_i} = \widehat{\boldsymbol{\theta}}_i'\mathbf{M}_i\widehat{\boldsymbol{\theta}}_i - \frac{M\widehat{\sigma}^2_{e_i}}{n}, \tag{4.29}$$

where $\widehat{\boldsymbol{\theta}}_i$ is the estimated vector of regression coefficients of the selected markers, $\mathbf{M}_i = \frac{2}{n}\mathbf{X}_i'\mathbf{X}_i$ is the covariance matrix $M \times M$ of the selected markers that are statistically linked to the *i*th trait marker loci; $\widehat{\sigma}^2_{e_i} = \frac{\mathbf{y}'(\mathbf{I}-\mathbf{H})\mathbf{y}}{n-M-1}$ is the unbiased estimated variance of the residuals, $\mathbf{H} = \mathbf{I} - \mathbf{X}_i\left(\mathbf{X}_i'\mathbf{X}_i\right)^{-1}\mathbf{X}_i'$, $\mathbf{I}$ is an identity matrix $n \times n$, M is the number of selected markers statistically linked to the QTL, and $\mathbf{X}_i$ is a matrix $n \times M$ with the coded values of the selected markers. According to Lande and Thompson (1990), Eq. (4.29) is an unbiased estimator of $\sigma^2_{s_i}$ and its variance can be written as

$$Var\left(\widehat{\sigma}^2_{\widehat{s}_i}\right) = \frac{4\sigma^2_{s_i}\sigma^2_{e_i}}{n} + \frac{2M\left(\sigma^2_{e_i}\right)^2}{n^2} + \frac{2M^2\left(\sigma^2_{e_i}\right)^2}{n^2(n-M)}, \tag{4.30}$$

which tends to zero when n, the number of genotypes or individuals, is very high.

From Eq. (4.29), it is possible to obtain an estimator of the covariance between the *i*th and *j*th marker scores when the number of selected markers statistically linked to the QTL is the same in the *i*th and *j*th traits. Thus, by Eq. (4.29), the covariance between the *i*th and *j*th marker scores can be estimated as

$$\widehat{\sigma}_{\widehat{s}_{ij}} = \widehat{\boldsymbol{\theta}}_i'\mathbf{M}_{ij}\widehat{\boldsymbol{\theta}}_j - \frac{M\widehat{\sigma}_{e_{ij}}}{n}, \tag{4.31}$$

where $\widehat{\boldsymbol{\theta}}_i$ and $\widehat{\boldsymbol{\theta}}_j$ are the estimated vectors of regression coefficients of the selected markers associated with the *i*th and *j*th trait loci respectively; $\mathbf{M}_{ij} = \frac{2}{n}\mathbf{X}_i'\mathbf{X}_j$ is the covariance matrix $M \times M$ of the markers statistically linked to the *i*th and *j*th trait marker loci; $\mathbf{X}_i$ and $\mathbf{X}_j$ are $n \times M$ matrices with the coded values of the selected markers associated with the *i*th and *j*th trait loci respectively; $\widehat{\sigma}_{e_{ij}} = \frac{\mathbf{y}_i'(\mathbf{I} - \mathbf{H}_{ij})\mathbf{y}_j}{n - M - 1}$ is the estimated covariance of the residuals between the *i*th ($\mathbf{y}_i$) and *j*th ($\mathbf{y}_j$) trait values, $\mathbf{H}_{ij} = \mathbf{I} - \mathbf{X}_i(\mathbf{X}_i'\mathbf{X}_j)^{-1}\mathbf{X}_j'$, $\mathbf{I}$ is an identity matrix $n \times n$, and M is the number of selected markers statistically linked to the QTL.

According to the PHT values described in Sect. 4.3.1 of this chapter, $M = 7$, $n = 247$, $\widehat{\sigma}^2_{e_i} = 180.80$ and $\widehat{\sigma}^2_{\widehat{s}_{PHT}} = 48.23$ (Eq. 4.29). Note that $\widehat{\sigma}^2_{\widehat{s}_{PHT}} \leq \widehat{\sigma}^2_{g_{PHT}}$, where $\widehat{\sigma}^2_{g_{PHT}} = 83.0$ is an estimate of the genetic variance of PHT. The estimated portion of the genetic variance attributable to $\widehat{\sigma}^2_{\widehat{s}_{PHT}} = 48.23$ was $\widehat{q}_{PHT} = \frac{48.23}{83} = 0.5811$; that is, the seven markers explain 58.11% of the genetic variance associated with PHT.

Charcosset and Gallais (1996) considered two possible methods of estimating $\sigma^2_{s_i}$ based on the coefficient of multiple determination or squared multiple correlation R^2 (note that in this case R^2 is not the square of the selection response). The coefficient R^2 gives the portion of the total variation in the phenotypic values that is "explained" by, or attributable to, the markers and can be written as

$$R^2 = \frac{\widehat{\boldsymbol{\theta}}\mathbf{X}'\mathbf{y} - n\bar{y}^2}{\mathbf{y}'\mathbf{y} - n\bar{y}^2} = \frac{\widehat{\sigma}^2_s}{\widehat{\sigma}^2_y}, \tag{4.32a}$$

where $\widehat{\boldsymbol{\theta}}\mathbf{X}'\mathbf{y} - n\bar{y}^2$ is the overall regression sum of squares adjusted for the intercept and $\mathbf{y}'\mathbf{y} - n\bar{y}^2$ is the total sum of squares adjusted for the mean. The coefficient R^2 is equal to 1 if the fitted equation $y_i = \theta_0 + \sum_{j \in M} \theta_j x_j + e_i$ passes through all the data points, so that all residuals are null; then, the markers explain all the phenotypic variance. At the other extreme, R^2 is zero if $\bar{y}_i = \widehat{\theta}_0$ and the estimated regression coefficients are null, i.e., $\widehat{\theta}_1 = \widehat{\theta}_2 = \cdots = \widehat{\theta}_M = 0$. In the latter case, markers do not affect the phenotypic observations and the variance of the marker score values is zero. Thus, the R^2 values are between 0 and 1, i.e., $0 \leq R^2 \leq 1.0$. Equation (4.32a) is useful for estimating $\sigma^2_{s_i}$ as $\widehat{\sigma}^2_{y_i}\sum_{j=1}^{M} R^2_j = \widehat{\sigma}^2_s$, where R^2_j is the estimated value of the *j*th marker and $\widehat{\sigma}_y^2$ is the phenotypic variance of the *i*th trait; however, this is a biased estimator of $\sigma_{s_i}^2$ (Hospital et al. 1997).

Charcosset and Gallais (1996) and Hospital et al. (1997) proposed an unbiased estimator of $\sigma_{s_i}^2$ based on all the selected markers using the adjusted coefficient of multiple determination, i.e.,

$$R_{Adj}^2 = 1 - \frac{n-1}{n-M-1}\left(1 - R^2\right) = \frac{\widehat{\sigma}_s^2}{\widehat{\sigma}_y^2}, \tag{4.32b}$$

whence we can obtain a unbiased estimator of $\sigma_{s_i}^2$ as $\widehat{\sigma}_y^2 R_{Adj}^2 = \widehat{\sigma}_{\widehat{s}_i}^2$ by jointly using all the markers that affect the phenotypic values. The problem with Eq. (4.32b) is that the R_{Adj}^2 values could be negative; in that case, the estimated value of $\sigma_{s_i}^2$ would also be negative. One additional problem with Eq. (4.32b) is that the R_{Adj}^2 values can produce $\widehat{\sigma}_s^2$ values that are higher than those of the estimated variance of the breeding values $\widehat{\sigma}_g^2$.

2

Using Eqs. (4.32a) and (4.32b), we can estimate σ_{s_i}, but from them it is not clear how we can estimate the covariance between two different estimated marker score values.

Consider the case of the PHT values described in Sect. 4.3.1 of this chapter, where $M = 7$, $n = 247$, and the estimated variance of PHT was $\widehat{\sigma}_{PHT}^2 = 191.81$. The estimated values of R^2 for each of the seven markers were 0.0038, 0.0005, 0.006, 0.0013, 0.0036, 0.0114, and 0.0298, whence, by multiplying each estimated R^2 value by $\widehat{\sigma}_{PHT}^2 = 191.81$ and summing the results, we found that the estimated value of $\sigma_{s_{PHT}}^2$ was $\widehat{\sigma}_{\widehat{s}_{PHT}}^2 = 9.78$. In this case, the estimated portion of the genetic variance attributable to $\widehat{\sigma}_{\widehat{s}_{PHT}}^2 = 9.78$ was $\widehat{q}_{PHT} = \frac{9.78}{83} = 0.1178$; thus, when we estimated $\sigma_{s_{PHT}}^2$ according to Eq. (4.32a), the seven markers explained only 11.78% of the genetic variance associated with PHT.

The estimated value of R_{Adj}^2 for the seven markers jointly was 0.06, whence $\widehat{\sigma}_{s_{PHT}}^2 = (191.81)(0.06) = 11.50$ is an estimate of $\sigma_{s_{PHT}}^2$. In the latter case, the estimated portion of the genetic variance attributable to $\widehat{\sigma}_{s_{PHT}}^2 = 11.50$ was $\widehat{q}_{PHT} = \frac{11.5}{83} = 0.1385$; that is, according to Eq. (4.32b), the seven markers explain 13.85% of the genetic variance associated with PHT.

One additional way of estimating the variance of the marker score $\sigma_{s_i}^2$ was proposed by Lange and Whittaker (2001) as

$$\frac{1}{n-1}\sum_{i=1}^{n}\left(\widehat{s}_i - \widehat{\mu}_{s_i}\right)^2, \tag{4.33}$$

where $\widehat{s}_i = \sum_{j=1}^{M} \widehat{\theta}_j x_j$ and $\widehat{\mu}_{s_i}$ is the mean of $\widehat{s}_i$ values. The covariance between the ith and jth marker scores can be estimated as the cross products of the marker score values divided by $n - 1$. Note that in this case, the number of markers associated with the ith and jth traits may be different.

For the PHT values described in Sect. 4.3.1 of this chapter, where $n = 247$, the estimated value of $\sigma^2_{s_i}$ was $\widehat{\sigma}^2_{s_{PHT}} = 15.75$ and the estimated portion of the genetic variance attributable to $\widehat{\sigma}^2_{s_{PHT}} = 15.75$ was $\widehat{q}_{PHT} = \frac{15.75}{83} = 0.1897$. That is, the seven markers jointly explain 18.97% of the genetic variance associated with PHT according to Eq. (4.33).

4.3.3 Estimating LMSI Selection Response and Efficiency

With the estimated phenotypic variances ($\widehat{\sigma}^2_{PHT} = 191.81$), the estimated genetic variance ($\widehat{\sigma}^2_{g_{PHT}} = 83.0$) and the estimated marker score variances: $\widehat{\sigma}^2_{\widehat{s}_{PHT}} = 48.23$ (Eq. 4.29), $\widehat{\sigma}^2_{\widehat{s}_{PHT}} = 9.78$ (Eq. 4.32a), $\widehat{\sigma}^2_{s_{PHT}} = 11.50$ (Eq. 4.32b), and $\widehat{\sigma}^2_{s_{PHT}} = 15.75$ (Eq. 4.33), we can estimate the LMSI coefficient, selection response, and efficiency.

Using the estimated value $\widehat{\sigma}^2_{\widehat{s}_{PHT}} = 48.23$ obtained with Eq. (4.29), it is possible to estimate the LMSI weight as $\widehat{\beta}_{PHT} = \frac{\widehat{\sigma}^2_{g_{PHT}} - \widehat{\sigma}^2_{s_{PHT}}}{\widehat{\sigma}^2_{PHT} - \widehat{\sigma}^2_{s_{PHT}}} = \frac{83.0 - 48.23}{191.81 - 48.23} = 0.242$,

whereas for $\widehat{\sigma}^2_{\widehat{s}_{PHT}} = 9.78$, $\widehat{\sigma}^2_{s_{PHT}} = 11.50$, and $\widehat{\sigma}^2_{s_{PHT}} = 15.75$, the estimated values of β_{PHT} were 0.402, 0.40, and 0.382 respectively. The latter results indicate that the estimated values of β_{PHT} associated with the phenotypic values tend to decrease when the estimated values of the variance of the marker score increase. This means that at the limit, when all the genetic variance is explained by the markers, the estimated values of β_{PHT} are zero and the estimated LMSI is equal to $\widehat{I}_M = \widehat{s}$. Thus, for trait PHT, when the estimated values of β_{PHT} are not zero, the estimated LMSI can be written as $\widehat{I}_{M_{PHT}} = \widehat{s}_{PHT} + \widehat{\beta}_{PHT}(PHT_i - \widehat{s}_{PHT})$. The $\widehat{I}_{M_{PHT}}$ values are used to predict, rank, and select the net genetic merit value of each individual candidate for selection.

Based on the result $\widehat{\sigma}^2_{\widehat{s}_{PHT}} = 48.23$ obtained with Eq. (4.29) and using a selection intensity of 10% (k_I= 1.755), the estimated LMSI selection response can be obtained as

$$\widehat{R}_M = k_I\sqrt{\frac{\widehat{\sigma}_g^2\left(\widehat{\sigma}_g^2 - \widehat{\sigma}_s^2\right) + \widehat{\sigma}_s^2\left(\widehat{\sigma}_y^2 - \widehat{\sigma}_g^2\right)}{\widehat{\sigma}_y^2 - \widehat{\sigma}_s^2}}$$

$$= 1.755\sqrt{\frac{83(83 - 48.23) + 48.23(191.81 - 83)}{191.81 - 48.23}}$$

$$= 1.755\sqrt{56.65} = 13.21.$$

In a similar manner, using the result $\widehat{\sigma}^2_{s_{PHT}} = 15.75$, the estimated selection response was $\widehat{R}_M = 1.755\sqrt{\frac{83(83 - 15.75) + 15.75(191.81 - 83)}{191.81 - 15.75}} = 1.755\sqrt{41.44}$ $= 11.30$. With $\widehat{\sigma}^2_{\widehat{s}_{PHT}} = 9.78$ and $\widehat{\sigma}^2_{s_{PHT}} = 11.50$, the estimated values of the LMSI selection responses were 10.99 and 11.10 respectively. The latter results indicate that the estimated values of the LMSI selection responses tend to increase when the estimated values of the variance of the marker score increase.

We can estimate LMSI versus phenotypic efficiency for one trait as $\widehat{\lambda}_M = \sqrt{\frac{\widehat{q}}{\widehat{h^2}} + \frac{(1-\widehat{q})^2}{1-\widehat{q}\widehat{h^2}}}$, where $\widehat{h^2}$ is the estimated trait heritability and $\widehat{q} = \frac{\widehat{\sigma}_s^2}{\widehat{\sigma}_g^2}$ is the estimated portion of additive genetic variance explained by the markers. When $\widehat{\sigma}^2_{\widehat{s}_{PHT}} = 48.23$, $\widehat{q}_{PHT} = \frac{48.23}{83} = 0.5811$, and $\widehat{h^2} = 0.433$, the estimated LMSI efficiency was $\widehat{\lambda}_M = \sqrt{1.58} = 1.25$. For $\widehat{\sigma}^2_{s_{PHT}} = 15.75$, $\widehat{\sigma}^2_{\widehat{s}_{PHT}} = 9.78$, and $\widehat{\sigma}^2_{s_{PHT}} = 11.50$, the estimated portions of the additive genetic variance explained by the markers were $\widehat{q}_{PHT} = \frac{15.75}{83} = 0.1897$, $\widehat{q}_{PHT} = \frac{9.78}{83} = 0.1178$, and $\widehat{q}_{PHT} = \frac{11.5}{83} = 0.1385$ respectively, whence the estimated LMSI efficiencies were 1.1, 1.04, and 1.05 respectively. The latter results indicate that the estimated values of LMSI efficiency tend to increase when the estimated values of the variance of the marker score increase (Fig. 4.1).

Figure 4.1 presents the change in LMSI efficiency with respect to phenotypic selection for different values of the variance of the marker score when the phenotypic (191.81) and genetic (83) variances are fixed. In a similar manner, Fig. 4.2 presents the change in the LMSI selection response for different values of the variance of the marker score when the phenotypic (191.81) and genetic (83) variances are fixed. In effect, LMSI efficiency and the selection response depend on the genetic variance explained by the markers.

4.3.4 Estimating the Variance of the Marker Score in the Multi-Trait Case

Equation (4.33) can be used in the multi-trait context when the numbers of markers associated with the *i*th and *j*th traits are different. Also, it is possible to adapt Eqs. (4.32a) and (4.32b) to the multi-trait case. However, in the latter case, in addition to the markers linked to the QTL that affect one specific trait, we need to find markers that affect more than one trait, which may be very difficult. For this reason, in the multi-trait context, Eqs. (4.32a) and (4.32b) could be used to estimate the variance of the marker score (**S**) without preselecting the markers that affect the phenotypic traits, only when the number of genotypes is higher than the number of markers.

Let $\mathbf{y}_1, \mathbf{y}_2, \ldots, \mathbf{y}_r$ be r independent multivariate normal vectors of observations, each with n observations, such that $\mathbf{Y} = \begin{bmatrix} y_{11} & y_{12} & \cdots & y_{1t} \\ y_{21} & y_{22} & \cdots & y_{2t} \\ \vdots & \vdots & \cdots & \vdots \\ y_{n1} & y_{n2} & \cdots & y_{nt} \end{bmatrix}$ is a matrix $n \times t$ of observations for t traits; then, the multivariate linear regression model can be written as $\mathbf{Y} = \mathbf{XB} + \mathbf{U}$, where $\mathbf{X}$ is a matrix $n \times m$ (m= number of markers and $m < n$) of known coded marker values, $\mathbf{B}$ is a matrix $m \times n$ of regression coefficients, and $\mathbf{U}$ is a matrix $n \times t$ of unobserved random disturbance whose rows for given $\mathbf{X}$ are uncorrelated, each with mean $\mathbf{0}$ and common covariance matrix $\mathbf{E}$ (Mardia et al. 1982; Rencher 2002). According to the least squares method of estimation, $\widehat{\mathbf{B}} = (\mathbf{X}'\mathbf{X})^{-1}\mathbf{X}'\mathbf{Y}$ is an estimator of $\mathbf{B}$ and $\widehat{\mathbf{E}} = \frac{(\mathbf{Y} - \widehat{\mathbf{B}}\mathbf{X})'(\mathbf{Y} - \widehat{\mathbf{B}}\mathbf{X})}{n - m - 1}$ is an estimator of the residual covariance matrix $\mathbf{E}$ assuming that $n > m$ (Johnson and Wichern 2007).

Note that $1 - R^2 = \frac{\widehat{\mathbf{e}}'\widehat{\mathbf{e}}}{\mathbf{y}'\mathbf{y}}$, where $\widehat{\mathbf{e}}$ is a vector of estimated residual values of the model $y_i = \theta_0 + \sum_{j \in M} \theta_j x_j + e_i$ and R^2 is the coefficient of multiple determination (Eq. 4.32a). In addition, as in the multi-trait context the estimated matrix of residuals is $\widehat{\mathbf{U}} = \mathbf{Y} - \widehat{\mathbf{B}}\mathbf{X}$, $1 - R^2$ can be written as $\mathbf{D} = (\mathbf{Y}'\mathbf{Y})^{-1}\widehat{\mathbf{U}}'\widehat{\mathbf{U}}$ (Mardia et al. 1982), whence R^2 in the multivariate context can written as

$$\mathbf{R}^2 = \mathbf{I} - \mathbf{D} = \widehat{\mathbf{P}}^{-1}\widehat{\mathbf{S}}, \tag{4.34a}$$

whereas R^2_{Adj} (Eq. 4.32b) can be written as

$$\mathbf{R}^2_{Adj} = \mathbf{I} - \frac{n-1}{n-m-1}\mathbf{D} = \widehat{\mathbf{P}}^{-1}\widehat{\mathbf{S}}, \tag{4.34b}$$

where $\mathbf{I}$ is an identity matrix $t \times t$, $\widehat{\mathbf{P}}^{-1}$ is the inverse of the estimated covariance matrix of phenotypic values ($\widehat{\mathbf{P}}$), and $\widehat{\mathbf{S}}$ is the estimated covariance matrix of marker score values. From Eq. (4.34b),

$$\widehat{\mathbf{P}}\mathbf{R}^2_{Adj} = \widehat{\mathbf{S}} \tag{4.34c}$$

is an unbiased estimator of matrix $\widehat{\mathbf{S}}$, whereas $\widehat{\mathbf{P}}\mathbf{R}^2 = \widehat{\mathbf{S}}$ (Eq. 4.34a) is a biased estimator of matrix $\widehat{\mathbf{S}}$. The main problem of Eq. (4.34c) is that the diagonal elements of $\widehat{\mathbf{S}}$ could be negative.

From the maize F_2 population including 247 genotypes (each one with two repetitions) and 195 molecular markers described in Sect. 4.3.1, we used two traits—PHT (cm) and EHT (cm)—to illustrate the multivariate method of estimating the LMSI parameters. The estimated phenotypic and genetic covariance matrices were $\widehat{\mathbf{P}} = \begin{bmatrix} 191.81 & 106.89 \\ 106.89 & 167.93 \end{bmatrix}$ and $\widehat{\mathbf{C}} = \begin{bmatrix} 83.00 & 57.44 \\ 57.44 & 59.80 \end{bmatrix}$, whereas the estimated covariance matrix of marker scores, using Eq. (4.33), was $\widehat{\mathbf{S}} = \begin{bmatrix} 15.750 & 0.983 \\ 0.983 & 28.083 \end{bmatrix}$. When we used Eq. (4.34a) and Eq. (4.34c), we obtained estimated values of the variance and covariance of the marker scores that were higher than the genetic values (data not presented). Equations (4.29) and (4.31) are used later to compare LMSI efficiency versus GW-LMSI efficiency using the simulated data described in Chap. 2, Sect. 2.8.1.

With matrices $\widehat{\mathbf{P}}$, $\widehat{\mathbf{C}}$, and $\widehat{\mathbf{S}}$, and the vector of economic weights $\mathbf{a}' = [\mathbf{w}' \quad \mathbf{0}']$, where $\mathbf{w}' = [-1 \quad -1]$ and $\mathbf{0}' = [0 \quad 0]$, we obtained the estimated matrices $\widehat{\mathbf{T}} = \begin{bmatrix} \widehat{\mathbf{P}} & \widehat{\mathbf{S}} \\ \widehat{\mathbf{S}} & \widehat{\mathbf{S}} \end{bmatrix}$ and $\mathbf{Z} = \begin{bmatrix} \widehat{\mathbf{C}} & \widehat{\mathbf{S}} \\ \widehat{\mathbf{S}} & \widehat{\mathbf{S}} \end{bmatrix}$, whence the estimated LMSI vector of coefficients was $\widehat{\boldsymbol{\beta}}' = \mathbf{a}'\widehat{\mathbf{Z}}_M\widehat{\mathbf{T}}_M^{-1} = [-0.59 \quad -0.18 \quad -0.41 \quad -0.82]$. Using a selection intensity of 10% ($k_I = 1.755$), the estimated LMSI selection response and the expected genetic gains per trait were $\widehat{R}_M = k_I\sqrt{\widehat{\boldsymbol{\beta}}'\widehat{\mathbf{T}}_M\widehat{\boldsymbol{\beta}}} = 20.41$ and $\widehat{\mathbf{E}}'_M = k_I\frac{\widehat{\boldsymbol{\beta}}'\widehat{\mathbf{Z}}_M}{\sqrt{\widehat{\boldsymbol{\beta}}'\widehat{\mathbf{T}}_M\widehat{\boldsymbol{\beta}}}} = [-10.09 \quad -10.31 \quad -2.53 \quad -4.39]$ respectively, whereas the estimated LMSI accuracy was $\widehat{\rho}_{H\hat{I}_M} = \frac{\widehat{\sigma}_{I_M}}{\widehat{\sigma}_H} = 0.72$.

The estimated LPSI parameters (see Chap. 2 for details) using the phenotypic information from the maize F_2 population for traits PHT and EHT are as follows. The estimated LPSI vector of coefficients was $\widehat{\mathbf{b}}' = \mathbf{w}'\widehat{\mathbf{C}}\widehat{\mathbf{P}}^{-1} = [-0.53 \quad -0.36]$, and, with a selection intensity of 10% ($k_I = 1.755$), the estimated LPSI selection response and the expected genetic gains per trait were $\widehat{R}_I = k_I\sqrt{\widehat{\mathbf{b}}'\widehat{\mathbf{P}}\widehat{\mathbf{b}}} = 18.97$ and $\widehat{\mathbf{E}}' = k_I\frac{\widehat{\mathbf{b}}'\widehat{\mathbf{C}}}{\widehat{\sigma}_I} = [-10.52 \quad -8.45]$ respectively, whereas the estimated LPSI accuracy was $\widehat{\rho}_{H\hat{I}} = \frac{\widehat{\sigma}_I}{\widehat{\sigma}_H} = 0.67$.

We can determine LMSI efficiency versus LPSI efficiency to predict the net genetic merit using the ratio of estimated accuracy values $\widehat{\rho}_{H\hat{I}_M} = 0.72$ and $\widehat{\rho}_{H\hat{I}} = 0.67$ of the LMSI and LPSI respectively, i.e., $\widehat{\lambda}_M = \frac{0.72}{0.67} = 1.075$, whence, according to Eq. (4.19), the estimated LMSI efficiency versus the LPSI efficiency, in percentage terms, was $\widehat{p}_M = 100(1.075 - 1) = 7.5$. That is, for these data, the estimated LMSI efficiency was only 7.5% greater than LPSI efficiency at predicting the net genetic merit.

4.4 Estimating the GW-LMSI Parameters in the Asymptotic Context

Lange and Whittaker (2001) proposed the GW-LMSI. However, these authors did not provide detailed procedures for estimating matrices **P**, **C**, **W**, and **M**. They indicated that matrix **C** can be estimated using the estimated matrix of covariance of marker scores ($\widehat{\mathbf{S}}$) and that matrices **P**, **W**, and **M** can be estimated *directly by their empirical variances and covariances*, but this assertion does not indicate a clear method for estimating those covariance matrices. In Chap. 2, we described the REML method of estimating **C** and **P**. Crossa and Cerón-Rojas (2011) described matrices **W** and **M** in a doubled haploid population. In this study, we describe and estimate matrices **W** and **M** for an F_2 population in the asymptotic context according to the Wright and Mowers (1994) approach, which is based on regressing phenotype values on marker coded values. We used this latter approach to estimate **W** and **M**, because it is a clearer estimation method than that of Lange and Whittaker (2001); however, the Wright and Mowers (1994) approach is an asymptotic method and should be regarded with precaution.

Matrix **M** is the covariance matrix of the molecular marker code values. All marker information used to construct matrix **M** is presented in Table 4.2. Based on this information, we found that the expectations ($E(X_1)$ and $E(X_2)$) and the variances ($V(X_1)$ and $V(X_2)$) of the marker coded values X_1 and X_2 are $E(X_1) = E(X_2) = 0$ and $V(X_1) = V(X_2) = 1$, whereas the covariance ($Cov(X_1, X_2)$) and correlation ($Corr(X_1, X_2)$), between X_1 and X_2 were

$$Cov(X_1, X_2) = Corr(X_1, X_2) = 1 - 2\delta. \tag{4.35}$$

Thus, as the variances of X_1 and X_2 are equal to 1, the correlation between X_1 and X_2 is $Corr(X_1, X_2) = \frac{Cov(X_1, X_2)}{\sqrt{V(X_1)V(X_2)}} = 1 - 2\delta$, i.e., the covariance and correlation between X_1 and X_2 are the same. Equation (4.35) results indicate that if we perform the same operation with many markers, we will obtain similar results; they also indicate that this is the way to construct matrix **M**.

Table 4.2 Marker genotypes, expected frequency, and coded values (X_1 and X_2) of the marker genotypes in an F_2 population

Marker genotype	Expected frequency	X_1	X_2
A_1B_1/A_1B_1	$(1-\delta)^2/4$	1	1
A_1B_1/A_1B_2	$2(\delta-\delta^2)/4$	1	0
A_1B_2/A_1B_2	$\delta^2/4$	1	−1
A_1B_1/A_2B_1	$2(\delta-\delta^2)/4$	0	1
A_1B_2/A_2B_1	$2(1-2\delta+2\delta^2)/4$	0	0
A_1B_2/A_2B_2	$2(\delta-\delta^2)/4$	0	−1
A_2B_1/A_2B_1	$\delta^2/4$	−1	1
A_2B_1/A_2B_2	$2(\delta-\delta^2)/4$	−1	0
A_2B_2/A_2B_2	$(1-\delta)^2/4$	−1	−1

Let **X** be a matrix of coded markers of size $n \times m$, where $n \geq m$ and $m=$ number of markers; then according to Wright and Mowers (1994), because all marker information is contained in matrix $\mathbf{X}'\mathbf{X}$, when the number of observations (n) tends to infinity, the product $\mathbf{x}_i'\mathbf{x}_j/n$ tends to the covariance between markers ith and jth, whence matrix $n^{-1}\mathbf{X}'\mathbf{X}$ should tend to the covariance matrix between the markers that conform matrix **X** with the ijth element equal to $(0.5 - \delta_{ij})$. Thus, matrix $2n^{-1}\mathbf{X}'\mathbf{X}$ should tend to a covariance matrix where the ijth entry is equal to $(1 - 2\delta_{ij})$. Based on the latter result, an estimator of matrix **M** in the asymptotic context is

$$\widehat{\mathbf{M}} = 2n^{-1}\mathbf{X}'\mathbf{X}. \tag{4.36}$$

Equation (4.36) is an asymptotic result and should be taken with caution. To date, there has been no clear method for estimating **M** in the non-asymptotic context; for this reason, Eq. (4.36) is used to estimate the GW-LMSI parameters.

Assume that a QTL is between the two markers in Table 4.2; then, δ can be written as $\delta = r_1 + r_2 - 2r_1r_2$, where r_1 and r_2 denote the recombination frequency between marker 1 and marker 2 respectively, with the QTL between them. When the number of genotypes or individuals tends to infinity, the covariance between the phenotypic trait values (y) and the marker 1 coded values (X_1) in an F_2 population can be written as

$$Cov(X_1, y) = \frac{1}{2}\alpha_1(1 - 2r_1), \tag{4.37}$$

where $\alpha_1(1 - 2r_1)$ is the portion of the additive effect (α_1) of the QTL linked to marker 1 (Edwards et al. 1987), and r_1 is the recombination frequency between the QTL and marker 1. We can assume that for many markers, the covariance of the phenotypic values is similar to Eq. (4.37), whence matrix **W** can be obtained.

Let **y** be a vector $n \times 1$ of recorded phenotypic values, where n denotes the number of observation or records, and **X** is a matrix of coded markers of size $n \times m$.

When n tends to infinity, $2n^{-1}\mathbf{X}'\mathbf{y}$ tends to be a vector with elements equal to $\alpha_i(1-2r_i)$, where α_i is the additive effect of the ith QTL linked to the ith marker, and r_i is the recombination frequency between the ith QTL and the ith marker. Now let $\mathbf{Y}=\begin{bmatrix} y_{11} & y_{12} & \cdots & y_{1t} \\ y_{21} & y_{22} & \cdots & y_{2t} \\ \vdots & \vdots & \cdots & \vdots \\ y_{n1} & y_{n2} & \cdots & y_{nt} \end{bmatrix}$ be a matrix of observations for t traits; then, an estimator of matrix $\mathbf{W}$ in the asymptotic context is

$$\widehat{\mathbf{W}}=2n^{-1}\mathbf{X}'\mathbf{Y}. \tag{4.38}$$

Once again, Eq. (4.38) is an asymptotic result and should be accepted with caution. But to date, there has been no clear method for estimating $\mathbf{W}$ in the non-asymptotic context; for this reason, Eq. (4.38) is used to estimate the GW-LMSI parameters.

4.5 Comparing LMSI Versus LPSI and GW-LMSI Efficiency

To compare LMSI efficiency versus GW-LMSI efficiency for predicting the net genetic merit, we use the simulated data set described in Chap. 2, Sect. 2.8.1.

Figure 4.4 presents the estimated accuracy values of the LPSI ($\widehat{\rho}_{H\hat{I}}=\frac{\widehat{\sigma}_I}{\widehat{\sigma}_H}$), the LMSI ($\widehat{\rho}_{H\hat{I}_M}=\frac{\widehat{\sigma}_{I_M}}{\widehat{\sigma}_H}$), and the GW-LMSI ($\widehat{\rho}_{H\hat{I}_W}=\frac{\widehat{\sigma}_{I_W}}{\widehat{\sigma}_H}$) for five simulated selection cycles. In addition, Table 4.3 presents the estimated LPSI, LMSI, and GW-LMSI selection responses, the estimated LPSI, LMSI, and GW-LMSI variances of the predicted error ($(1-\widehat{\rho}^2_{H\hat{I}})\widehat{\sigma}^2_H$, $(1-\widehat{\rho}^2_{H\hat{I}_M})\widehat{\sigma}^2_H$ and $(1-\widehat{\rho}^2_{H\hat{I}_W})\widehat{\sigma}^2_H$ respectively), the ratios of the estimated LMSI accuracy to the estimated LPSI accuracy and the estimated LMSI accuracy to the estimated GW-LMSI accuracy, expressed as percentages (Eq. 4.19), for five simulated selection cycles.

According to Fig. 4.4, for this data set the estimated LMSI accuracy ($\widehat{\rho}_{H\hat{I}_M}$) was higher than the estimated LPSI and GW-LMSI accuracy ($\widehat{\rho}_{H\hat{I}}$ and $\widehat{\rho}_{H\hat{I}_W}$ respectively), for the five simulated selection cycles, that is, $\widehat{\rho}_{H\hat{I}_M}>\widehat{\rho}_{H\hat{I}}>\widehat{\rho}_{H\hat{I}_W}$. In a similar manner, Table 4.3 results indicate that the estimated LMSI selection response ($\widehat{R}_M$) was higher than the estimated LPSI and GW-LMSI selection responses ($\widehat{R}_I$ and $\widehat{R}_W$ respectively): $\widehat{R}_M>\widehat{R}_I>\widehat{R}_W$.

Note that the estimated LPSI, LMSI, and GW-LMSI variances of the predicted error, and the estimated LMSI efficiency versus LPSI efficiency and versus GW-LMSI efficiency (expressed in percentages) are related to the estimated

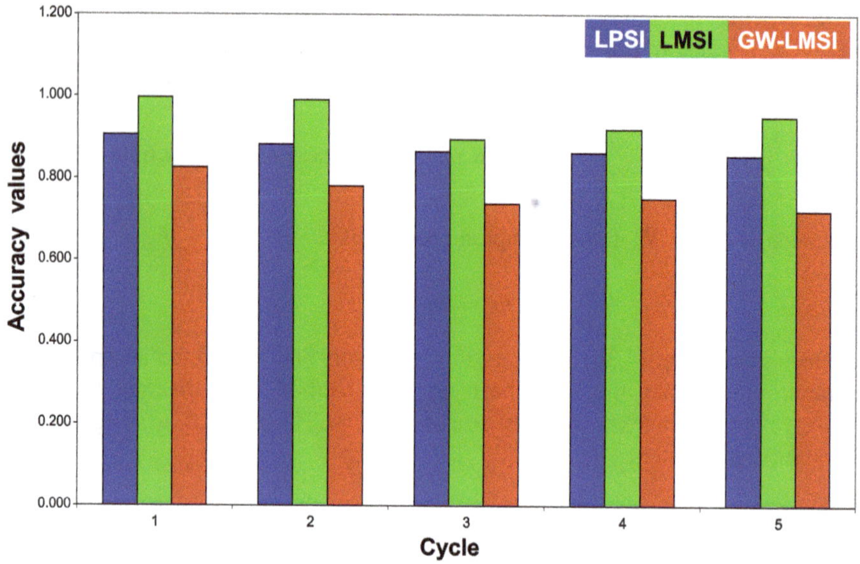

Fig. 4.4 Estimated correlation values of the linear phenotypic selection index (LPSI), the linear molecular selection index (LMSI), and the genome-wide LMSI (GW-LMSI) with the net genetic merit for four traits, 2500 markers and 500 genotypes (each with four repetitions) in one environment for five simulated selection cycles

Table 4.3 Estimated linear phenotypic, molecular, and genome-wide selection indices (LPSI, LMSI, and GW-LMSI respectively), selection responses and variance of the predicted error, and estimated ratio of LMSI accuracy to LPSI and GW-LMSI accuracy expressed in percentages for 4 traits, 2500 markers and 500 genotypes (each with four repetitions) in one environment for five simulated selection cycles

	Selection response			Variance of the predicted error			Efficiency of LMSI versus	
Cycle	LPSI	LMSI	GW-LMSI	LPSI	LMSI	GW-LMSI	LPSI	GW-LMSI
1	17.84	19.60	16.24	22.53	0.07	39.84	10.07	20.67
2	15.66	24.36	13.88	22.66	0.07	40.06	12.14	26.81
3	14.44	14.70	12.13	21.95	1.86	39.86	3.43	21.27
4	14.29	15.29	12.48	22.84	1.46	39.09	6.57	22.50
5	13.86	15.15	11.49	22.13	0.88	39.65	11.11	31.88
Average	15.22	17.82	13.24	22.42	0.87	39.70	8.66	24.63

LMSI, LPSI, and GW-LMSI accuracies, and that in all five selection cycles, $\widehat{\rho}_{H\hat{I}_M} > \widehat{\rho}_{H\hat{I}} > \widehat{\rho}_{H\hat{I}_W}$. This implies that the estimated LMSI variance of the predicted error was lower than the estimated LPSI and GW-LMSI variance of the predicted error. In a similar manner, because $\widehat{\rho}_{H\hat{I}_M} > \widehat{\rho}_{H\hat{I}} > \widehat{\rho}_{H\hat{I}_W}$, the estimated LMSI efficiency was higher than the estimated LPSI efficiency and the estimated GW-LMSI efficiency.

Based on Fig. 4.4 and Table 4.3 results, we conclude that the LMSI was a better predictor of the net genetic merit than the LPSI, and that the LPSI is a better predictor of the net genetic merit than the GW-LMSI for this simulated data set.

References

Bulmer MG (1980) The mathematical theory of quantitative genetics. Lectures in biomathematics. University of Oxford, Clarendon Press, Oxford

Charcosset A, Gallais A (1996) Estimation of the contribution of quantitative trait loci (QTL) to the variance of a quantitative trait by means of genetic markers. Theor Appl Genet 93:1193–1201

Crossa J, Cerón-Rojas JJ (2011) Multi-trait multi-environment genome-wide molecular marker selection indices. J Indian Soc Agric Stat 62(2):125–142

Dekkers JCM, Settar P (2004) Long-term selection with known quantitative trait loci. Plant Breed Rev 24:311–335

Edwards MD, Stuber CW, Wendel JF (1987) Molecular-marker-facilitated investigations of quantitative-trait loci in maize. I. Numbers, genomic distribution and types of gene action. Genetics 116:113–125

Hospital F, Moreau L, Lacoudre F, Charcosset A, Gallais A (1997) More on the efficiency of marker-assisted selection. Theor Appl Genet 95:1181–1189

Johnson RA, Wichern DW (2007) Applied multivariate statistical analysis, 6th edn. Pearson Prentice Hall, Upper Saddle River, NJ

Knapp SJ (1998) Marker-assisted selection as a strategy for increasing the probability of selecting superior genotypes. Crop Sci 38:1164–1174

Lande R, Thompson R (1990) Efficiency of marker-assisted selection in the improvement of quantitative traits. Genetics 124:743–756

Lange C, Whittaker JC (2001) On prediction of genetic values in marker-assisted selection. Genetics 159:1375–1381

Mardia KV, Kent JT, Bibby JM (1982) Multivariate analysis. Academic Press, New York

Moreau L, Charcosset A, Hospital F, Gallais A (1998) Marker-assisted selection efficiency in populations of finite size. Genetics 148:1353–1365

Moreau L, Hospital F, Whittaker J (2007) Marker-assisted selection and introgression. In: Balding DJ, Bishop M, Cannings C (eds) Handbook of statistical genetics, vol 1, 3rd edn. Wiley, New York, pp 718–751

Rencher AC (2002) Methods of multivariate analysis. Wiley, New York

Searle S, Casella G, McCulloch CE (2006) Variance components. Wiley, Hoboken, NJ

Whittaker JC (2003) Marker-assisted selection and introgression. In: Balding DJ, Bishop M, Cannings C (eds) Handbook of statistical genetics, vol 1, 2nd edn. Wiley, New York, pp 554–574

Wright AJ, Mowers RP (1994) Multiple regression for molecular marker, quantitative trait data from large F_2 population. Theor Appl Genet 89:305–312

Zhang W, Smith C (1992) Computer simulation of marker-assisted selection utilizing linkage disequilibrium. Theor Appl Genet 83:813–820

Zhang W, Smith C (1993) Simulation of marker-assisted selection utilizing linkage disequilibrium: the effects of several additional factors. Theor Appl Genet 86:492–496

5

LGSI and Combined LGSI

Abstract The linear genomic selection index (LGSI) is a linear combination of genomic estimated breeding values (GEBVs) used to predict the individual net genetic merit and select individual candidates from a nonphenotyped testing population as parents of the next selection cycle. In the LGSI, phenotypic and marker data from the training population are fitted into a statistical model to estimate all individual available genome marker effects; these estimates can then be used in subsequent selection cycles to obtain GEBVs that are predictors of breeding values in a testing population for which there is only marker information. The GEBVs are obtained by multiplying the estimated marker effects in the training population by the coded marker values obtained in the testing population in each selection cycle. Applying the LGSI in plant or animal breeding requires the candidates to be genotyped for selection to obtain the GEBV, and predicting and ranking the net genetic merit of the candidates for selection using the LGSI. We describe the LGSI and show that it is a direct application of the linear phenotypic selection index theory in the genomic selection context; next, we present the combined LGSI (CLGSI), which uses phenotypic and GEBV information jointly to predict the net genetic merit. The CLGSI can be used only in training populations when there are phenotypic and maker information, whereas the LGSI is used in testing populations where there is only marker information. We validate the theoretical results of the LGSI and CLGSI using real and simulated data.

5.1 The Linear Genomic Selection Index

5.1.1 Basic Conditions for Constructing the LGSI

Conditions described in Chap. 4 (Sect. 4.1.1) for constructing a valid linear molecular selection index (LMSI), are also necessary for the linear genomic selection index (LGSI); however, in addition to those conditions, the LGSI also requires:

1. All marker effects to be estimated simultaneously in the training population.
2. The estimated marker effects to be used in subsequent selection cycles to obtain GEBVs that are predictors of the individual breeding values in the testing population (candidates for selection) for which there is only marker information.

3. The GEBV values to be composed entirely of the additive genetic effects.
4. Phenotypes to be used to estimate all marker effects in the training population, not to make selections in the testing population (Heffner et al. 2009; Lorenz et al. 2011).

5.1.2 *Genomic Breeding Values and Marker Effects*

The breeding value (g_i) is the average additive effects of the genes an individual receives from both parents; thus, it is a function of the genes transmitted from parents to progeny and is the only component that can be selected and, therefore, the main component of interest in breeding programs (Mrode 2005). The *i*th phenotypic value (y_i) can be denoted as $y_i = g_i + e_i$, where g_i is the *breeding value* and e_i the residual. Basic assumptions for g_i and e_i are: both g_i and e_i have normal distribution with expectation equal to zero and variance $\sigma^2_{g_i}$ and $\sigma^2_{e_i}$ respectively. This means that $y_i = \mu_i + g_i + e_i$ is a linear mixed model (Mrode 2005; Searle et al. 2006), where μ_i is the mean of y_i.

Let $\mathbf{y}'_i = [y_{i1} \;\; y_{i2} \;\; \cdots \;\; y_{in}]$ be a vector $1 \times n$ of observations in the *i*th trait and let $\mathbf{g}'_i = [g_{i1} \;\; g_{i2} \;\; \cdots \;\; g_{in}]$ be a vector $1 \times n$ of unobservable breeding values associated with $\mathbf{y}_i$; then $\mathbf{y}_i$ can be written as

$$\mathbf{y}_i = \mathbf{1}\mu_i + \mathbf{Z}\mathbf{g}_i + \mathbf{e}_i, \qquad (5.1)$$

where μ_i is the mean of the *i*th trait, $\mathbf{1}$ is a vector $n \times 1$ of 1s, $\mathbf{Z}$ is a design matrix of 0s and 1s, $\mathbf{g}_i \sim$ MVN $(\mathbf{0}, \mathbf{A}\sigma^2_{g_i})$ is a vector of breeding values, and $\mathbf{e}_i \sim$ MVN $(\mathbf{0}, \mathbf{I}_n\sigma^2_{e_i})$ is a vector of residuals; $\mathbf{0}$ is the mean and $\mathbf{A}\sigma^2_{g_i}$ and $\mathbf{I}_n\sigma^2_{e_i}$ the covariance matrix of $\mathbf{g}_i$ and $\mathbf{e}_i$ respectively; $\mathbf{A}$ is the numerical relationship matrix (Mrode 2005) and $\mathbf{I}_n$ an identity matrix $n \times n$; $\sigma^2_{g_i}$ and $\sigma^2_{e_i}$ are the additive and residual variances associated with g_i and e_i; and MVN stands for multivariate normal distribution.

Suppose that $\mathbf{A}$, $\mathbf{Z}$, μ_i, $\sigma^2_{g_i}$, and $\sigma^2_{e_i}$ are known; then, according to Mrode (2005), the best linear unbiased predictor (BLUP) of $\mathbf{g}_i$ can be written as

$$\widehat{\mathbf{g}}_i = \sigma^2_{g_i}\mathbf{A}\mathbf{Z}'\mathbf{V}^{-1}(\mathbf{y}_i - \mathbf{1}\mu_i), \qquad (5.2)$$

where $\mathbf{V}^{-1}$ is the inverse matrix of the variance of $\mathbf{y}_i$, i.e., $Var(\mathbf{y}_i) = \sigma^2_{g_i}\mathbf{Z}\mathbf{A}\mathbf{Z}' + \mathbf{I}_n\sigma^2_{e_i} = \mathbf{V}$. In the context of animal breeding, Eq. (5.2) is considered a univariate linear phenotypic selection index (LPSI) (Mrode 2005) and is used to rank and select individuals as parents of the next generation in the context of one trait. Equation (5.2) can be extended to the multi-trait phenotypic selection index case, but to predict the net genetic merit ($H = \mathbf{w}'\mathbf{g}$, see Chap. 2 for details) it would be necessary to construct linear combinations of the predicted values of $\mathbf{g}_i$ associated with the traits of interest as was described in the Foreword of this book.

The vector of the individual genomic breeding values ($\boldsymbol{\gamma}_i$) associated with the ith characteristic ($i = 1, 2, \ldots, t$; t = number of traits) of the candidates for selection can be written as

$$\boldsymbol{\gamma}_i = \mathbf{X}\mathbf{u}_i, \tag{5.3}$$

where $\mathbf{X}$ is an $n \times m$ matrix (n = number of observations and m = number of markers in the population) of coded marker values ($2 - 2p$, $1 - 2p$, and $-2p$ for genotypes AA, Aa, and aa respectively) associated with the additive effects of the quantitative trait loci (QTL) and $\mathbf{u}_i$ is an $m \times 1$ vector of the additive effects of the QTL associated with markers that affect the ith trait. It is assumed that $\boldsymbol{\gamma}_i$ has MVN with mean $\mathbf{0}$ and variance $\mathbf{G}\sigma_{\gamma}^2$, i.e., $\boldsymbol{\gamma}_i \sim$ MVN ($\mathbf{0}$, $\mathbf{G}\sigma_{\gamma_i}^2$), where $\sigma_{\gamma_i}^2$ is the additive genomic variance of $\boldsymbol{\gamma}_i$ and $\mathbf{G} = \mathbf{X}\mathbf{X}'/c$ is the $n \times n$ additive genomic relationship matrix between genotypes; $c = \sum_{j=1}^{m} 2p_j(1 - p_j)$ in an F_2 population, and $c = \sum_{j=1}^{m} 4p_j(1 - p_j)$ in a double haploid population; p is the frequency of allele A and $1 - p$ is the frequency of allele a in the jth marker ($j = 1, 2, \ldots, m$).

The additive genomic relationship matrix $\mathbf{G} = \mathbf{X}\mathbf{X}'/c$ has special properties. For example, in the asymptotic context, the expectation of matrix $\mathbf{G}$ is equal to the numerical relationship matrix $\mathbf{A}$, i.e., $E(\mathbf{G}) = \mathbf{A}$ (Habier et al. 2007; Van Raden 2008); this means that $\mathbf{G}$ is a particular realization of $\mathbf{A}$ and when the number of markers and genotypes increases in the training population, the value of $\mathbf{G}$ tends to concentrate around $\mathbf{A}$. Thus, it can be assumed that at the limit, when the number of markers and genotypes is very high, $\mathbf{G} = \mathbf{A}$ (Cerón-Rojas and Sahagún-Castellanos 2016).

The vector of genomic breeding values (Eq. 5.3) has a similar function in genomic selection as $\mathbf{g}_i$ in the phenotypic selection context. In addition, $\mathbf{g}_i$ can be written as $\mathbf{g}_i = \boldsymbol{\gamma}_i + \eta_i$, where $\eta_i = \mathbf{g}_i - \boldsymbol{\gamma}_i$ (Gianola et al. 2003). Also, note that

$$Cov(\mathbf{g}_i, \boldsymbol{\gamma}_i) = \sigma_{\gamma_i}^2, \tag{5.4}$$

i.e., the covariance between $\boldsymbol{\gamma}_i$ and $\mathbf{g}_i$ is equal to the variance of $\boldsymbol{\gamma}_i$ (Dekkers 2007).

Let $\mathbf{y}_i' = [y_{i1} \quad y_{i2} \quad \cdots \quad y_{in}]$ be a vector $1 \times n$ of observation of the ith trait in the training population and let $\boldsymbol{\gamma}_i' = [\gamma_{i1} \quad \gamma_{i2} \quad \cdots \quad \gamma_{in}]$ be a vector $1 \times n$ of unobservable genomic breeding values associated with $\mathbf{y}_i$; then, $\mathbf{y}_i$ can also be written as

$$\mathbf{y}_i = \mathbf{1}\mu_i + \mathbf{Z}\boldsymbol{\gamma}_i + \boldsymbol{\varepsilon}_i, \tag{5.5}$$

where μ_i is the mean of the ith trait, $\mathbf{1}$ is a vector $n \times 1$ of 1s, $\mathbf{Z}$ is a design matrix, $\boldsymbol{\gamma}_i \sim$ MVN ($\mathbf{0}$, $\mathbf{G}\sigma_{\gamma}^2$) and $\boldsymbol{\varepsilon}_i \sim$ MVN ($\mathbf{0}$, $\mathbf{I}_n\sigma_{\varepsilon_i}^2$) are vectors of genomic breeding values and of residuals respectively, and $\sigma_{\varepsilon_i}^2$ is the residual variance. $\mathbf{I}_n$, $\mathbf{G}$, and σ_{γ}^2 were defined in Eqs. (5.2) and (5.3).

According to Eqs. (5.2) and (5.3), when μ_i, σ_{γ}^2 and $\sigma_{\varepsilon_i}^2$ are known, the vector of GEBVs for the individuals with the ith trait can be obtained as

$$\widehat{\boldsymbol{\gamma}}_i = \sigma^2_{\gamma_i}\mathbf{GZ}'\mathbf{V}^{-1}(\mathbf{y}_i - \mathbf{1}\mu_i), \tag{5.6}$$

where the variance of $\mathbf{y}_i$ should now be written as $\mathbf{V} = \sigma^2_{\gamma_i}\mathbf{ZGZ}' + \mathbf{I}_n\sigma^2_{\varepsilon_i}$. In the context of genomic selection, Eq. (5.6) is considered a univariate LGSI and is used to rank and select individuals as parents of the next generation (Van Raden 2008; Togashi et al. 2011). Equation (5.6) is the BLUP of $\boldsymbol{\gamma}_i$ and can be extended to a multi-trait genomic selection index, but to predict the net genetic merit ($H = \mathbf{w}'\mathbf{g}$), it is necessary to construct an LGSI, which is a linear combination of $\boldsymbol{\gamma}_i$.

Although Eq. (5.6) is theoretically very important in LGSI, in practice we need to estimate the marker effects associated with all the traits of interest and to use these estimates in the testing population to obtain the GEBV of the candidates for selection. Let $\mathbf{u}' = [\mathbf{u}_1' \quad \mathbf{u}_2' \quad \cdots \quad \mathbf{u}_t']$ be a vector $1 \times nt$ associated with t traits. In the univariate context, Van Raden (2008) showed that the ith vector $\mathbf{u}_i$ of marker effects in the training population can be estimated as

$$\widehat{\mathbf{u}}_i = c^{-1}\mathbf{X}'[\mathbf{G} + \upsilon\mathbf{I}_n]^{-1}(\mathbf{y}_i - \mathbf{1}\mu_i), \tag{5.7}$$

where $\upsilon = \frac{\sigma^2_{e_i}}{\sigma^2_{g_i}}$; $\sigma^2_{g_i}$, $\sigma^2_{e_i}$ and the other parameters were defined earlier. According to Ceron-Rojas et al. (2015), to estimate the vector $\mathbf{u}' = [\mathbf{u}_1' \quad \mathbf{u}_2' \quad \cdots \quad \mathbf{u}_t']$ in the multi-trait context, Eq. (5.7) can be written as

$$\widehat{\mathbf{u}} = c^{-1}\mathbf{W}_t'[(\mathbf{I}_t \otimes \mathbf{G}) + (\mathbf{N} \otimes \mathbf{I}_n)]^{-1}(\mathbf{y} - \boldsymbol{\mu} \otimes \mathbf{1}), \tag{5.8}$$

where $\mathbf{W}_t = \mathbf{I}_t \otimes \mathbf{X}$, "$\otimes$" denotes the Kronecker product (Schott 2005), c and $\mathbf{X}$ were defined in Eq. (5.3); $\mathbf{N} = \mathbf{RC}^{-1}$, where$\mathbf{R}$ and $\mathbf{C}$ are the residual and breeding value covariance matrices for t traits respectively; $\mathbf{y}' = [\mathbf{y}_1' \quad \mathbf{y}_2' \quad \cdots \quad \mathbf{y}_t'] \sim \text{MVN}(\boldsymbol{\mu}, \mathbf{V})$ is a vector of size $1 \times tn$, with covariance matrix $\mathbf{V} = \mathbf{C} \otimes \mathbf{G} + \mathbf{R} \otimes \mathbf{I}_n$; $\mathbf{I}_t$ is an identity matrix of size $t \times t$ and $\mathbf{I}_n$ was defined earlier; $\boldsymbol{\mu}' = [\mu_1 \quad \mu_2 \quad \cdots \quad \mu_t]$ is a vector $1 \times t$ of means associated with vector $\mathbf{y}$, and $\mathbf{1}$ is a vector $n \times 1$ of 1s. In this case, the estimator of the vector of sub-vectors of genomic breeding values $\boldsymbol{\gamma}' = [\boldsymbol{\gamma}_1 \quad \boldsymbol{\gamma}_2 \quad \ldots \quad \boldsymbol{\gamma}_t]$ in the testing population can be obtained as

$$\widehat{\boldsymbol{\gamma}} = \mathbf{W}_t\widehat{\mathbf{u}}. \tag{5.9}$$

Equation (5.9) is the vector of GEBVs for the multi-trait case. Thus, in the testing population, in Eq. (5.9), only the coded values in matrix $\mathbf{X}$ change, whereas $\widehat{\mathbf{u}}$ is the same in each selection cycle. Note that to obtain Eqs. (5.7) and (5.8), we assumed that $\boldsymbol{\mu}$, $\mathbf{C}$, and $\mathbf{R}$ are known.

We indicated that the genomic breeding values have normal distribution (Eq. 5.5). Using the simulated data described in Chap. 2, Sect. 2.8.1, in Fig. 5.1 we present the distribution of the GEBVs (Eq. 5.9) associated with traits T1 in the first (Fig. 5.1a) and the fifth (Fig. 5.1b) selection cycles in the testing population. In effect, the frequency distribution of the GEBVs approaches normal distribution in both selection cycles.

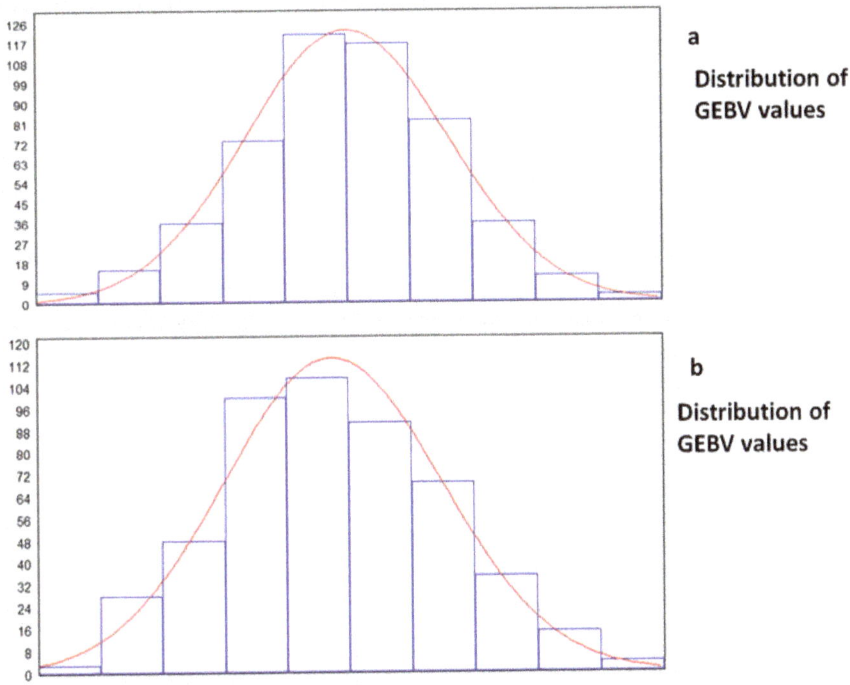

Fig. 5.1 Distribution of the genomic estimated breeding values (GEBVs) associated with traits T1 in (**a**) the first and (**b**) the fifth selection cycles in the testing population

5.1.3 The LGSI and Its Parameters

Similar to the LPSI (Chap. 2), the objective of the LGSI is to predict the net genetic merit $H = \mathbf{w}'\mathbf{g}$, where $\mathbf{g}' = [g_1 \quad g_2 \quad \dots \quad g_t]$ (t = number of traits) is a vector of unobservable true breeding values and $\mathbf{w}' = [w_1 \quad w_2 \quad \dots \quad w_t]$ is a vector of economic weights. Suppose that the genomic breeding values $\boldsymbol{\gamma}_i = \mathbf{X}\mathbf{u}_i$ are known; then, the LGSI can be written as

$$I_G = \boldsymbol{\beta}'\boldsymbol{\gamma}, \tag{5.10}$$

where $\boldsymbol{\beta}$ is an unknown vector of weights.

The main advantage of the LGSI over the LPSI lies in the possibility of reducing the intervals between selection cycles (L_G) by more than two thirds (Lorenz et al. 2011); thus, this parameter should be incorporated into the LGSI selection response and the expected genetic gain per trait to reflect the main advantage of the LGSI over the LPSI and the other indices. Assuming that $L_G = 1$, in the LPSI context we wrote the selection response as $R_I = k_I \sigma_H \rho_{HI}$; however, if $L_G \neq 1$, the LGSI selection response can be written as

$$R_{I_G} = \frac{k_I}{L_G}\frac{\sigma_{HI_G}}{\sigma_{I_G}^2} = \frac{k_I}{L_G}\sigma_H \rho_{HI_G}, \tag{5.11}$$

where k_I is the standardized selection differential (or selection intensity) associated with the LGSI, σ_{HI_G} is the covariance between $H = \mathbf{w}'\mathbf{g}$ and the LGSI, $\sigma_{I_G}^2$ is the variance of the LGSI, σ_H is the standard deviation of H, ρ_{HI_G} is the correlation between H and the LGSI, and L_G denotes the intervals between selection cycles.

Let $\mathbf{C}$ and $\boldsymbol{\Gamma}$ be matrices of covariance of the breeding values ($\mathbf{g}$) and of the genomic breeding values ($\boldsymbol{\gamma}$) respectively; then, the correlation between $H = \mathbf{w}'\mathbf{g}$ and $I_G = \boldsymbol{\beta}'\boldsymbol{\gamma}$ can be written as

$$\rho_{HI_G} = \frac{\mathbf{w}'\boldsymbol{\Gamma}\boldsymbol{\beta}}{\sqrt{\mathbf{w}'\mathbf{C}\mathbf{w}}\sqrt{\boldsymbol{\beta}'\boldsymbol{\Gamma}\boldsymbol{\beta}}}, \tag{5.12}$$

where $\mathbf{w}'\boldsymbol{\Gamma}\boldsymbol{\beta} = \sigma_{HI_G}$ is the covariance between $H = \mathbf{w}'\mathbf{g}$ and $I_G = \boldsymbol{\beta}'\boldsymbol{\gamma}$, $\sigma_H = \sqrt{\mathbf{w}'\mathbf{C}\mathbf{w}}$ is the standard deviation of the variance of $H = \mathbf{w}'\mathbf{g}$, and $\sigma_{I_G} = \sqrt{\boldsymbol{\beta}'\boldsymbol{\Gamma}\boldsymbol{\beta}}$ is the standard deviation of the variance of $I_G = \boldsymbol{\beta}'\boldsymbol{\gamma}$.

5.1.4 Maximizing LGSI Parameters

To maximize the genomic selection response (Eq. 5.11), suppose that k_I, σ_H and L_G are fixed and take the derivative of the natural logarithm (ln) of the correlation between H and I_G (Eq. 5.12) with respect to vector $\boldsymbol{\beta}$, equate the result of the derivative to the null vector, and isolate $\boldsymbol{\beta}$, i.e.,

$$\frac{\partial}{\partial \boldsymbol{\beta}} \ln \rho_{HI_g} = \frac{\partial}{\partial \boldsymbol{\beta}} \ln\left(\frac{\mathbf{w}'\boldsymbol{\Gamma}\boldsymbol{\beta}}{\sqrt{\mathbf{w}'\mathbf{C}\mathbf{w}}\sqrt{\boldsymbol{\beta}'\boldsymbol{\Gamma}\boldsymbol{\beta}}}\right) = \mathbf{0}. \tag{5.13}$$

The result is $\boldsymbol{\beta} = s\mathbf{w}$, where $s = \boldsymbol{\beta}'\boldsymbol{\Gamma}\boldsymbol{\beta}/\mathbf{w}'\boldsymbol{\Gamma}\boldsymbol{\beta}$ is a proportional constant that does not affect the maximum value of ρ_{HI_G}, because this is invariant to the scale change; then, assuming that $\boldsymbol{\beta} = \mathbf{w}$, the maximized LGSI selection response can be written as

$$R_{I_G} = \frac{k_I}{L_G}\sqrt{\mathbf{w}'\boldsymbol{\Gamma}\mathbf{w}}. \tag{5.14}$$

Hereafter, we refer to the LGSI genomic selection response as that of Eq. (5.14). Also, because $\boldsymbol{\beta} = \mathbf{w}$, Eq. (5.12) can be written as

$$\rho_{HI_G} = \frac{\sqrt{\mathbf{w}'\boldsymbol{\Gamma}\mathbf{w}}}{\sqrt{\mathbf{w}'\mathbf{C}\mathbf{w}}} = \frac{\sigma_{I_G}}{\sigma_H}, \tag{5.15}$$

which is the maximized correlation between $H = \mathbf{w}'\mathbf{g}$ and $I_G = \boldsymbol{\beta}'\boldsymbol{\gamma}$, or LGSI accuracy; $\sigma_H = \sqrt{\mathbf{w}'\mathbf{C}\mathbf{w}}$ is the standard deviation of the variance of H, and $\sigma_{I_G} = \sqrt{\boldsymbol{\beta}'\boldsymbol{\Gamma}\boldsymbol{\beta}}$ is the standard deviation of the variance of I_G.

The LGSI expected genetic gain per trait ($\mathbf{E}_{I_G}$) can be written as

$$\mathbf{E}_{I_G} = \frac{k_I}{L_G}\frac{\boldsymbol{\Gamma}\mathbf{w}}{\sqrt{\mathbf{w}'\boldsymbol{\Gamma}\mathbf{w}}}. \tag{5.16}$$

All the terms in Eq. (5.16) were previously defined.

Let $\lambda_G = \frac{\rho_{HI_G}}{\rho_{HI}}$ be LGSI efficiency versus LPSI efficiency to predict the net genetic merit, where ρ_{HI_G} is the LGSI accuracy and ρ_{HI} the LPSI accuracy; in percentage terms, LGSI efficiency versus LPSI efficiency for each selection cycle can be written as

$$p_G = 100(\lambda_G - 1). \tag{5.17}$$

According to Eq. (5.17), if $p_G > 0$, LGSI efficiency is greater than LPSI efficiency; if $p_G = 0$, the efficiency of both selection indices is equal, and if $p_G < 0$, the LPSI is more efficient than the LGSI at predicting $H = \mathbf{w}'\mathbf{g}$.

Equation (5.17) is useful for measuring LGSI efficiency in terms of accuracy when predicting the net genetic merit ($H = \mathbf{w}'\mathbf{g}$), whereas the Technow et al. (2013) inequality measures LGSI efficiency in terms of the time needed to complete one selection cycle. In the context of the LGSI and the LPSI, the Technow inequality can be written as

$$L_G < \frac{\rho_{HI_G}}{h_I} L_P, \tag{5.18}$$

where L_G and L_P denote the time required to complete one selection cycle for the LGSI and the LPSI respectively, ρ_{HI_G} is the LGSI accuracy, and h_I is the square root of the heritability (Lin and Allaire 1977; Nordskog 1978) of the LPSI, which can be denoted as $h_I = \sqrt{\frac{\mathbf{b}'\mathbf{C}\mathbf{b}}{\mathbf{b}'\mathbf{P}\mathbf{b}}}$ (see Chap. 2 for details). Then, assuming that the selection intensity is the same for both selection indices, if Eq. (5.18) is true, the LGSI is more efficient than the LPSI per unit of time.

5.1.5 Relationship Between the LGSI and LPSI Selection Responses

To obtain the relationship between R_{I_G} and R_I in the asymptotic context, we omitted the intervals between selection cycles (L_G and L_I respectively) to simplify the algebra. Consider a population where the number of genotypes and markers tends to infinity; in this case, markers explain most of the true additive genetic variances and covariances. Thus, we can assume that matrices $\mathbf{\Gamma}$ and $\mathbf{C}$ are very similar, and at the limit, $\mathbf{\Gamma} = \mathbf{C}$. Now suppose that in this population the phenotypic variance–covariance matrix ($\mathbf{P}$) is known and comprises matrix $\mathbf{\Gamma}$ and the variance–covariance residual matrix ($\mathbf{R}$). In this case, the inverse of $\mathbf{P}$ can be written as $\mathbf{P}^{-1} = (\mathbf{\Gamma} + \mathbf{R})^{-1} = \mathbf{\Gamma}^{-1} - \mathbf{\Gamma}^{-1}(\mathbf{\Gamma}^{-1} + \mathbf{R}^{-1})^{-1}\mathbf{\Gamma}^{-1}$, where $\mathbf{\Gamma}^{-1}$ and $\mathbf{R}^{-1}$ are the inverses of matrices $\mathbf{\Gamma}$ and $\mathbf{R}$ respectively. Thus, the LPSI selection response is given by

$$R_I = k_I\sqrt{\mathbf{b}'\mathbf{P}\mathbf{b}} = k_I\sqrt{\mathbf{w}'\mathbf{\Gamma}\mathbf{P}^{-1}\mathbf{\Gamma}\mathbf{w}} = k_I\sqrt{\mathbf{w}'\mathbf{\Gamma}\mathbf{w} - \mathbf{w}'\left(\mathbf{\Gamma}^{-1} + \mathbf{R}^{-1}\right)^{-1}\mathbf{w}}, \quad (5.19)$$

where $\mathbf{b} = \mathbf{P}^{-1}\mathbf{\Gamma}\mathbf{w}$ is the vector of coefficients of the LPSI in the asymptotic context. Note that $\mathbf{b}'\mathbf{P}\mathbf{b} \geq 0$ and $\mathbf{w}'\mathbf{\Gamma}\mathbf{w} \geq 0$, i.e., $\mathbf{b}'\mathbf{P}\mathbf{b}$ and $\mathbf{w}'\mathbf{\Gamma}\mathbf{w}$ are positive semi-definite, meaning that $\mathbf{w}'\mathbf{\Gamma}\mathbf{w} \geq \mathbf{w}'(\mathbf{\Gamma}^{-1} + \mathbf{R}^{-1})^{-1}\mathbf{w} \geq 0$; then, in the asymptotic context, $R_{I_G} \geq R_I$. This result is not common when the number of genotypes and markers is small; however, it gives an idea of the theoretical behavior of R_{I_G} with respect to R_I when the number of markers and genotypes is very large.

Because $\mathbf{g}_q$ can be written as $\mathbf{g}_q = \boldsymbol{\gamma}_q + \eta_q$, where $\eta_q = \mathbf{g}_q - \boldsymbol{\gamma}_q$ $(q = 1, 2, \cdots, t)$, for low numbers of markers and genotypes, the covariance genotypic matrix $\mathbf{C}$ can be written as $\mathbf{C} = \mathbf{\Gamma} + \mathbf{E}$, where $\mathbf{E} = \mathbf{C} - \mathbf{\Gamma}$; then, the inverse of matrix $\mathbf{P}$ can be written as $\mathbf{P}^{-1} = [(\mathbf{\Gamma} + \mathbf{E}) + \mathbf{R}]^{-1} = (\mathbf{\Gamma} + \mathbf{E})^{-1} - (\mathbf{\Gamma} + \mathbf{E})^{-1}[(\mathbf{\Gamma} + \mathbf{E})^{-1} + \mathbf{R}^{-1}]^{-1}(\mathbf{\Gamma} + \mathbf{E})^{-1}$. In the latter case, the LPSI selection response R_I can be written as

$$\begin{aligned} R_I &= k_I\sqrt{\mathbf{w}'(\mathbf{\Gamma} + \mathbf{E})\mathbf{P}^{-1}(\mathbf{\Gamma} + \mathbf{E})\mathbf{w}} \\ &= k_I\sqrt{\mathbf{w}'\mathbf{\Gamma}\mathbf{w} + \mathbf{w}'\mathbf{E}\mathbf{w} - \mathbf{w}'\left[(\mathbf{\Gamma} + \mathbf{E})^{-1} + \mathbf{R}^{-1}\right]^{-1}\mathbf{w}}. \end{aligned} \quad (5.20)$$

Equation (5.20) indicates that in the non-asymptotic context (low numbers of markers and genotypes), R_{I_G} and R_I are related in three possible ways:

1. $R_I > R_{I_G}$ if $\mathbf{w}'\mathbf{E}\mathbf{w} > \mathbf{w}'[(\mathbf{\Gamma} + \mathbf{E})^{-1} + \mathbf{R}^{-1}]^{-1}\mathbf{w}$
2. $R_I = R_{I_G}$ if $\mathbf{w}'\mathbf{E}\mathbf{w} = \mathbf{w}'[(\mathbf{\Gamma} + \mathbf{E})^{-1} + \mathbf{R}^{-1}]^{-1}\mathbf{w}$
3. $R_{I_G} > R_I$ if $\mathbf{w}'\mathbf{E}\mathbf{w} < \mathbf{w}'[(\mathbf{\Gamma} + \mathbf{E})^{-1} + \mathbf{R}^{-1}]^{-1}\mathbf{w}$

The second and third points indicate that R_{I_G} may be equal to or larger than R_I, even under a small number of markers, depending on the size of $\mathbf{w}'\mathbf{E}\mathbf{w}$ and $\mathbf{w}'[(\mathbf{\Gamma} + \mathbf{E})^{-1} + \mathbf{R}^{-1}]^{-1}\mathbf{w}$. These three points explain the theoretical relationship between R_I and R_{I_G} for a low number of markers and genotypes. When $\mathbf{\Gamma} = \mathbf{C}$, $\mathbf{E} = \mathbf{0}$, and $R_I = k_I\sqrt{\mathbf{w}'\mathbf{\Gamma}\mathbf{w} - \mathbf{w}'\left(\mathbf{\Gamma}^{-1} + \mathbf{R}^{-1}\right)^{-1}\mathbf{w}}$, then $R_{I_G} \geq R_I$.

5.1.6 Statistical LGSI Properties

Assuming that H and I_G have joint bivariate normal distribution and that $\boldsymbol{\Gamma}$, $\mathbf{C}$, and $\mathbf{w}$ are known, the LGSI has the following properties:

1. The variance of I_G ($\sigma^2_{I_G}$) and the covariance between H and I_G (σ_{HI_G}) are equal, i.e., $\sigma^2_{I_G} = \sigma_{HI_G}$.
2. The maximized correlation between H and I_G (or LGSI accuracy) is equal to $\rho_{HI_G} = \frac{\sigma_{I_G}}{\sigma_H}$, where σ_{I_G} is the standard deviation of $\sigma^2_{I_G}$ and σ_H is the standard deviation of the variance of H (σ^2_H).
3. The variance of the predicted error, $Var(H - I_G) = \left(1 - \rho^2_{HI_G}\right)\sigma^2_H$, is minimal. Note that $Var(H - I_G) = \sigma^2_{I_G} + \sigma^2_H - 2\sigma_{HI_G}$, and when $\boldsymbol{\beta} = \mathbf{w}$, $\sigma^2_{I_G} = \sigma_{HI_G}$, whence $Var(H - I_G) = \sigma^2_H - \sigma^2_{I_G} = \left(1 - \rho^2_{HI_G}\right)\sigma^2_H$ is minimal.
4. The total variance of H explained by I_G is $\sigma^2_{I_G} = \rho^2_{HI_G}\sigma^2_H$. It is evident that if $\rho_{HI_G} = 1, \sigma^2_{I_G} = \sigma^2_H$, and if $\rho_{HI_G} = 0, \sigma^2_{I_G} = 0$. That is, the variance of H explained by I_G is proportional to ρ_{HI_G}, and when ρ_{HI_G} is close to $1, \sigma^2_{I_G}$ is close to σ^2_H; if ρ_{HI_G} is close to 0, $\sigma^2_{I_G}$ is close to 0.

The LGSI properties described in points 1–4 of this subsection are the same as the LPSI properties described in Chap. 2. This corroborates the LGSI as an application of the LPSI theory to the genomic selection context.

5.1.7 Genomic Covariance Matrix in the Training and Testing Population

To derive the LGSI theory, we assumed that the true genomic additive variance–covariance matrix $\boldsymbol{\Gamma}$ was known. However, in practice, we need to estimate it. In the training population, matrix $\boldsymbol{\Gamma}$ can be estimated by restricted maximum likelihood (REML) using phenotypic and genomic information, as described by Vattikuti et al. (2012) and Su et al. (2012). In Eqs. (2.22) to (2.24) of Chap. 2, we presented the formulas for estimating the genotypic and residual variance and covariance based on the formulas described by Lynch and Walsh (1998). Here, we present a brief description of how we can estimate the qth component ($\sigma_{\gamma qq}$) of $\boldsymbol{\Gamma}$ in the training population using the REML method.

We estimated $\sigma_{\gamma qq} = \sigma^2_{\gamma q}$ (q, $q' = t$ = number of traits) in the absence of dominance and epistatic effects, using the model $\mathbf{y}_q = \mathbf{1}\mu_q + \mathbf{Z}\boldsymbol{\gamma}_q + \boldsymbol{\varepsilon}_q$, where the vector $\mathbf{y}_q \sim \text{NMV}(\mathbf{1}\mu_q, \mathbf{V}_q)$ $g \times 1$ (g = number of genotypes in the population) had a multivariate normal distribution; $\mathbf{1}$ was a $g \times 1$ vector of 1s, μ_q was the mean of the qth trait, $\mathbf{Z}$ was an identity matrix $g \times g$; $\boldsymbol{\gamma}_q \sim \text{NMV}(\mathbf{0}, \mathbf{G}\sigma^2_{\gamma q})$ was a vector of genomic breeding values, and $\boldsymbol{\varepsilon}_q \sim NMV(\mathbf{0}, \mathbf{I}\sigma^2_{\varepsilon_q})$ was a $g \times 1$ vector of residuals. Matrix $\mathbf{G} = \mathbf{XX}'/c$ was the genomic relationship matrix, and in an F_2 population, $c = \sum_{j=1}^{N} 2p_jq_j$; $\mathbf{X}$ was a $g \times m$ matrix (m = number of markers) of the coded marker values ($2-2p$ for AA, $1-2p$ for Aa, and $-2p$ for aa) for the additive effects of the markers; p and q denote the frequency of allele A and the frequency of allele a in the jth marker ($j = 1, 2, \ldots, m$), and $\mathbf{V}_q = \mathbf{G}\sigma^2_{\gamma q} + \mathbf{I}\sigma^2_{\varepsilon_q}$.

The expectation–maximization algorithm allowed the REML for the variance components $\sigma^2_{\gamma q}$ and $\sigma^2_{\varepsilon_q}$ to be computed by iterating the following equations:

$$\sigma^{2(n+1)}_{\gamma q} = \sigma^{2(n)}_{\gamma q} + \frac{\left(\sigma^{2(n)}_{\gamma q}\right)^2}{g}\left[\mathbf{y}'_q\left(\mathbf{T}^{(n)}\mathbf{G}\mathbf{T}^{(n)}\right)\mathbf{y}_q - tr\left(\mathbf{T}^{(n)}\mathbf{G}\right)\right] \quad (5.21)$$

and

$$\sigma^{2(n+1)}_{\varepsilon_q} = \sigma^{2(n)}_{\varepsilon_q} + \frac{\left(\sigma^{2(n)}_{\varepsilon_q}\right)^2}{g}\left[\mathbf{y}'_q\left(\mathbf{T}^{(n)}\mathbf{T}^{(n)}\right)\mathbf{y}_q - tr\left(\mathbf{T}^{(n)}\right)\right], \quad (5.22)$$

where g is the number of genotypes. After n iterations, when $\sigma^{2(n+1)}_{\gamma q}$ was very similar to $\sigma^{2(n)}_{\gamma q}$ and $\sigma^{2(n+1)}_{\varepsilon_q}$ was very similar to $\sigma^{2(n)}_{\varepsilon_q}$, $\sigma^{2(n+1)}_{\gamma q}$ and $\sigma^{2(n+1)}_{\varepsilon_q}$ were the estimated variance components of $\sigma^2_{\gamma q}$ and $\sigma^2_{\varepsilon_q}$ respectively. In Eqs. (5.21) and (5.22) $tr(.)$ denoted the trace of the matrices within brackets; $\mathbf{T} = \mathbf{V}_q^{-1} - \mathbf{V}_q^{-1}\mathbf{1}\left(\mathbf{1}'\mathbf{V}_q^{-1}\mathbf{1}\right)^{-1}\mathbf{1}'\mathbf{V}_q^{-1}$, and $\mathbf{V}_q^{-1}$ was the inverse of $\mathbf{V}_q = \mathbf{G}\sigma^2_{\gamma q} + \mathbf{I}\sigma^2_{\varepsilon_q}$. In matrix $\mathbf{T}^{(n)}$, $\mathbf{V}_q^{-1(n)}$ was the inverse of matrix $\mathbf{V}_q^{(n)} = \mathbf{G}\sigma^{2(n)}_{\gamma q} + \mathbf{I}\sigma^{2(n)}_{\varepsilon_q}$.

The genomic additive genetic covariance between the observations of the qth and ith traits, $\mathbf{y}_q$ and $\mathbf{y}_i$ ($\sigma_{\gamma_{qi}}$, $q, i = 1, 2, \ldots, t$), can be estimated by REML. Here, we adapted Eqs. (5.21) and (5.22) using the variance of the sum of $\mathbf{y}_q$ and $\mathbf{y}_i$, i.e., $Var(\mathbf{y}_i + \mathbf{y}_q) = \mathbf{V}_i + \mathbf{V}_q + 2\mathbf{C}_{iq}$, where $\mathbf{V}_i = \mathbf{G}\sigma^2_{\gamma_i} + \mathbf{I}\sigma^2_{\varepsilon_i} = Var(\mathbf{y}_i)$ is the variance of $\mathbf{y}_i$ and $\mathbf{V}_q = \mathbf{G}\sigma^2_{\gamma_q} + \mathbf{I}\sigma^2_{\varepsilon_q} = Var(\mathbf{y}_q)$ is the variance of $\mathbf{y}_q$; $2\mathbf{C}_{iq} = 2\mathbf{G}\sigma_{\gamma_{iq}} + 2\mathbf{I}\sigma_{\varepsilon_{iq}} = 2Cov(\mathbf{y}_i, \mathbf{y}_q)$ is the covariance of $\mathbf{y}_q$ and $\mathbf{y}_i$, and $\sigma_{\gamma_{iq}}$ and $\sigma_{\varepsilon_{iq}}$ are the genomic and residual covariance respectively, associated with $\mathbf{y}_i$ and $\mathbf{y}_q$. Thus, one way of estimating $\sigma_{\gamma_{iq}}$ and $\sigma_{e_{iq}}$ is by using the following equation:

$$0.5Var(\mathbf{y}_i + \mathbf{y}_q) - 0.5Var(\mathbf{y}_i) - 0.5Var(\mathbf{y}_q), \quad (5.23)$$

for which Eqs. (5.21) and (5.22) can be adapted.

If there is only marker information on the testing population, then it is not possible to estimate $\boldsymbol{\Gamma}$ using Eqs. (5.21) to (5.23). Another way of estimating $\boldsymbol{\Gamma}$ is to use the method proposed by Ceron-Rojas et al. (2015), which requires the estimated values of $\boldsymbol{\gamma}_q$ ($\widehat{\boldsymbol{\gamma}}_q$) in the cycle of interest. Let $\widehat{\mathbf{u}}$ be the estimator of the vector of marker effects $\mathbf{u}' = [\mathbf{u}_1' \quad \mathbf{u}_2' \quad \cdots \quad \mathbf{u}_t']$ for t traits obtained in the training population. We obtained the qth GEBVs ($q = 1, 2, \ldots, t$) in the lth selection cycle ($l = 1, 2, \ldots$, number of cycles) as

$$\widehat{\boldsymbol{\gamma}}_{ql} = \mathbf{X}_l \widehat{\mathbf{u}}_q \tag{5.24}$$

where $\widehat{\mathbf{u}}_q$ is the vector of size $m \times 1$ of the estimated marker effects of the qth trait in the training population and $\mathbf{X}_l$ is a matrix of size $n \times m$ of the coded values of marker genotypes in the lth selection cycle of the testing population.

Now suppose that $\boldsymbol{\gamma}_q$ and $\boldsymbol{\gamma}_{q'}$ have multivariate normal distribution jointly, with mean $\mathbf{1}\mu_{\gamma_q}$ and $\mathbf{1}\mu_{\gamma_{q'}}$ respectively, and covariance matrix $\mathbf{G}\sigma_{\gamma_{qq'}}$, where $\mathbf{1}$ is an $n \times 1$ vector of 1s and $\mathbf{G} = \mathbf{XX}'/c$ is the additive genomic relationship matrix. Then, $\boldsymbol{\Gamma} = \left\{\sigma_{\gamma_{qq'}}\right\}$ can be estimated as

$$\widehat{\boldsymbol{\Gamma}}_l = \left\{\widehat{\sigma}_{\gamma_{qq'}}\right\}, \tag{5.25}$$

where $\widehat{\sigma}_{\gamma_{qq'}} = \frac{1}{g}\left(\widehat{\boldsymbol{\gamma}}_{ql} - \mathbf{1}\widehat{\mu}_{\gamma_{ql}}\right)'\mathbf{G}_l^{-1}\left(\widehat{\boldsymbol{\gamma}}_{q'l} - \mathbf{1}\widehat{\mu}_{\gamma_{q'l}}\right)$ is the estimated covariance between $\boldsymbol{\gamma}_q$ and $\boldsymbol{\gamma}_{q'}$ in the lth selection cycle of the testing population; g is the number of genotypes; $\widehat{\boldsymbol{\gamma}}_{ql}$ was defined in Eq. (5.24); $\widehat{\mu}_{\gamma_{ql}}$ and $\widehat{\mu}_{\gamma_{q'l}}$ are the estimated arithmetic means of the values of $\widehat{\boldsymbol{\gamma}}_{ql}$ and $\widehat{\boldsymbol{\gamma}}_{q'l}$; $\mathbf{1}$ is a $g \times 1$ vector of 1s and $\mathbf{G}_l = c^{-1}\mathbf{X}_l\mathbf{X}_l'$ is the additive genomic relationship matrix in the lth selection cycle ($l = 1, 2, \ldots$, number of cycles) in the testing population.

From Eq. (5.25) we can estimate the LGSI response and expected genetic gain per trait in the testing population as

$$\widehat{R}_{I_G} = \frac{k_I}{L_G}\sqrt{\mathbf{w}'\widehat{\boldsymbol{\Gamma}}\mathbf{w}} \quad \text{and} \quad \widehat{\mathbf{E}}_{I_G} = \frac{k_I}{L_G}\frac{\widehat{\boldsymbol{\Gamma}}\mathbf{w}}{\sqrt{\mathbf{w}'\widehat{\boldsymbol{\Gamma}}\mathbf{w}}}, \tag{5.26}$$

respectively. The estimated LGSI ($\widehat{I}_{\mathrm{G}}$) values in the lth selection cycle can be obtained as

$$\widehat{I}_{\mathrm{G}} = \sum_{q=1}^{t} w_q \widehat{\boldsymbol{\gamma}}_{ql}, \tag{5.27}$$

where w_q is the qth economic weight and $\widehat{\boldsymbol{\gamma}}_{ql}$ was defined in Eq. (5.24). Equation (5.27) is a vector of size $g \times 1$ (g= number of genotypes). In practice, $\widehat{I}_{\mathrm{G}}$ values are ranked to select individual genotypes with optimal GEBVs.

5.1.8 Numerical Examples

To estimate matrices **C** and **R** and the marker effects in the training population, we used a real maize (*Zea mays*) F_2 population with 248 genotypes (each with two repetitions), 233 molecular markers, and three traits—grain yield (GY, ton ha^{-1}), ear height (EHT, cm), and plant height (PHT, cm)—evaluated in one environment. The estimated matrices were $\widehat{\mathbf{C}} = \begin{bmatrix} 0.07 & 0.61 & 1.06 \\ 0.61 & 17.93 & 22.75 \\ 1.06 & 22.75 & 44.53 \end{bmatrix}$ and $\widehat{\mathbf{R}} = \begin{bmatrix} 0.38 & 0.72 & 1.27 \\ 0.72 & 47.14 & 60.96 \\ 1.27 & 60.96 & 121.46 \end{bmatrix}$, which were estimated by Eqs. (5.21) to (5.23) using the numerical relationship **A** instead of the genomic relationship matrix ($\mathbf{G} = \mathbf{XX}'/c$).

Table 5.1 presents the first 20 BLUPs of the estimated marker effects (Eq. 5.8) in the training population and the first 20 marker coded values and GEBVs (Eq. 5.9) obtained in the testing population associated with trait GY. In the

Table 5.1 The 20 best linear unbiased predictors (BLUPs) of the estimated marker effects in the training population and the first 20 marker coded values and genomic estimated breeding values (GEBVs) obtained in the testing population associated with grain yield

Training population	Testing population					
	Marker coded values					
BLUPs	M1	M2	M3	...	M233	GEBVs
−0.0003	1	1	0	...	−1	0.195
−0.0038	0	0	0	...	−1	0.221
−0.0085	−1	1	0	...	−1	−0.643
0.0069	0	1	0	...	1	0.525
−0.0042	0	0	0	...	0	−0.603
0.0038	−1	0	0	...	0	0.062
0.0008	0	1	1	...	0	−0.226
0.0012	0	1	1	...	1	0.023
−0.0004	0	−1	0	...	0	0.444
0.0062	0	0	1	...	−1	−0.286
0.0121	−1	1	0	...	1	−0.196
0.0077	−1	−1	−1	...	0	−0.566
0.0033	−1	0	0	...	0	0.073
0.0102	−1	1	0	...	1	0.058
0.0054	0	1	0	...	0	0.874
0.0002	0	0	0	...	0	0.102
0.0171	0	1	0	...	−1	−0.342
0.0159	−1	0	1	...	−1	−0.428
0.0117	−1	0	0	...	−1	0.072
0.0121	0	−1	0	...	−1	−0.428

testing population, there were 380 genotypes and 233 molecular markers. In this population, the estimated genomic covariance matrix $\boldsymbol{\Gamma} = \left\{\sigma_{\gamma_{qq'}}\right\}$ was $\widehat{\boldsymbol{\Gamma}} = \begin{bmatrix} 0.21 & 2.95 & 5.00 \\ 2.95 & 42.41 & 71.11 \\ 5.00 & 71.11 & 121.53 \end{bmatrix}$. The first GEBV (0.195) related to GY in Table 5.1 was obtained as $0.195 = -0.0003(1) - 0.0038(1) - 0.0085(0) + \cdots - 0.03(-1)$. The other GEBVs can be obtained in a similar manner.

Suppose a selection intensity of 10% ($k_I = 1.755$) and a vector of economic weights of $\mathbf{w}' = [5 \quad -0.1 \quad -0.1]$; then, the estimated LGSI selection response and the expected genetic gain per trait without including the interval between selection cycle is $\widehat{R}_{I_G} = (1.755)\sqrt{\mathbf{w}'\widehat{\boldsymbol{\Gamma}}\mathbf{w}} = 0.92$ and $\widehat{\mathbf{E}}'_{I_G} = (1.755)\frac{\mathbf{w}'\widehat{\boldsymbol{\Gamma}}}{\sqrt{\mathbf{w}'\widehat{\boldsymbol{\Gamma}}\mathbf{w}}} = [0.80 \quad 11.41 \quad 19.28]$ respectively, whereas the estimated LGSI accuracy was $\widehat{\rho}_{HI_G} = 0.48$.

Chapter 11 presents RIndSel, a graphical unit interface that uses selection index theory to select individual candidates as parents for the next selection cycle, which can be used to obtain the results of the real numerical example described in this subsection.

To compare LGSI efficiency versus LPSI efficiency we used the simulated data described in Chap. 2, Sect. 2.8.1. According to Beyene et al. (2015), at least 4 years are required to complete one phenotypic selection cycle in maize, whereas genomic selection requires only 1.5 years. Thus, to compare LGSI efficiency versus LPSI efficiency in terms of time, we can use the Technow et al. (2013) inequality described in Eq. (5.18).

Table 5.2 presents the estimated value of Eq. (5.18) for five simulated selection cycles. The LGSI efficiency was higher than LPSI efficiency in terms of time, because the Technow et al. (2013) inequality was true in the five selection cycles. An additional result obtained by Ceron-Rojas et al. (2015) is presented in Fig. 5.2, which shows the correlation among the LGSI, the LPSI, and the true net genetic

Table 5.2 Five simulated selection cycles

Cycle	L_G	L_P	$\widehat{\rho}_{HI_G}$	$\widehat{h}_I$	$\frac{\widehat{\rho}_{HI_G}}{\widehat{h}_I}L_P$
1	1.5	4.0	0.73	0.92	3.17
2	1.5	4.0	0.78	0.89	3.50
3	1.5	4.0	0.83	0.88	3.77
4	1.5	4.0	0.74	0.87	3.40
5	1.5	4.0	0.71	0.87	3.30

Time required for the linear genomic selection index (L_G) and linear phenotypic selection index (L_P) to complete one selection cycle; estimated accuracy ($\widehat{\rho}_{HI_G}$) of the linear genomic selection index and the square root of the estimated heritability of the linear phenotypic selection index ($\widehat{h}_I$); estimated right-hand side ($\frac{\widehat{\rho}_{HI_G}}{\widehat{h}_I}L_P$) of the inequality formula ($L_G < \frac{\rho_{H,I_G}}{h_I}L_P$)

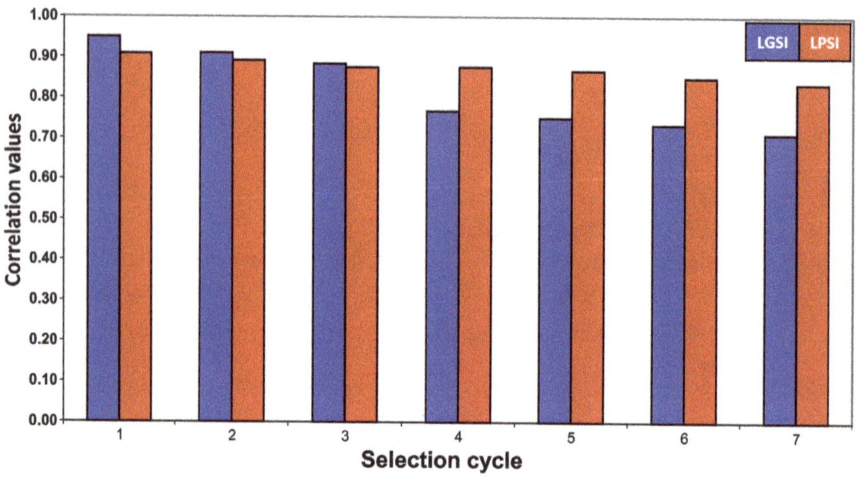

Fig. 5.2 Correlation between the linear genomic selection index (LGSI), the linear phenotypic selection index (LPSI), and true net genetic merit (H) values in seven selection cycles. For each selection cycle, the first column indicates the correlation between the LGSI estimated values and the *H* true values, whereas the second column shows the correlation between the LPSI estimated values and the *H* true values

merit values in seven selection cycles. According to Fig. 5.2, the correlation between the LGSI and the true net genetic merit values was higher than the correlation between the LPSI and the true net genetic merit values for the first three selection cycles; after this cycle, the correlation between LGSI and the true net genetic merit values tended to decrease.

5.2 The Combined Linear Genomic Selection Index

The combined LGSI (CLGSI) developed by Dekkers (2007) is a slightly modified version of the LMSI (see Chap. 4 for details), which, instead of using the marker scores, uses the GEBVs and the phenotypic information jointly to predict the net genetic merit. The main difference between the CLGSI and the LGSI is that the CLGSI can only be used in training populations, whereas the LGSI is used in testing populations. The basic conditions for constructing a valid CLGSI include conditions for constructing the LPSI, the LMSI, and the LGSI, because the CLGSI uses GEBVs and phenotypic information jointly to predict the net genetic merit.

5.2.1 The CLGSI Parameters

The net genetic merit can be written in a similar manner to that in the LMSI context, that is, as

$$H = \mathbf{w}'\mathbf{g} + \mathbf{w}_2'\boldsymbol{\gamma} = \begin{bmatrix} \mathbf{w}' & \mathbf{w}_2' \end{bmatrix} \begin{bmatrix} \mathbf{g} \\ \boldsymbol{\gamma} \end{bmatrix} = \mathbf{a}_G'\mathbf{z}_G, \tag{5.28}$$

where $\mathbf{g}' = [g_1 \quad \cdots \quad g_t]$ is the vector of breeding values, $\mathbf{w}' = [w_1 \quad \cdots \quad w_t]$ is the vector of economic weights associated with breeding values, $\mathbf{w}_2' = [0_1 \quad \cdots \quad 0_t]$ is a null vector associated with the vector of genomic breeding values $\boldsymbol{\gamma}' = [\gamma_1 \quad \gamma_2 \quad \cdots \quad \gamma_t]$, $\mathbf{a}_G' = [\mathbf{w}' \quad \mathbf{w}_2']$ and $\mathbf{z}_G = [\mathbf{g}' \quad \boldsymbol{\gamma}']$.

The CLGSI can be written as

$$I_C = \boldsymbol{\beta}_y'\mathbf{y} + \boldsymbol{\beta}_G'\boldsymbol{\gamma} = \begin{bmatrix} \boldsymbol{\beta}_y' & \boldsymbol{\beta}_G' \end{bmatrix} \begin{bmatrix} \mathbf{y} \\ \boldsymbol{\gamma} \end{bmatrix} = \boldsymbol{\beta}_C'\mathbf{t}_C, \tag{5.29}$$

where $\mathbf{y}' = [y_1 \quad \cdots \quad y_t]$ (t = number of traits) is the vector of phenotypic values; $\boldsymbol{\gamma}$ was defined earlier; $\boldsymbol{\beta}_y'$ and $\boldsymbol{\beta}_G$ are vectors of coefficients of phenotypic and genomic weight values respectively; $\boldsymbol{\beta}_C' = \begin{bmatrix} \boldsymbol{\beta}_y' & \boldsymbol{\beta}_G' \end{bmatrix}$ and $\mathbf{t}_G' = [\mathbf{y}' \quad \boldsymbol{\gamma}']$.

The CLGSI selection response can be written as

$$R_C = k_I\sigma_H\rho_{HI_C} = k_I\sigma_H \frac{\mathbf{a}_C'\boldsymbol{\Psi}_C\boldsymbol{\beta}_C}{\sqrt{\mathbf{a}_C'\boldsymbol{\Psi}_C\mathbf{a}_C}\sqrt{\boldsymbol{\beta}_C'\mathbf{T}_C\boldsymbol{\beta}_C}}, \tag{5.30}$$

where k_I is the standardized selection differential of the CLGSI, $\sigma_H^2 = \mathbf{a}_C'\boldsymbol{\Psi}_C\mathbf{a}_C$ and $Var(I_C) = \boldsymbol{\beta}_C'\mathbf{T}_C\boldsymbol{\beta}_C$ are the variances of H and I_C, whereas $\mathbf{a}_C'\boldsymbol{\Psi}_C\boldsymbol{\beta}_C$ and ρ_{HI_C} are the covariance and the correlation between H and I_C respectively; $\mathbf{T}_C = Var\begin{bmatrix} \mathbf{y} \\ \boldsymbol{\gamma} \end{bmatrix} = \begin{bmatrix} \mathbf{P} & \boldsymbol{\Gamma} \\ \boldsymbol{\Gamma} & \boldsymbol{\Gamma} \end{bmatrix}$ and $\boldsymbol{\Psi}_C = Var\begin{bmatrix} \mathbf{g} \\ \boldsymbol{\gamma} \end{bmatrix} = \begin{bmatrix} \mathbf{C} & \boldsymbol{\Gamma} \\ \boldsymbol{\Gamma} & \boldsymbol{\Gamma} \end{bmatrix}$ are block matrices of the phenotypic covariance matrix, $\mathbf{P} = Var(\mathbf{y})$, the genomic covariance matrix, $\boldsymbol{\Gamma} = Var(\boldsymbol{\gamma})$, and the genetic breeding values covariance matrix, $\mathbf{C} = Var(\mathbf{g})$.

Suppose that matrices $\boldsymbol{\Psi}_C$ and $\mathbf{T}_C$ are known; then the CLGSI vector of coefficients that simultaneously maximizes ρ_{HI_C} and R_C can be written as

$$\boldsymbol{\beta}_C = \mathbf{T}_C^{-1}\boldsymbol{\Psi}_C\mathbf{a}_C, \tag{5.31}$$

whence the optimized CLGSI is

$$I_C = \boldsymbol{\beta}_C'\mathbf{t}_C, \tag{5.32}$$

Equations (5.31) and (5.32) indicate that the CLGSI is an application of the LPSI to the genomic selection context.

From Eq. (5.31), the maximized CLGSI selection response, expected genetic gain per trait and accuracy can be written as

$$R_C = k_I\sqrt{\boldsymbol{\beta}'_C\mathbf{T}_C\boldsymbol{\beta}_C}, \tag{5.33}$$

$$\mathbf{E}_C = k_I\frac{\boldsymbol{\Psi}_C\boldsymbol{\beta}_C}{\sqrt{\boldsymbol{\beta}'_C\mathbf{T}_C\boldsymbol{\beta}_C}} \tag{5.34}$$

and

$$\rho_{HI_C} = \frac{\sqrt{\boldsymbol{\beta}'_C\mathbf{T}_C\boldsymbol{\beta}_C}}{\sqrt{\mathbf{w}'\mathbf{C}\mathbf{w}}}, \tag{5.35}$$

respectively. Note that the maximized LPSI accuracy is $\rho_{HI} = \frac{\sqrt{\mathbf{b}'\mathbf{P}\mathbf{b}}}{\sqrt{\mathbf{w}'\mathbf{C}\mathbf{w}}}$ (see Chap. 2). The denominator of the accuracy of the CLGSI and $\rho_{HI} = \frac{\sqrt{\mathbf{b}'\mathbf{P}\mathbf{b}}}{\sqrt{\mathbf{w}'\mathbf{C}\mathbf{w}}}$ is the same; however, the numerator of the two indices accuracy is different. We would expect that $\sqrt{\boldsymbol{\beta}'_C\mathbf{T}_C\boldsymbol{\beta}_C} \geq \sqrt{\mathbf{b}'\mathbf{P}\mathbf{b}}$, and then $\rho_{HI_C} \geq \rho_{HI}$. Similar results can be observed when we compared the maximized LPSI selection response and expected genetic gain per trait with the maximized CLGSI selection response and expected genetic gain per trait.

5.2.2 *Relationship Between the CLGSI and the LGSI*

As we have indicated, the CLGSI is mathematically equivalent to the LMSI; thus, it has similar statistical properties to those of the LMSI, some of which are described in this section. The rest can be seen in Chap. 4. Let $\mathbf{Q}_C = \mathbf{T}_C^{-1}\boldsymbol{\Psi}_C$, then matrix $\mathbf{Q}_C$ can be written as

$$\mathbf{Q}_C = \begin{bmatrix} (\mathbf{P}-\boldsymbol{\Gamma})^{-1}(\mathbf{C}-\boldsymbol{\Gamma}) & \mathbf{0} \\ \mathbf{I}-(\mathbf{P}-\boldsymbol{\Gamma})^{-1}(\mathbf{C}-\boldsymbol{\Gamma}) & \mathbf{I} \end{bmatrix}, \tag{5.36}$$

whence as $\mathbf{w}'_2 = [0_1 \quad \cdots \quad 0_t]$, the two sub-vectors that conform vector $\boldsymbol{\beta}_C = \mathbf{Q}_C\mathbf{a}_C$ or $\boldsymbol{\beta}'_C = [\boldsymbol{\beta}'_\mathbf{y} \quad \boldsymbol{\beta}'_G]$ can be written as

$$\boldsymbol{\beta}_\mathbf{y} = (\mathbf{P}-\boldsymbol{\Gamma})^{-1}(\mathbf{C}-\boldsymbol{\Gamma})\mathbf{w}, \tag{5.37}$$

and

$$\boldsymbol{\beta}_G = \left[\mathbf{I}-(\mathbf{P}-\boldsymbol{\Gamma})^{-1}(\mathbf{C}-\boldsymbol{\Gamma})\right]\mathbf{w} = \mathbf{w} - \boldsymbol{\beta}_\mathbf{y}. \tag{5.38}$$

When $\mathbf{\Gamma}$ is equal to the null matrix (no genomic information), Eq. (5.37) is equal to $\boldsymbol{\beta}_{\mathbf{y}} = \mathbf{P}^{-1}\mathbf{C}\mathbf{w} = \mathbf{b}$ and $R_C = k_I\sqrt{\mathbf{b}'\mathbf{P}\mathbf{b}} = R_I$, which are the LPSI vector of coefficients and the selection response.

By Eqs. (5.37) and (5.38), the maximized CLGSI selection response and the optimized CLGSI can be written as

$$R_C = k_I\sqrt{\mathbf{w}'\mathbf{C}(\mathbf{P}-\mathbf{\Gamma})^{-1}(\mathbf{C}-\mathbf{\Gamma})\mathbf{w} + \mathbf{w}'\mathbf{\Gamma}\left[\mathbf{I}-(\mathbf{P}-\mathbf{\Gamma})^{-1}(\mathbf{C}-\mathbf{\Gamma})\right]\mathbf{w}} \quad (5.39)$$

and

$$I_C = \boldsymbol{\beta}_{\mathbf{y}}\mathbf{y} + \boldsymbol{\beta}_G\boldsymbol{\gamma} = \mathbf{w}'\boldsymbol{\gamma} + \boldsymbol{\beta}_{\mathbf{y}}(\mathbf{y}-\boldsymbol{\gamma}), \quad (5.40)$$

respectively.

Assume that when the number of markers and genotypes increases, matrix $\mathbf{\Gamma}$ tends to matrix $\mathbf{C}$ and that, at the limit, $\mathbf{\Gamma} = \mathbf{C}$; then, Eq. (5.39) can be written as $R_C = k_I\sqrt{\mathbf{w}'\mathbf{\Gamma}\mathbf{w}} = R_G$ (except by L_G); in addition, $\boldsymbol{\beta}_{\mathbf{y}} = \mathbf{0}$ and $\boldsymbol{\beta}_G = \mathbf{w}$, the weights of the LGSI, and, in this latter case, the CLGSI is equal to the LGSI, as we would expect. Thus, in the asymptotic context, the LGSI and the CLGSI are the same.

An additional interesting result of the relationship between the CLGSI and the LGSI is as follows. The maximized correlation between H and I_C (or CLGSI accuracy) can be written as

$$\rho_{HI_C} = \frac{\mathbf{a}_C'\mathbf{\Psi}_C\boldsymbol{\beta}_C}{\sqrt{\mathbf{a}_C'\mathbf{\Psi}_C\mathbf{a}_C}\sqrt{\boldsymbol{\beta}_C'\mathbf{T}_C\boldsymbol{\beta}_C}}; \quad (5.41)$$

However, when $\mathbf{\Gamma} = \mathbf{C}$, $\mathbf{\Psi}_C = \begin{bmatrix}\mathbf{\Gamma} & \mathbf{\Gamma}\\ \mathbf{\Gamma} & \mathbf{\Gamma}\end{bmatrix}$, $\boldsymbol{\beta}_{\mathbf{y}} = \mathbf{0}$, $\boldsymbol{\beta}_G = \mathbf{w}$, and $\boldsymbol{\beta}_C' = \begin{bmatrix}\boldsymbol{\beta}_{\mathbf{y}}' & \boldsymbol{\beta}_G'\end{bmatrix} = \begin{bmatrix}\mathbf{0} & \mathbf{w}'\end{bmatrix}$, whence $\mathbf{a}_C'\mathbf{\Psi}_C\boldsymbol{\beta}_C = \mathbf{a}_C'\mathbf{\Psi}_C\mathbf{a}_C = \boldsymbol{\beta}_C'\mathbf{T}_C\boldsymbol{\beta}_C = \mathbf{w}'\mathbf{\Gamma}\mathbf{w}$, and Eq. (5.41) is equal to 1. That is, the maximum correlation between H and I_C in the asymptotic context is equal to the maximum correlation between H and the LGSI, and that value will be equal to 1.

The asymptotic relationship between the CLGSI expected genetic gain per trait, $\mathbf{E}_C$ (Eq. 5.34), and the LGSI expected genetic gain per trait, $\mathbf{E}_{I_G}$ (Eq. 5.16), is as follows. When $\mathbf{\Gamma} = \mathbf{C}$, $\mathbf{\Psi}_C = \begin{bmatrix}\mathbf{\Gamma} & \mathbf{\Gamma}\\ \mathbf{\Gamma} & \mathbf{\Gamma}\end{bmatrix}$ and $\boldsymbol{\beta}_C' = \begin{bmatrix}\mathbf{0} & \mathbf{w}'\end{bmatrix}$, whence

$$\mathbf{E}_C = k_I\frac{\mathbf{\Psi}_C\boldsymbol{\beta}_C}{\sqrt{\boldsymbol{\beta}_C'\mathbf{T}_C\boldsymbol{\beta}_C}} = k_I\frac{2\mathbf{\Gamma}\mathbf{w}}{\sqrt{\mathbf{w}'\mathbf{\Gamma}\mathbf{w}}} = 2\mathbf{E}_{I_G}. \quad (5.42)$$

This means that in the asymptotic context, the CLGSI expected genetic gain per trait is twice the LGSI expected genetic gain per trait. Of course, 2 is only a proportionality constant; thus, in reality, $\mathbf{E}_C = \mathbf{E}_{I_G}$.

5.2.3 Statistical Properties of the CLGSI

Assume that H and I_C have bivariate joint normal distribution; $\mathbf{P}$, $\mathbf{C}$, $\mathbf{\Gamma}$, and $\mathbf{w}$ are known, and $\boldsymbol{\beta}_C = \mathbf{T}_C^{-1}\mathbf{\Psi}_C\mathbf{a}_C$; then, the CLGSI properties are as follow:

1. $\sigma_{I_C}^2 = \sigma_{HI_C}$, i.e., the variance of I_C ($\sigma_{I_C}^2$) and the covariance between H and I_C (σ_{HI_C}) are the same.
2. The maximized correlation between H and I_C is $\rho_{HI_C} = \frac{\sigma_{I_C}}{\sigma_H}$, where σ_{I_C} is the standard deviation of the variance of I_C ($\sigma_{I_C}^2$) and σ_H is the standard deviation of the variance of $H(\sigma_H^2)$.
3. The variance of the predicted error, $Var(H - I_C) = \left(1 - \rho_{HI_C}^2\right)\sigma_H^2$, is minimal.
4. The total variance of H explained by I_C is $\sigma_{I_C}^2 = \rho_{HI_C}^2\sigma_H^2$.

Note that CLGSI properties 1 to 4 are the same as LMSI properties 1 to 4 and that both indices jointly incorporate phenotypic and marker information to predict the net genetic merit; however, the LMSI incorporates the marker information by the marker score values, whereas the CLGSI uses the GEBVs.

5.2.4 Estimating the CLGSI Parameters

Using the real maize (*Zea mays*) F_2 population with 248 genotypes (each with two repetitions), 233 molecular markers and three traits—GY (ton ha^{-1}), EHT (cm), and PHT (cm)—described in Sect. 5.1.8 of this chapter, we estimated matrices $\mathbf{P}$ and $\mathbf{C}$ using Eqs. (2.22) to (2.24) described in Chap. 2 of this book. The estimated matrices were $\widehat{\mathbf{P}} = \begin{bmatrix} 0.45 & 1.33 & 2.33 \\ 1.33 & 65.07 & 83.71 \\ 2.33 & 83.71 & 165.99 \end{bmatrix}$ and $\widehat{\mathbf{C}} = \begin{bmatrix} 0.07 & 0.61 & 1.06 \\ 0.61 & 17.93 & 22.75 \\ 1.06 & 22.75 & 44.53 \end{bmatrix}$.

In a similar manner, we estimated matrix $\mathbf{\Gamma}$ using Eqs. (5.21) to (5.23). The estimated matrix was $\widehat{\mathbf{\Gamma}} = \begin{bmatrix} 0.07 & 0.65 & 1.05 \\ 0.65 & 10.62 & 14.25 \\ 1.05 & 14.25 & 26.37 \end{bmatrix}$. Note that matrices $\widehat{\mathbf{C}}$ and $\widehat{\mathbf{\Gamma}}$ have similar values. This means that, in the asymptotic context, we can assume that matrix $\mathbf{\Gamma}$ tends to matrix $\mathbf{C}$.

To estimate the CLMSI and its associated parameters (selection response, expected genetic gain per trait, etc.), we need to estimate the vector of coefficients $\boldsymbol{\beta}_C = \mathbf{T}_C^{-1}\mathbf{\Psi}_C\mathbf{a}_C$ as $\widehat{\boldsymbol{\beta}}_C = \widehat{\mathbf{T}}_C^{-1}\widehat{\mathbf{\Psi}}_C\mathbf{a}_C$, where $\widehat{\mathbf{T}}_C = \begin{bmatrix} \widehat{\mathbf{P}} & \widehat{\mathbf{\Gamma}} \\ \widehat{\mathbf{\Gamma}} & \widehat{\mathbf{\Gamma}} \end{bmatrix}$ and $\widehat{\mathbf{\Psi}}_C = \begin{bmatrix} \widehat{\mathbf{C}} & \widehat{\mathbf{\Gamma}} \\ \widehat{\mathbf{\Gamma}} & \widehat{\mathbf{\Gamma}} \end{bmatrix}$ are estimates of matrices $\mathbf{T}_C = \begin{bmatrix} \mathbf{P} & \mathbf{\Gamma} \\ \mathbf{\Gamma} & \mathbf{\Gamma} \end{bmatrix}$ and $\mathbf{\Psi}_C = \begin{bmatrix} \mathbf{C} & \mathbf{\Gamma} \\ \mathbf{\Gamma} & \mathbf{\Gamma} \end{bmatrix}$ respectively. The estimated CLGSI vector of coefficients $\widehat{\boldsymbol{\beta}}_C = \widehat{\mathbf{T}}_C^{-1}\widehat{\mathbf{\Psi}}_C\mathbf{a}_C$ is conformed by the vector of phenotypic weights, $\widehat{\boldsymbol{\beta}}_{\mathbf{y}} = \left(\widehat{\mathbf{P}} - \widehat{\boldsymbol{\Gamma}}\right)^{-1}\left(\widehat{\mathbf{C}} - \widehat{\boldsymbol{\Gamma}}\right)\mathbf{w}$, and by the vector of genomic weights, $\widehat{\boldsymbol{\beta}}_G = \left[\mathbf{I} - \left(\widehat{\mathbf{P}} - \widehat{\boldsymbol{\Gamma}}\right)^{-1}\left(\widehat{\mathbf{C}} - \widehat{\boldsymbol{\Gamma}}\right)\right]\mathbf{w}$.

Let $\mathbf{w}' = [5 \quad -0.1 \quad -0.1]$ be the vector of economic weights; then, according to the estimated matrices $\widehat{\mathbf{P}}$, $\widehat{\mathbf{C}}$, and $\widehat{\boldsymbol{\Gamma}}$, $\widehat{\boldsymbol{\beta}}'_{\mathbf{y}} = [0.08 \quad -0.02 \quad -0.01]$ and $\widehat{\boldsymbol{\beta}}'_G = [4.92 \quad -0.08 \quad -0.09]$, whence the estimated CLGSI in the training population can be written as

$$\widehat{I}_C = \widehat{\boldsymbol{\beta}}_{\mathbf{y}}\mathbf{y} + \widehat{\boldsymbol{\beta}}_G\widehat{\boldsymbol{\gamma}}. \tag{5.43}$$

Suppose a selection intensity of 10% ($k_I = 1.755$); then, the estimated CLGSI selection response and expected genetic gain per trait were $\widehat{R}_C = k_I\sqrt{\widehat{\boldsymbol{\beta}}'_C\widehat{\mathbf{T}}_C\widehat{\boldsymbol{\beta}}_C} = 1.54$ and $\widehat{\mathbf{E}}'_C = k_I\frac{\widehat{\boldsymbol{\beta}}'_C\widehat{\boldsymbol{\Psi}}_C}{\sqrt{\widehat{\boldsymbol{\beta}}'_C\widehat{\mathbf{T}}_C\widehat{\boldsymbol{\beta}}_C}} = [0.36 \quad 1.04 \quad 1.70 \quad 0.36 \quad 1.53 \quad 2.38]$ respectively, whereas the estimated CLGSI accuracy was $\widehat{\rho}_{HI_C} = \frac{\widehat{\sigma}_{I_C}}{\widehat{\sigma}_H} = 0.814$.

The estimated LPSI selection response, expected genetic gain per trait, and accuracy were 0.601, $[0.09 \quad -0.81 \quad -0.89]$, and 0.32 respectively; thus, the CLGSI was more efficient to predict the net genetic merit than the LPSI because the CLGSI accuracy and selection response were 0.814 and 1.54 respectively.

5.2.5 *LGSI and CLGSI Efficiency Vs LMSI, GW-LMSI and LPSI Efficiency*

In this subsection, we compare the accuracy, selection response, and efficiency of the LGSI and CLGSI with the LMSI, the GW-LMSI, and the LPSI using the simulated data for a maize (*Zea mays*) population described in Chap. 2, Sect. 2.8.1.

Figure 5.3 presents the estimated accuracy values of the LMSI, the LGSI, the CLGSI, the LPSI, and the GW-LMSI for five simulated selection cycles. According to these results, for the first three selection cycles, the estimated accuracies of the indices, in decreasing order, were LMSI > LGSI > CLGSI > LPSI > GW-LMSI. That is, the highest estimated accuracy was obtained with the LMSI, whereas the lowest was obtained with the GW-LMSI. For the fourth and fifth selection cycles, the estimated accuracies, in decreasing order, were LMSI > LPSI > CLGSI > LGSI > GW-LMSI. This means that in all five selection cycles, the LMSI had the highest accuracy and the GW-LMSI had the lowest accuracy, whereas the estimated LGSI accuracy was reduced to fourth place. Thus, the accuracy of the LGSI tended to decrease after the first three selection cycles whereas LPSI accuracy was a constant.

To compare LGSI efficiency versus the efficiency of the other selection indices, we assumed that the interval between selection cycles in the LGSI is 1.5 years, whereas for CLGSI, LMSI, GW-LMSI, and LPSI, the interval was 4.0 years. Table 5.3 presents the estimated selection response of the LPSI, the LMSI, the

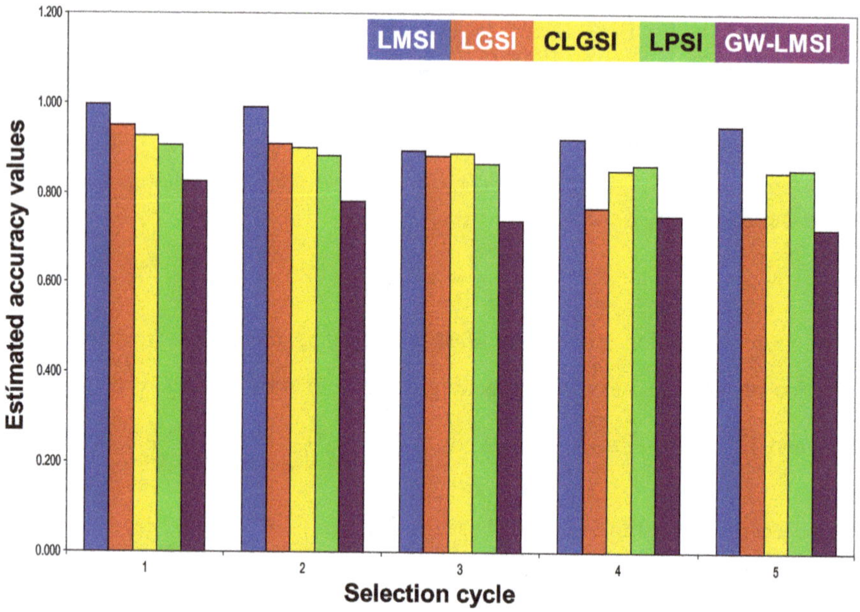

Fig. 5.3 Estimated accuracy values of the linear molecular selection index (LMSI), the LGSI, the combined LGSI (CLGSI), the LPSI, and the genome-wide LMSI (GW-LMSI) with the net genetic merit for four traits, 2500 markers, and 500 genotypes (each with four repetitions) in one environment for five simulated selection cycles

Table 5.3 Estimated selection response of the linear phenotypic selection index (LPSI), the linear molecular selection index (LMSI), the genome-wide LMSI (GW-LMSI), the linear genomic selection index (LGSI), and the combined LGSI (CLGSI), not including (first part of the Table) and including (second part of the Table) the interval length between selection cycles, obtained using five simulated selection cycles

Cycle	LPSI	LMSI	GW-LMSI	LGSI	C-LGSI
Estimated selection response not including the interval length					
1	17.84	19.60	16.24	14.36	18.24
2	15.66	24.36	13.88	13.90	16.02
3	14.44	14.70	12.13	13.59	14.61
4	14.29	15.29	12.48	12.30	14.14
5	13.86	15.15	11.49	11.38	13.51
Average	15.22	17.82	13.24	13.11	15.30
Estimated selection response including the interval length[a]					
1	4.46	4.90	4.06	9.58	4.56
2	3.92	6.09	3.47	9.27	4.00
3	3.61	3.68	3.03	9.06	3.65
4	3.57	3.82	3.12	8.20	3.53
5	3.47	3.79	2.87	7.59	3.38
Average	3.80	4.46	3.31	8.74	3.83

[a]The interval length for the LPSI, LMSI, GW-LMSI, and C-LGSI was 4 years, whereas the interval length for the LGSI was 1.5 years

Table 5.4 Estimated accuracy of the LMSI, the LGSI, the CLGSI, the LPSI, and the GW-LMSI; LMSI efficiency compared with LGSI, CLGSI, LPSI, and GW-LMSI efficiencies, expressed in percentages, for five simulated selection cycles

	Estimated accuracy					LMSI efficiency compared with			
Cycle	LMSI	LGSI	CLGSI	LPSI	GW-LMSI	LGSI	CLGSI	LPSI	GW-LMSI
1	1.00	0.95	0.93	0.91	0.83	4.93	7.45	10.07	20.67
2	0.99	0.91	0.90	0.88	0.78	8.78	9.88	12.14	26.81
3	0.90	0.88	0.89	0.87	0.74	1.26	0.64	3.43	21.27
4	0.92	0.77	0.85	0.86	0.75	19.99	8.12	6.57	22.5
5	0.95	0.75	0.85	0.86	0.72	26.71	12.2	11.11	31.88
Average	0.95	0.85	0.88	0.87	0.76	12.33	7.66	8.66	24.63

GW-LMSI, the LGSI, and the CLGSI, including and not including the interval between selection cycles (first and second parts of Table 5.3 respectively), obtained using five simulated selection cycles. According to the first part of Table 5.3, the average estimated selection responses, in decreasing order, of the LMSI, CLGSI, LPSI, GW-LMSI, and LGSI for the five simulated selection cycles were 17.82, 15.30, 15.22, 13.24, and 13.11 respectively, when the length of the interval between selection was not included. If the length of the interval between selection cycles is included when comparing the selection response of the indices in terms of time, the estimated selection response of LMSI, CLGSI, LPSI, GW-LMSI must be divided by 4 in each selection cycle, and the estimated LGSI selection response should be divided by 1.5. Thus, according to the second part of Table 5.3, if we include the length of the interval between selection cycles, the average estimated selection responses, in decreasing order, of LGSI, LMSI, CLGSI, LPSI, and GW-LMSI for the five simulated selection cycles were 8.74, 4.46, 3.83, 3.80, and 3.31. This means that in terms of time, the efficiency of the LGSI was higher than the efficiency of the other four selection indices.

Table 5.4 presents the estimated accuracy of the LMSI, LGSI, CLGSI, LPSI, and the GW-LMSI. In addition, Table 5.4 presents the efficiency when predicting the net genetic merit of the LMSI with respect to the LGSI, CLGSI, LPSI, and GW-LMSI as percentages, for five simulated selection cycles. Note that in this case, LMSI efficiency was higher than the efficiency of the other four selection indices, because the LMSI had the highest correlation with the net genetic merit.

References

Beyene Y, Semagn K, Mugo S, Tarekegne A, Babu R et al (2015) Genetic gains in grain yield through genomic selection in eight bi-parental maize populations under drought stress. Crop Sci 55:154–163

Ceron-Rojas JJ, Crossa J, Arief VN, Basford K, Rutkoski J, Jarquín D, Alvarado G, Beyene Y, Semagn K, DeLacy I (2015) A genomic selection index applied to simulated and real data. G3 (Bethesda) 5:2155–2164
Cerón-Rojas JJ, Sahagún-Castellanos J (2016) Multivariate empirical Bayes to predict plant breeding values. Agrociencia 50:633–648
Dekkers JCM (2007) Prediction of response to marker-assisted and genomic selection using selection index theory. J Anim Breed Genet 124:331–341
Gianola D, Perez-Enciso M, Toro MA (2003) On marker-assisted prediction of genetic value: beyond the ridge. Genetics 163:347–365
Habier D, Fernando RL, Dekkers JCM (2007) The impact of genetic relationship information on genome-assisted breeding values. Genetics 177:2389–2397
Heffner EL, Sorrells ME, Jannink JL (2009) Genomic selection for crop improvement. Crop Sci 49 (1):12
Lin CY, Allaire FR (1977) Heritability of a linear combination of traits. Theor Appl Genet 51:1–3
Lorenz AJ, Chao S, Asoro FG, Heffner EL, Hayashi T et al (2011) Genomic selection in plant breeding: knowledge and prospects. Adv Agron 110:77–123
Lynch M, Walsh B (1998) Genetics and analysis of quantitative traits. Sinauer Associates, Sunderland, ME
Mrode RA (2005) Linear models for the prediction of animal breeding values, 2nd edn. CABI Publishing, Cambridge, MA
Nordskog AW (1978) Some statistical properties of an index of multiple traits. Theor Appl Genet 52:91–94
Schott JR (2005) Matrix analysis for statistics, 2nd edn. Wiley, Hoboken, NJ
Searle S, Casella G, McCulloch CE (2006) Variance components. Wiley, Hoboken, NJ
Su G, Christensen OF, Ostersen T, Henryon M, Lund MS (2012) Estimating additive and non-additive genetic variances and predicting genetic merits using genome-wide dense single nucleotide polymorphism markers. PLoS One 7(9):e45293
Technow F, Bürger A, Melchinger AE (2013) Genomic prediction of northern corn leaf blight resistance in maize with combined or separate training sets for heterotic groups. G3 (Bethesda) 3:197–203
Togashi K, Lin CY, Yamazaki T (2011) The efficiency of genome-wide selection for genetic improvement of net merit. J Anim Sci 89:2972–2980
Van Raden PM (2008) Efficient methods to compute genomic predictions. J Dairy Sci 91:4414–4423
Vattikuti S, Guo J, Chow CC (2012) Heritability and genetic correlations explained by common SNPs for metabolic syndrome traits. PLoS Genet 8(3):e1002637

6

Description and Theoretical Validation of Constrained LGSI

Abstract The constrained linear genomic selection indices are null restricted and predetermined proportional gain linear genomic selection indices (RLGSI and PPG-LGSI respectively), which are a linear combination of genomic estimated breeding values (GEBVs) to predict the net genetic merit. They are the results of a direct application of the restricted and the predetermined proportional gain linear phenotypic selection index theory to the genomic selection context. The RLGSI can be extended to a combined RLGSI (CRLGSI) and the PPG-LGSI can be extended to a combined PPG-LGSI (CPPG-LGSI); the latter indices use phenotypic and GEBV information jointly in the prediction of net genetic merit. The main difference between the RLGSI and PPG-LGSI with respect to the CRLGSI and the CPPG-LGSI is that although the RLGSI and PPG-LGSI are useful in a testing population where there is only marker information, the CRLGSI and CPPG-LGSI can be used only in training populations when there are joint phenotypic and marker information. The RLGSI and CRLGSI allow restrictions equal to zero to be imposed on the expected genetic advance of some traits, whereas the PPG-LGSI and CPPG-LGSI allow predetermined proportional restriction values to be imposed on the expected trait genetic gains to make some traits change their mean values based on a predetermined level. We describe the foregoing four indices and we validated their theoretical results using real and simulated data.

6.1 The Restricted Linear Genomic Selection Index

Let $H = \mathbf{w}'\mathbf{g}$ be the net genetic merit and $I_G = \boldsymbol{\beta}'\boldsymbol{\gamma}$ the linear genomic selection index (LGSI, see Chap. 5 for details), where $\mathbf{g}$, $\boldsymbol{\gamma}$, $\mathbf{w}$, and $\boldsymbol{\beta}$ are vectors $t \times 1$ (t= number of traits) of breeding values, genomic breeding values, economic weights, and LGSI coefficients respectively. It can be shown that $Cov(I_G, \mathbf{g}) = \boldsymbol{\Gamma}\boldsymbol{\beta}$ is the covariance between $\mathbf{g}$ and $I_G = \boldsymbol{\beta}'\boldsymbol{\gamma}$, and that $Var(\boldsymbol{\gamma}) = \boldsymbol{\Gamma}$ is the genomic covariance matrix of size $t \times t$ (see Chap. 5 for details). The objective of the restricted linear genomic selection index (RLGSI) is to improve only $(t - r)$ of t $(r < t)$ traits (leaving r of them fixed) in a testing population using only genomic estimated breeding values

(GEBVs). The RLGSI minimizes the mean squared difference between I_G and H, $E[(H - I_G)^2]$, with respect to $\boldsymbol{\beta}$ under the restriction $Cov(I_G, \mathbf{U}'\mathbf{g}) = \mathbf{U}'\boldsymbol{\Gamma\beta} = \mathbf{0}$, where $\mathbf{U}'$ is a matrix $(t - 1) \times t$ of 1s and 0s, in a similar manner to the restricted linear phenotypic selection index (RLPSI) described in Chap. 3 in the phenotypic selection context.

6.1.1 The Maximized RLGSI Parameters

Let $Var(I_G) = \boldsymbol{\beta}'\boldsymbol{\Gamma\beta}$ be the variance of $I_G = \boldsymbol{\beta}'\boldsymbol{\gamma}$, $\mathbf{w}'\mathbf{Cw}$ the variance of $H = \mathbf{w}'\mathbf{g}$, and $Cov(I_G, H) = \mathbf{w}'\boldsymbol{\Gamma\beta}$ the covariance between $H = \mathbf{w}'\mathbf{g}$ and $I_G = \boldsymbol{\beta}'\boldsymbol{\gamma}$. The mean squared difference between H and I_G can be written as $E[(H - I_G)^2]$, which should be minimized under the restriction $\mathbf{U}'\boldsymbol{\Gamma\beta} = \mathbf{0}$ assuming that $\boldsymbol{\Gamma}$, $\mathbf{C}$, $\mathbf{U}'$, and $\mathbf{w}$ are known, i.e., it is necessary to minimize the function

$$f_R(\boldsymbol{\beta}, \mathbf{v}) = \mathbf{w}'\mathbf{Cw} + \boldsymbol{\beta}'\boldsymbol{\Gamma\beta} - 2\mathbf{w}'\boldsymbol{\Gamma\beta} + 2\mathbf{v}'\mathbf{U}'\boldsymbol{\Gamma\beta} \tag{6.1}$$

with respect to vectors $\boldsymbol{\beta}$ and $\mathbf{v}' = [v_1 \ v_2 \ \cdots \ v_{r-1}]$, where $\mathbf{v}$ is a vector of Lagrange multipliers. In matrix notation, the derivative results of Eq. (6.1) are

$$\begin{bmatrix} \boldsymbol{\beta} \\ \mathbf{v} \end{bmatrix} = \begin{bmatrix} \boldsymbol{\Gamma} & \boldsymbol{\Gamma}\mathbf{U} \\ \mathbf{U}'\boldsymbol{\Gamma} & \mathbf{0} \end{bmatrix}^{-1} \begin{bmatrix} \boldsymbol{\Gamma}\mathbf{w} \\ \mathbf{0} \end{bmatrix}. \tag{6.2}$$

Following the procedure described in Chap. 3 (Eqs. 3.2 to 3.5), it can be shown that the RLGSI vector of coefficients that minimizes $E[(H - I_G)^2]$ under the restriction $\mathbf{U}'\boldsymbol{\Gamma\beta} = \mathbf{0}$ is

$$\boldsymbol{\beta}_{RG} = \mathbf{K}_G\mathbf{w}, \tag{6.3}$$

where $\mathbf{K}_G = [\mathbf{I}_t - \mathbf{Q}_G]$, $\mathbf{Q}_G = \mathbf{U}(\mathbf{U}'\boldsymbol{\Gamma}\mathbf{U})^{-1}\mathbf{U}'\boldsymbol{\Gamma}$, $\mathbf{w}$ is a vector of economic weights, and $\mathbf{I}_t$ is an identity matrix $t \times t$. When no restrictions are imposed on any of the traits, $\mathbf{U}'$ is a null matrix and $\boldsymbol{\beta}_{RG} = \mathbf{w}$, the optimized LGSI vector of coefficients (see Chap. 5 for details).

By Eq. (6.3), the RLGSI, and the maximized RLGSI selection response and expected genetic gain per trait can be written as

$$I_{RG} = \boldsymbol{\beta}'_{RG}\boldsymbol{\gamma}, \tag{6.4}$$

$$R_{RG} = \frac{k_I}{L_G}\sqrt{\boldsymbol{\beta}'_{RG}\boldsymbol{\Gamma}\boldsymbol{\beta}_{RG}} \tag{6.5}$$

and

$$\mathbf{E}_{RG} = \frac{k_I}{L_G} \frac{\boldsymbol{\Gamma\beta}_{RG}}{\sqrt{\boldsymbol{\beta}'_{RG}\boldsymbol{\Gamma\beta}_{RG}}}, \tag{6.6}$$

respectively, where k_I is the standardized selection differential (or selection intensity) associated with the RLGSI, and L_G is the interval between selection cycles or the time required to complete a selection cycle using the RLGSI. Equations (6.4) to (6.6) depend only on GEBV information; thus, they are useful in testing populations.

6.1.2 Statistical Properties of RLGSI

Assuming that $H = \mathbf{w}'\mathbf{g}$ and $I_{RG} = \boldsymbol{\beta}'_{RG}\boldsymbol{\gamma}$ have bivariate joint normal distribution, $\boldsymbol{\beta}_{RG} = \mathbf{K}_G\mathbf{w}$, and $\boldsymbol{\Gamma}$, $\mathbf{C}$, and $\mathbf{w}$ are known, it can be shown that the RLGSI has the following properties:

1. Matrices $\mathbf{K}_G$ and $\mathbf{Q}_G$ are idempotent ($\mathbf{K}_G = \mathbf{K}_G^2$ and $\mathbf{Q}_G = \mathbf{Q}_G^2$) and orthogonal ($\mathbf{K}_G\mathbf{Q}_G = \mathbf{Q}_G\mathbf{K}_G = \mathbf{0}$), that is, they are projectors. Matrix $\mathbf{Q}_G$ projects vector $\boldsymbol{\beta} = \mathbf{w}$ into a space generated by the columns of matrix $\mathbf{U}'\boldsymbol{\Gamma}$ due to the restriction $\mathbf{U}'\boldsymbol{\Gamma\beta} = \mathbf{0}$ used when $f_R(\boldsymbol{\beta}, \mathbf{v})$ (Eq. 6.1) is minimized with respect to vectors $\boldsymbol{\beta}$ and $\mathbf{v}$, whereas matrix $\mathbf{K}_G$ projects $\mathbf{w}$ into a space perpendicular to that generated by the $\mathbf{U}'\boldsymbol{\Gamma}$ matrix columns.
2. Because of the restriction $\mathbf{U}'\boldsymbol{\Gamma\beta} = \mathbf{0}$, matrix $\mathbf{K}_G$ projects vector $\mathbf{w}$ into a space smaller than the original space of $\mathbf{w}$. The space reduction into which matrix $\mathbf{K}_G$ projects $\mathbf{w}$ is equal to the number of zeros that appears in Eq. (6.6).
3. Vector $\boldsymbol{\beta}_{RG} = \mathbf{K}_G\mathbf{w}$ minimizes the mean square error under the restriction $\mathbf{U}'\boldsymbol{\Gamma\beta} = \mathbf{0}$.
4. The variance of $I_{RG} = \boldsymbol{\beta}'_{RG}\boldsymbol{\gamma}$ ($\sigma^2_{I_{RG}} = \boldsymbol{\beta}'_{RG}\boldsymbol{\Gamma\beta}_{RG}$) is equal to the covariance between $I_{RG} = \boldsymbol{\beta}'_{RG}\boldsymbol{\gamma}$ and $H = \mathbf{w}'\mathbf{g}$ ($\sigma_{HI_{RG}} = \mathbf{w}'\boldsymbol{\Gamma\beta}_{RG}$).
5. The maximized correlation between H and I_{RG} is equal to $\rho_{HI_{RG}} = \frac{\sigma_{I_{RG}}}{\sigma_H}$, where $\sigma_{I_{RG}} = \sqrt{\boldsymbol{\beta}'_{RG}\boldsymbol{\Gamma\beta}_{RG}}$ and $\sigma_H = \sqrt{\mathbf{w}'\mathbf{C}\mathbf{w}}$ are the standard deviations of $I_{RG} = \boldsymbol{\beta}'_{RG}\boldsymbol{\gamma}$ and $H = \mathbf{w}'\mathbf{g}$ respectively.
6. The variance of the predicted error, $Var(H - I_{RG}) = \left(1 - \rho^2_{HI_{RG}}\right)\sigma^2_H$, is minimal. Note that $Var(H - I_{RG}) = \sigma^2_{I_{RG}} + \sigma^2_H - 2\sigma_{HI_{RG}}$, and when $\boldsymbol{\beta}_{RG} = \mathbf{K}_G\mathbf{w}$, $\sigma^2_{I_{RG}} = \sigma_{HI_{RG}}$, whence $Var(H - I_{RG}) = \sigma^2_H - \sigma^2_{I_{RG}} = \left(1 - \rho^2_{HI_{RG}}\right)\sigma^2_H$ is minimal.

The statistical RLGSI properties are equal to the statistical RLPSI properties. Thus the RLGSI is an application of the RLPSI to the genomic selection context.

6.1.3 Numerical Examples

To estimate the parameters associated with the RLGSI, we use the real data set described in Chap. 5, Sect. 5.1.8, where we found that, in the testing population, the estimate of matrix $\mathbf{\Gamma}$ was $\widehat{\mathbf{\Gamma}} = \begin{bmatrix} 0.21 & 2.95 & 5.00 \\ 2.95 & 42.41 & 71.11 \\ 5.00 & 71.11 & 121.53 \end{bmatrix}$. We use this matrix and the GEBVs associated with the traits grain yield (GY, ton ha^{-1}), ear height (EHT, cm), and plant height (PHT, cm) to illustrate the RLGSI theoretical results.

Suppose that on the RLGSI expected genetic gain per trait we impose one and two null restrictions using matrices $\mathbf{U}_1' = [1 \quad 0 \quad 0]$ and $\mathbf{U}_2' = \begin{bmatrix} 1 & 0 & 0 \\ 0 & 1 & 0 \end{bmatrix}$ (see Chap. 3, Sect. 3.1.3, for details about matrix $\mathbf{U}'$). We need to estimate the RLGSI vector of coefficients ($\boldsymbol{\beta}_{RG} = \mathbf{K}_G\mathbf{w}$) as $\widehat{\boldsymbol{\beta}}_{RG} = \widehat{\mathbf{K}}_G\mathbf{w}$, where $\widehat{\mathbf{K}}_G = [\mathbf{I}_3 - \widehat{\mathbf{Q}}_G]$ and $\widehat{\mathbf{Q}}_G = \mathbf{U}(\mathbf{U}'\widehat{\mathbf{\Gamma}}\mathbf{U})^{-1}\mathbf{U}'\widehat{\mathbf{\Gamma}}$ are estimates of matrices $\mathbf{K}_G = [\mathbf{I}_3 - \mathbf{Q}_G]$ and $\mathbf{Q}_G = \mathbf{U}(\mathbf{U}'\mathbf{\Gamma}\mathbf{U})^{-1}\mathbf{U}'\mathbf{\Gamma}$ respectively, and $\mathbf{I}_3$ is an identity matrix 3 × 3. The estimated $\mathbf{Q}_G$ matrices for restrictions $\mathbf{U}_1' = [1\ 0\ 0]$ and $\mathbf{U}_2' = \begin{bmatrix} 1\ 0\ 0 \\ 0\ 1\ 0 \end{bmatrix}$ were $\widehat{\mathbf{Q}}_{G_1} = \mathbf{U}_1(\mathbf{U}_1'\widehat{\mathbf{\Gamma}}\mathbf{U}_1)^{-1}$ $\mathbf{U}_1'\widehat{\mathbf{\Gamma}} = \begin{bmatrix} 1.0 & 14.05 & 23.81 \\ 0 & 0 & 0 \\ 0 & 0 & 0 \end{bmatrix}$ and $\widehat{\mathbf{Q}}_{G_2} = \mathbf{U}_2(\mathbf{U}_2'\widehat{\mathbf{\Gamma}}\mathbf{U}_2)^{-1}\mathbf{U}_2'\widehat{\mathbf{\Gamma}} = \begin{bmatrix} 1.0 & 0 & 11.18 \\ 0 & 1.0 & 0.90 \\ 0 & 0 & 0 \end{bmatrix}$ respectively, whereas the estimated $\mathbf{K}_G$ matrices for both restrictions were $\widehat{\mathbf{K}}_{G_1} = [\mathbf{I}_3 - \widehat{\mathbf{Q}}_{G_1}]$ $= \begin{bmatrix} 0 & -14.05 & -23.81 \\ 0 & 1.0 & 0 \\ 0 & 0 & 1.0 \end{bmatrix}$ and $\widehat{\mathbf{K}}_{G_2} = [\mathbf{I}_3 - \widehat{\mathbf{Q}}_{G_2}] = \begin{bmatrix} 0 & 0 & -11.18 \\ 0 & 0 & -0.90 \\ 0 & 0 & 1.0 \end{bmatrix}$.

Let $\mathbf{w}' = [5 \quad -0.1 \quad -0.1]$ be the vector of economic weights; then the estimated RLGSI vector of coefficients for one and two null restrictions were $\widehat{\boldsymbol{\beta}}_{RG_1}' = \mathbf{w}'\widehat{\mathbf{K}}_{G_1}' = [3.78 \quad -0.1 \quad -0.1]$ and $\widehat{\boldsymbol{\beta}}_{RG_2}' = \mathbf{w}'\widehat{\mathbf{K}}_{G_2}' = [1.12 \quad 0.09 \quad -0.1]$ respectively, and the estimated RLGSI for both restrictions can be written as $\widehat{I}_{RG_1} = 3.78\text{GEBV}_1 - 0.1\text{GEBV}_2 - 0.1\text{GEBV}_3$ and $\widehat{I}_{RG_2} = 1.12\text{GEBV}_1 + 0.09\text{GEBV}_2 - 0.1\text{GEBV}_3$, where GEBV_1, GEBV_2, and GEBV_3 are the genomic estimated breeding values associated with traits GY, EHT, and PHT respectively in the testing population.

Table 6.1 presents 20 genotypes selected from a population of 380 genotypes and the GEBVs in the testing population ranked according to the estimated RLGSI values for one restriction, where $\mathbf{U}_1' = [1 \quad 0 \quad 0]$. The estimated RLGSI values for genotypes 5 and 306 can be obtained as follows: $\widehat{I}_{RG_5} = 3.78(-0.6) - 0.1(-8.67) - 0.1(15.97) = 0.196$ and $\widehat{I}_{RG_{306}} = 3.78(0.13) - 0.1(1.31) - 0.1(1.66) = 0.194$ respectively. This procedure is valid for any number of genotypes and GEBVs in the testing population.

Assume a selection intensity of 10% ($k_{I_G} = 1.755$); then the estimated RLGSI selection response and expected genetic gain per trait not including the interval length were $\widehat{R}_{RG_1} = k_{I_G}\sqrt{\widehat{\boldsymbol{\beta}}_{RG_1}'\widehat{\mathbf{\Gamma}}\widehat{\boldsymbol{\beta}}_{RG_1}} = 0.40$ and $\widehat{\mathbf{E}}_{RG_1}' = k_I\dfrac{\widehat{\boldsymbol{\beta}}_{RG_1}'\widehat{\mathbf{\Gamma}}}{\sqrt{\widehat{\boldsymbol{\beta}}_{RG_1}'\widehat{\mathbf{\Gamma}}\widehat{\boldsymbol{\beta}}_{RG_1}}}$

Table 6.1 Number of genotypes selected from 380 genotypes of a real testing population; genomic estimated breeding values (GEBVs) associated with three traits: grain yield (GY, ton ha^{-1}), ear height (EHT, cm), and plant height (PHT, cm) in the testing population, and estimated and ranked restricted linear genomic selection index (RLGSI) values obtained in the testing population for one null restriction

	Estimated GEBVs in the testing population			
Number of genotypes	GEBV-GY	GEBV-EHT	GEBV-PHT	Estimated RLGSI
5	−0.6	−8.67	−15.97	0.196
306	0.13	1.31	1.66	0.194
6	0.06	1.83	−1.13	0.157
349	0.37	4.34	8.12	0.153
142	−0.26	−5.47	−5.85	0.149
69	−0.11	−3.43	−2.16	0.143
24	0.03	−0.43	0.19	0.137
192	−0.8	−13.91	−17.7	0.137
33	−0.18	−1.44	−6.71	0.135
18	−0.43	−5.48	−12.08	0.131
21	−1.00	−16.11	−22.96	0.127
41	0.17	1.09	4.08	0.126
351	0.16	2.64	2.15	0.126
323	0.04	−0.79	1.04	0.126
158	−0.49	−8.95	−10.83	0.126
25	−0.24	−3.46	−6.86	0.125
338	0.37	3.88	8.89	0.122
316	−0.01	−0.51	−1.09	0.122
32	−0.19	−3.97	−4.43	0.122
204	−0.46	−7.41	−11.19	0.121

$= [0 \quad -1.42 \quad -2.58]$ respectively. For two restrictions, with $\mathbf{U}_2' = \begin{bmatrix} 1 & 0 & 0 \\ 0 & 1 & 0 \end{bmatrix}$, the estimated RLGSI selection response and expected genetic gains not including the interval length were $\widehat{R}_{RG_2} = k_{I_G}\sqrt{\widehat{\boldsymbol{\beta}}'_{RG_2}\widehat{\boldsymbol{\Gamma}}\widehat{\boldsymbol{\beta}}_{RG_2}} = 0.23$ and $\widehat{\mathbf{E}}'_{RG_2} = k_I \frac{\widehat{\boldsymbol{\beta}}'_{RG_2}\widehat{\boldsymbol{\Gamma}}}{\sqrt{\widehat{\boldsymbol{\beta}}'_{RG_2}\widehat{\boldsymbol{\Gamma}}\widehat{\boldsymbol{\beta}}_{RG_2}}} = [0 \quad 0 \quad -2.29]$ respectively. When the number of restrictions increases, the estimated RLGSI selection response value decreases, whereas the number of zeros increases in the estimated RLGSI expected genetic gain per trait. The number of zeros in the estimated RLGSI expected genetic gain per trait is equal to the number of restrictions imposed on RLGSI by matrix $\mathbf{U}'$, where each restriction appears as 1.

Figure 6.1 presents the frequency distribution of the estimated RLGSI values for one (Fig. 6.1a) and two null restrictions (Fig. 6.1b). For both restrictions the frequency distribution of the estimated RLGSI values approaches the normal distribution.

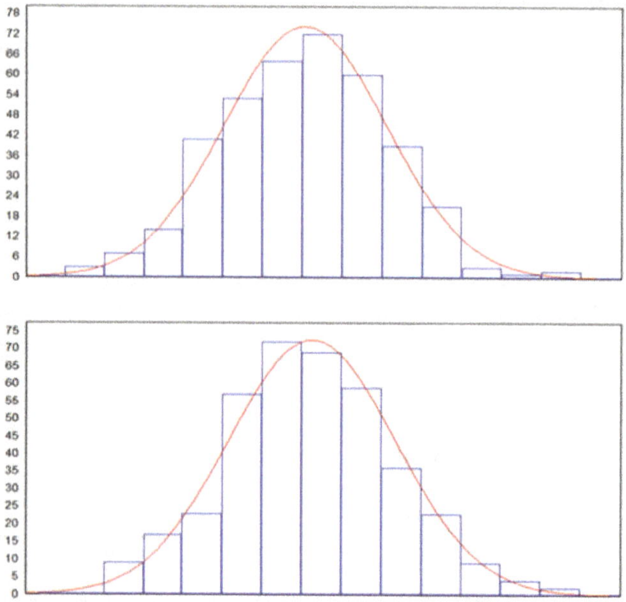

Fig. 6.1 Distribution of 380 estimated restricted linear genomic selection index (RLGSI) values with one (**a**) and two (**b**) null restrictions respectively obtained in a real testing population for one selection cycle in one environment

Now we use the simulated data set described in Chap. 2, Sect. 2.8.1, to compare RLPSI (restricted linear phenotypic selection index, Chap. 3 for details) efficiency versus RLGSI efficiency. Table 6.2 presents the estimated RLPSI and RLGSI selection response for one, two, and three null restrictions imposed by matrices $\mathbf{U}_1' = [1 \quad 0 \quad 0]$, $\mathbf{U}_2' = \begin{bmatrix} 1 & 0 & 0 \\ 0 & 1 & 0 \end{bmatrix}$, and $\mathbf{U}_3' = \begin{bmatrix} 1 & 0 & 0 & 0 \\ 0 & 1 & 0 & 0 \\ 0 & 0 & 1 & 0 \end{bmatrix}$ for five simulated selection cycles including and not including the interval between selection cycles. In each selection cycle, the sample size was equal to 500 genotypes, each with four repetitions and four traits, whereas the selection intensity was 10% ($k_I = 1.755$); the interval lengths for the RLPSI and RLGSI were 4 and 1.5 years (Beyene et al. 2015) respectively.

Table 6.2 was divided in two parts. The first part presents the estimated RLPSI whereas the second part presents the estimated RLGSI selection responses. Columns 2, 3, and 4 in Table 6.2 present the estimated RLPSI and RLGSI selection responses not including the interval length, whereas columns 5, 6, and 7 present the estimated RLPSI and RLGSI selection response, including the interval length. The averages of the estimated RLPSI selection response not including the interval length for one, two, and three restrictions were 7.04, 5.50, and 3.90, whereas when the interval length was included, the averages were 1.76, 1.38, and 0.98 respectively. The averages of the estimated RLGSI selection response not including the interval length

Table 6.2 Estimated restricted linear phenotypic selection index (RLPSI) and RLGSI selection responses for 1, 2, and 3 null restrictions for 5 simulated selection cycles including and not including the interval between selection cycles. The interval lengths for the RLPSI and the RLGSI were 4 and 1.5 years respectively

	Estimated RLPSI selection response					
	Not including interval length			Including interval length[a]		
Cycle	1	2	3	1	2	3
1	6.87	5.54	4.13	1.72	1.39	1.03
2	8.45	5.94	4.27	2.11	1.49	1.07
3	7.17	5.79	4.16	1.79	1.45	1.04
4	6.68	5.06	3.72	1.67	1.27	0.93
5	6.02	5.16	3.24	1.51	1.29	0.81
Average	7.04	5.50	3.90	1.76	1.38	0.98
	Estimated RLGSI selection response					
	Not including interval length			Including interval length[b]		
Cycle	1	2	3	1	2	3
1	6.41	5.58	4.71	4.28	3.72	3.14
2	5.04	3.47	2.47	3.36	2.32	1.65
3	4.76	3.36	2.22	3.17	2.24	1.48
4	4.51	3.07	2.28	3.01	2.05	1.52
5	4.46	3.10	2.26	2.97	2.07	1.51
Average	5.04	3.72	2.79	3.36	2.48	1.86

[a]The estimated RLPSI selection response was divided by 4
[b]The estimated RLGSI selection response was divided by 1.5

for one, two, and three restrictions were 5.04, 3.72, and 2.79, whereas when the interval length was included the averages were 3.36, 2.48, and 1.86 respectively. These results indicated that when the interval length was included in the estimation of the RLPSI and RLGSI selection response, RLGSI efficiency was greater than RLPSI efficiency, and vice versa, when the interval length was not included the RLPSI efficiency was greater than RLGSI efficiency.

Table 6.3 presents the estimated RLPSI (first part) and RLGSI (second part) expected genetic gain per trait not including the interval between selection cycles for one, two, and three null restrictions in five simulated selection cycles. In this case, RLPSI efficiency is greater than RLGSI efficiency because the averages of the estimated RLPSI expected genetic gain per trait were −2.52, 2.26, and 2.26 for one null restriction; 2.84 and 2.65 for two null restrictions; and 3.90 for three null restrictions. For the same set of restrictions, the averages of the estimated RLGSI expected genetic gain per trait were: −1.85, 1.13, and 2.06 for one null restriction; 1.52 and 2.19 for two null restrictions, and 2.79 for three null restrictions. However, divided by the interval length (4 years in the RLPSI), the averages of the estimated RLPSI expected genetic gain per trait were −0.63, 0.57, and 0.57 for one null restriction; 0.71 and 0.66 for two null restrictions, and 0.98 for three null restrictions. In a similar manner, dividing by the interval length (1.5 years in this case), the averages of the estimated RLGSI expected genetic gain per trait were −1.23, 0.75,

Table 6.3 Estimated RLPSI and RLGSI expected genetic gain per trait for 1, 2, and 3 null restrictions for 5 simulated selection cycles (each with 4 traits) not including the interval length between selection cycles

	Estimated RLPSI expected genetic gain for one, two, and three null restrictions											
	1				2				3			
Cycle	T1	T2	T3	T4	T1	T2	T3	T4	T1	T2	T3	T4
1	0	−2.18	2.03	2.66	0	0	2.77	2.77	0	0	0	4.13
2	0	−3.41	2.33	2.71	0	0	2.87	3.07	0	0	0	4.27
3	0	−2.30	3.12	1.74	0	0	3.11	2.68	0	0	0	4.16
4	0	−2.88	1.42	2.38	0	0	2.35	2.70	0	0	0	3.72
5	0	−1.83	2.38	1.81	0	0	3.12	2.04	0	0	0	3.24
Average	0	−2.52	2.26	2.26	0	0	2.84	2.65	0	0	0	3.90
	Estimated RLGSI expected genetic gain for 1, 2, and 3 null restrictions											
	1				2				3			
Cycle	T1	T2	T3	T4	T1	T2	T3	T4	T1	T2	T3	T4
1	0	−1.41	1.29	3.72	0	0	1.89	3.70	0	0	0	4.71
2	0	−2.16	1.07	1.81	0	0	1.49	1.98	0	0	0	2.47
3	0	−1.94	1.24	1.57	0	0	1.58	1.78	0	0	0	2.22
4	0	−1.90	1.02	1.60	0	0	1.34	1.73	0	0	0	2.28
5	0	−1.83	1.02	1.61	0	0	1.33	1.77	0	0	0	2.26
Average	0	−1.85	1.13	2.06	0	0	1.52	2.19	0	0	0	2.79

and 1.37 for one restriction; 1.01 and 1.46 for two restrictions; and 1.86 for three restrictions.

Table 6.4 presents the estimated RLPSI heritability ($\widehat{h}^2_{I_R}$) values, the estimated restricted linear genomic selection index (RLGSI) accuracy ($\widehat{\rho}_{HI_{RG}}$) values, the values of $W = \frac{\widehat{\rho}_{HI_{RG}}}{\widehat{h}_{I_R}} L_{RP}$ ($L_{RP} = 4$), and the values of $\widehat{p} = 100\left(\widehat{\lambda}_R - 1\right)$, where $\widehat{\lambda}_R = \widehat{\rho}_{HI_R}/\widehat{\rho}_{HI_{RG}}$ and $\widehat{\rho}_{HI_R}$ is the estimated RLPSI accuracy, for one, two, and three restrictions for five simulated selection cycles. The RLGSI interval length was $L_{RG} = 1.5$ whereas the averages of the values of $W = \frac{\widehat{\rho}_{HI_{RG}}}{\widehat{h}_{I_R}} L_{RP}$ for each restriction were 1.22, 0.85, and 0.60; this means that the estimated Technow inequality (Technow et al. 2013), $L_{RG} < \frac{\widehat{\rho}_{HI_{RG}}}{\widehat{h}_{I_R}} L_{RP}$ (Chap. 5, Eq. 5.18), was not true. Thus, according to the Technow inequality results, for this data set, RLGSI efficiency in terms of time was not greater than RLPSI efficiency. The inequality $L_{RG} < \frac{\widehat{\rho}_{HI_G}}{\widehat{h}_{I_R}} L_{I_R}$ was not true because the estimated RLGSI accuracy was very low, whereas RLPSI heritability was high. Thus, note that the averages of the estimated RLGSI accuracy for one, two, and three null restrictions were 0.25, 0.19, and 0.14 respectively, and the averages of the estimated RLPSI heritability values were 0.70, 0.78 and 0.88, respectively. Thus, according to these results, because the estimated RLGSI accuracy is very low and

Table 6.4 Estimated RLPSI heritability ($\widehat{h}^2_{I_R}$), estimated RLGSI accuracy ($\widehat{\rho}_{HI_{RG}}$), estimated values of $W = \frac{\widehat{\rho}_{HI_{RG}}}{\widehat{h}_{I_R}} L_{RP}$ ($L_{RP} = 4$), and values of $\widehat{p} = 100(\widehat{\lambda}_R - 1)$, where $\widehat{\lambda}_R = \widehat{\rho}_{HI_R}/\widehat{\rho}_{HI_{RG}}$, and $\widehat{\rho}_{HI_R}$ are the estimated RLPSI accuracy values, for 1, 2, and 3 restrictions for five simulated selection cycles

	RLPSI heritability			RLGSI accuracy			Estimated values of W			Estimated values of $\widehat{p}$		
Cycle	1	2	3	1	2	3	1	2	3	1	2	3
1	0.65	0.77	0.89	0.33	0.28	0.24	1.62	1.29	1.02	7.27	−1.40	−12.28
2	0.76	0.80	0.90	0.26	0.18	0.13	1.20	0.80	0.54	84.12	83.76	87.74
3	0.71	0.80	0.88	0.24	0.17	0.11	1.16	0.77	0.49	80.34	103.03	119.72
4	0.71	0.79	0.89	0.22	0.15	0.11	1.06	0.68	0.48	79.02	97.29	94.65
5	0.67	0.76	0.86	0.22	0.15	0.11	1.07	0.70	0.48	74.31	110.97	80.61
Average	0.70	0.78	0.88	0.25	0.19	0.14	1.22	0.85	0.60	65.01	78.73	74.09

RLPSI heritability is high, RLGSI efficiency was lower than RLPSI efficiency in terms of time.

The last three columns of Table 6.4, from left to right, present the estimated p values, $\widehat{p} = 100(\widehat{\lambda}_R - 1)$, for one, two, and three null restrictions in five simulated selection cycles. The average of the $\widehat{p}$ values indicates that for each of the three restrictions the RLPSI efficiency was 65.05%, 78.73%, and 74.09%, greater than RLGSI efficiency at predicting the net genetic merit. Thus, for this data set, the RLPSI was a better predictor of the net genetic merit than the RLGSI in each cycle.

6.2 The Predetermined Proportional Gain Linear Genomic Selection Index

6.2.1 *Objective of the PPG-LGSI*

Let $\mathbf{d}' = [d_1 \quad d_2 \quad \dots \quad d_r]$ be a vector $1 \times r$ (r is the number of predetermined proportional gains) of the predetermined proportional gains imposed by the breeder, and assume that μ_q is the population mean of the qth trait before selection. The objective of the predetermined proportional gain linear genomic selection index (PPG-LGSI) is to change μ_q to $\mu_q + d_q$ in the testing population, where d_q is a predetermined change in μ_q. It is possible to solve this problem minimizing the mean squared difference between $I_G = \boldsymbol{\beta}'\boldsymbol{\gamma}$ and $H = \mathbf{w}'\mathbf{g}$, $E[(H - I_G)^2]$, under the restriction $\mathbf{U}'\boldsymbol{\Gamma}\boldsymbol{\beta} = \theta_G\mathbf{d}$, where θ_G is a proportionality constant, or under the restriction $\mathbf{D}'\mathbf{U}'\boldsymbol{\Gamma}\boldsymbol{\beta} = \mathbf{0}$, where $\mathbf{D}' = \begin{bmatrix} d_r & 0 & \dots & 0 & -d_1 \\ 0 & d_r & \dots & 0 & -d_2 \\ \vdots & \vdots & \ddots & \vdots & \vdots \\ 0 & 0 & \dots & d_r & -d_{r-1} \end{bmatrix}$ is a matrix $(r - 1) \times r$ (see Chap. 3 for details), and d_q ($q = 1, 2\dots, r$) is the qth element of vector $\mathbf{d}' = [d_1 \quad d_2 \quad \dots \quad d_r]$; $\mathbf{U}'$ is a matrix $(t - 1) \times t$ of 1s and 0s, and $\boldsymbol{\Gamma} = \left\{\sigma_{\gamma_{qq'}}\right\}$ (q, $q' = 1, 2, \dots, t$, t = number of traits) is a covariance matrix of additive genomic breeding values, $\boldsymbol{\gamma}' = [\gamma_1\, \gamma_2 \cdots \gamma_t]$.

6.2.2 *The Maximized PPG-LGSI Parameters*

In this subsection, we minimize $E[(H - I_G)^2]$ under the restriction $\mathbf{D}'\mathbf{U}'\boldsymbol{\Gamma}\boldsymbol{\beta} = \mathbf{0}$ and later under the restriction $\mathbf{U}'\boldsymbol{\Gamma}\mathbf{b} = \theta_G\mathbf{d}$. Under the restriction $\mathbf{D}'\mathbf{U}'\boldsymbol{\Gamma}\boldsymbol{\beta} = \mathbf{0}$, it is necessary to minimize the function

$$f_P(\boldsymbol{\beta}, \mathbf{v}) = \boldsymbol{\beta}'\boldsymbol{\Gamma}\boldsymbol{\beta} + \mathbf{w}'\mathbf{C}\mathbf{w} - 2\mathbf{w}'\boldsymbol{\Gamma}\boldsymbol{\beta} + 2\mathbf{v}'\mathbf{D}'\mathbf{U}'\boldsymbol{\Gamma}\boldsymbol{\beta} \tag{6.7}$$

with respect to $\boldsymbol{\beta}$ and $\mathbf{v}' = [v_1 \quad v_2 \quad \dots \quad v_{r-1}]$, where $\mathbf{v}'$ is a vector of Lagrange multipliers. From a mathematical point of view, Eq. (6.7) is equal to Eq. (6.1); thus, the vector of coefficients $\boldsymbol{\beta}$ of the PPG-LGSI should be similar to the vector of coefficients of the RLGSI (Eq. 6.3), i.e., the PPG-LGSI vector of coefficients is equal to

$$\boldsymbol{\beta}_{PG} = \mathbf{K}_P\mathbf{w}, \tag{6.8}$$

where now $\mathbf{K}_P = [\mathbf{I}_t - \mathbf{Q}_P]$, $\mathbf{Q}_P = \mathbf{UD}(\mathbf{D}'\mathbf{U}'\boldsymbol{\Gamma}\mathbf{UD})^{-1}\mathbf{D}'\mathbf{U}'\boldsymbol{\Gamma}$, $\mathbf{w}$ is a vector of economic weights, and $\mathbf{I}_t$ is an identity matrix $t \times t$. When $\mathbf{D}' = \mathbf{U}'$, $\boldsymbol{\beta}_{PG} = \boldsymbol{\beta}_{RG}$ (the RLGSI vector of coefficients), and when $\mathbf{U}'$ is a null matrix, $\boldsymbol{\beta}_{PG} = \mathbf{w}$ (the LGSI vector of coefficients). This means that the PPG-LGSI includes the RLGSI and the LGSI as particular cases.

Under the restriction $\mathbf{U}'\boldsymbol{\Gamma}\boldsymbol{\beta} = \theta_G\mathbf{d}$ (see Chap. 3 for details) the vector of coefficients of the PPG-LGSI can be written as

$$\boldsymbol{\beta}_{PG} = \boldsymbol{\beta}_{RG} + \theta_G\mathbf{U}(\mathbf{U}'\boldsymbol{\Gamma}\mathbf{U})^{-1}\mathbf{d}, \tag{6.9}$$

where $\boldsymbol{\beta}_{RG} = \mathbf{K}_G\mathbf{w}$ (Eq. 6.3), $\mathbf{K}_G = [\mathbf{I} - \mathbf{Q}_G]$, $\mathbf{Q}_G = \mathbf{U}(\mathbf{U}'\boldsymbol{\Gamma}\mathbf{U})^{-1}\mathbf{U}'\boldsymbol{\Gamma}$, and $\mathbf{d}' = [d_1 \quad d_2 \quad \dots \quad d_r]$ is the vector of the predetermined proportional gains imposed by the breeder. It can be shown that θ_G, the proportionality constant, can be written as

$$\theta_G = \frac{\mathbf{d}'(\mathbf{U}'\boldsymbol{\Gamma}\mathbf{U})^{-1}\mathbf{U}'\boldsymbol{\Gamma}\mathbf{w}}{\mathbf{d}'(\mathbf{U}'\boldsymbol{\Gamma}\mathbf{U})^{-1}\mathbf{d}}. \tag{6.10}$$

When $\theta_G = 0$, $\boldsymbol{\beta}_{PG} = \boldsymbol{\beta}_{RG}$, and when $\mathbf{U}'$ is a null matrix, $\boldsymbol{\beta}_{PG} = \mathbf{w}$. Equations (6.8) and (6.9) give the same results, that is, both equations express the same result in a different mathematical way.

The maximized selection response and expected genetic gain per trait of the PPG-LGSI can be written as

$$R_{PG} = \frac{k_I}{L_G}\sqrt{\boldsymbol{\beta}'_{PG}\boldsymbol{\Gamma}\boldsymbol{\beta}_{PG}} \tag{6.11}$$

and

$$\mathbf{E}_{PG} = \frac{k_I}{L_G}\frac{\boldsymbol{\Gamma}\boldsymbol{\beta}_{PG}}{\sqrt{\boldsymbol{\beta}'_{PG}\boldsymbol{\Gamma}\boldsymbol{\beta}_{PG}}}, \tag{6.12}$$

respectively, where L_G is the time required to complete a selection cycle using the PPG-LGSI. Equations (6.11) and (6.12) depend only on GEBV information.

6.2.3 Statistical Properties of the PPG-LGSI

Assuming that $H = \mathbf{w}'\mathbf{g}$ and the PPG-LGSI ($I_{PG} = \boldsymbol{\beta}_{PG}'\boldsymbol{\gamma}$) have bivariate joint normal distribution, $\boldsymbol{\beta}_{PG} = \mathbf{K}_P\mathbf{w}$; $\boldsymbol{\Gamma}$, $\mathbf{C}$, and $\mathbf{w}$ are known, it can be shown that PPG-LGSI has the following statistical properties:

1. The vector $\boldsymbol{\beta}_{PG} = \mathbf{K}_P\mathbf{w}$ minimizes the mean square error under the restriction $\mathbf{D}'\mathbf{U}'\boldsymbol{\Gamma}\boldsymbol{\beta} = \mathbf{0}$.
2. The variance of $I_{PG} = \boldsymbol{\beta}_{PG}'\boldsymbol{\gamma}$ ($\sigma^2_{I_{PG}} = \boldsymbol{\beta}_{PG}'\boldsymbol{\Gamma}\boldsymbol{\beta}_{PG}$) is equal to the covariance between $I_{PG} = \boldsymbol{\beta}_{PG}'\boldsymbol{\gamma}$ and $H = \mathbf{w}'\mathbf{g}$ ($\sigma_{HI_{PG}} = \mathbf{w}'\boldsymbol{\Gamma}\boldsymbol{\beta}_{PG}$).
3. The maximized correlation between H and I_{PG} (also called PPG-LGSI accuracy) is equal to $\rho_{HI_{PG}} = \frac{\sigma_{I_{PG}}}{\sigma_H}$, where $\sigma_{I_{PG}} = \sqrt{\boldsymbol{\beta}_{PG}'\boldsymbol{\Gamma}\boldsymbol{\beta}_{PG}}$ and $\sigma_H = \sqrt{\mathbf{w}'\mathbf{C}\mathbf{w}}$ are the standard deviations of $I_{PG} = \boldsymbol{\beta}_{PG}'\boldsymbol{\gamma}$ and $H = \mathbf{w}'\mathbf{g}$ respectively.
4. The variance of the predicted error, $Var(H - I_{PG}) = \left(1 - \rho^2_{HI_{PG}}\right)\sigma^2_H$, is minimal.

The statistical PPG-LGSI properties are equal to the statistical PPG-LPSI properties, then, the PPG-LGSI is an application of the PPG-LPSI to the genomic selection context.

6.2.4 Numerical Example

To illustrate the PPG-LGSI theory, we use the estimated matrix $\widehat{\boldsymbol{\Gamma}} = \begin{bmatrix} 0.21 & 2.95 & 5.00 \\ 2.95 & 42.41 & 71.11 \\ 5.00 & 71.11 & 121.53 \end{bmatrix}$ and the GEBVs associated with the traits GY (ton ha^{-1}), EHT (cm), and PHT (cm), described in Sect. 6.1.3.

It is necessary to estimate the PPG-LGSI vector of coefficients $\boldsymbol{\beta}_{PG} = \boldsymbol{\beta}_{RG} + \theta_g\mathbf{U}(\mathbf{U}'\boldsymbol{\Gamma}\mathbf{U})^{-1}\mathbf{d}$ (Eqs. 6.9 and 6.10). In Sect. 6.1.3, we showed that the estimated vectors of coefficients of $\boldsymbol{\beta}_{RG} = \mathbf{K}_G\mathbf{w}$ for the null restrictions $\mathbf{U}_1' = [1 \quad 0 \quad 0]$ and $\mathbf{U}_2' = \begin{bmatrix} 1 & 0 & 0 \\ 0 & 1 & 0 \end{bmatrix}$ were $\widehat{\boldsymbol{\beta}}_{RG1}' = \mathbf{w}'\widehat{\mathbf{K}}_{G1}' = [3.78 \quad -0.1 \quad -0.1]$ and $\widehat{\boldsymbol{\beta}}_{RG2}' = \mathbf{w}'\widehat{\mathbf{K}}_{G2}' = [1.12 \quad 0.09 \quad -0.1]$ respectively, where $\mathbf{w}' = [5 \quad -0.1 \quad -0.1]$. This means that to estimate $\boldsymbol{\beta}_{PG} = \boldsymbol{\beta}_{RG} + \theta_G\mathbf{U}(\mathbf{U}'\boldsymbol{\Gamma}\mathbf{U})^{-1}\mathbf{d}$, we need only to estimate $\theta_G\mathbf{U}(\mathbf{U}'\boldsymbol{\Gamma}\mathbf{U})^{-1}\mathbf{d}$ for both sets of restrictions.

Consider matrix $\mathbf{U}_1' = [1 \quad 0 \quad 0]$ and let $d_1 = 7.0$ be the predetermined proportional gain restriction for trait 1. We can estimate θ_G and $\mathbf{U}(\mathbf{U}'\boldsymbol{\Gamma}\mathbf{U})^{-1}\mathbf{d}$ as

$$\widehat{\theta}_{G1} = \frac{7.0\left(\mathbf{U}_1'\widehat{\boldsymbol{\Gamma}}\mathbf{U}_1\right)^{-1}\mathbf{U}_1'\widehat{\boldsymbol{\Gamma}}\mathbf{w}}{7.0\left(\mathbf{U}_1'\widehat{\boldsymbol{\Gamma}}\mathbf{U}_1\right)^{-1}7.0} = 0.036 \quad \text{and} \quad \mathbf{U}_1\left(\mathbf{U}_1'\widehat{\boldsymbol{\Gamma}}\mathbf{U}_1\right)^{-1}7.0 = \begin{bmatrix} 33.333 \\ 0 \\ 0 \end{bmatrix},$$

whence the PPG-LGSI vector of coefficients was $\widehat{\boldsymbol{\beta}}_{PG_1} = \widehat{\boldsymbol{\beta}}_{RG_1} + \widehat{\theta}_{G_1}\mathbf{U}_1\left(\mathbf{U}_1'\widehat{\boldsymbol{\Gamma}}\mathbf{U}_1\right)^{-1}7.0 = \begin{bmatrix} 5.0 \\ -0.1 \\ -0.1 \end{bmatrix}$, and the estimated PPG-LGSI was $\widehat{I}_{PG_1} = 5.0\text{GEBV}_1 - 0.1\text{GEBV}_2 - 0.1\text{GEBV}_3$. In a similar manner, we can estimate the PPG-LGSI vector of coefficients under restrictions $\mathbf{U}_2' = \begin{bmatrix} 1 & 0 & 0 \\ 0 & 1 & 0 \end{bmatrix}$ and $\mathbf{d}_2' = [7 \quad -3]$. In this case, $\widehat{\boldsymbol{\beta}}_{PG_2} = \widehat{\boldsymbol{\beta}}_{RG_2} + \widehat{\theta}_{G_2}\mathbf{U}_2(\mathbf{U}_2'\widehat{\boldsymbol{\Gamma}}\mathbf{U}_2)^{-1}\mathbf{d}_2 = \begin{bmatrix} 4.97 \\ -0.18 \\ -0.10 \end{bmatrix}$ and the estimated PPG-LGSI was $\widehat{I}_{PG_2} = 4.97\text{GEBV}_1 - 0.18\text{GEBV}_2 - 0.1\text{GEBV}_3$.

Figure 6.2 presents the frequency distribution of the estimated PPG-LGSI values for one (Fig. 6.2a) and two (Fig. 6.2b) predetermined restrictions, $d = 7$ and $\mathbf{d}' = [7 \quad -3]$ respectively, obtained in a real testing population for one selection cycle in one environment. For both restrictions, the frequency distribution of the estimated PPG-LGSI values approaches the normal distribution.

Assume a selection intensity of 10% ($k_{I_G} = 1.755$); then, for one predetermined restriction, where $\mathbf{U}_1' = [1 \quad 0 \quad 0]$ and $d_1 = 7.0$, the estimated PPG-LGSI selection response and expected genetic gain per trait, not including the interval length, were $\widehat{R}_{PG_1} = k_{I_G}\sqrt{\widehat{\boldsymbol{\beta}}_{PG_1}'\widehat{\boldsymbol{\Gamma}}\widehat{\boldsymbol{\beta}}_{PG_1}} = 1.05$ and $\widehat{\mathbf{E}}_{PG_1}' = k_I \dfrac{\widehat{\boldsymbol{\beta}}_{PG_1}'\widehat{\boldsymbol{\Gamma}}}{\sqrt{\widehat{\boldsymbol{\beta}}_{PG_1}'\widehat{\boldsymbol{\Gamma}}\widehat{\boldsymbol{\beta}}_{PG_1}}} = [0.74 \ 9.92 \ 16.54]$

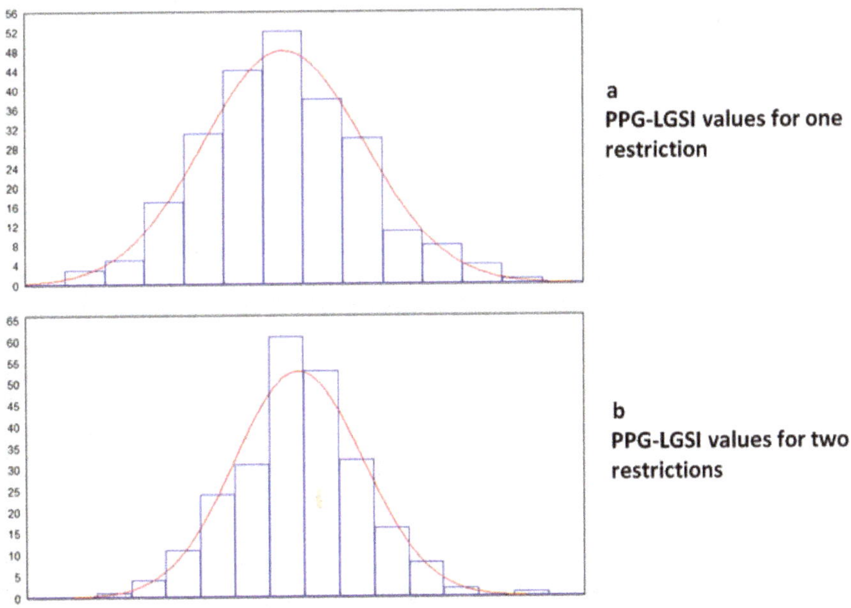

Fig. 6.2 Distribution of 380 estimated predetermined proportional gain linear genomic selection index (PPG-LGSI) values with one (**a**) and two (**b**) predetermined restrictions, $d = 7$ and $\mathbf{d}' = [7 \quad -3]$ respectively, obtained in a real testing population for one selection cycle in one environment

respectively. For two restrictions, with $\mathbf{U}_2' = \begin{bmatrix} 1 & 0 & 0 \\ 0 & 1 & 0 \end{bmatrix}$ and $\mathbf{d}' = [7 \ -3]$, the estimated RLGSI selection response and expected genetic gains, not including the interval length, were $\widehat{R}_{PG_2} = k_{I_G}\sqrt{\widehat{\boldsymbol{\beta}}'_{PG_2}\widehat{\boldsymbol{\Gamma}}\widehat{\boldsymbol{\beta}}_{G_2}} = 0.52$ and $\widehat{\mathbf{E}}'_{PG_2} = k_I \frac{\widehat{\boldsymbol{\beta}}'_{PG_2}\widehat{\boldsymbol{\Gamma}}}{\sqrt{\widehat{\boldsymbol{\beta}}'_{PG_2}\widehat{\boldsymbol{\Gamma}}\widehat{\boldsymbol{\beta}}_{PG_2}}} = [0.11 \ -0.05 \ 0.14]$ respectively.

Now, we use the simulated data set described in Chap. 2, Sect. 2.8.1 to compare PPG-LGSI efficiency versus predetermined proportional gain linear phenotypic selection index (PPG-LPSI) efficiency. Let $\mathbf{U}_1' = [1 \quad 0 \quad 0]$, $\mathbf{U}_2' = \begin{bmatrix} 1 & 0 & 0 \\ 0 & 1 & 0 \end{bmatrix}$, and $\mathbf{U}_3' = \begin{bmatrix} 1 & 0 & 0 & 0 \\ 0 & 1 & 0 & 0 \\ 0 & 0 & 1 & 0 \end{bmatrix}$ be the matrices and $d_1 = 7$, $\mathbf{d}_2' = [7 \quad -3]$, and $\mathbf{d}_3' = [7 \quad -3 \quad 5]$ the vectors for one, two, and three predetermined restrictions respectively. Table 6.5 presents the estimated PPG-LPSI and PPG-LGSI selection response for each predetermined restriction in five simulated selection cycles including and not including the interval between selection cycles (4 years for the PPG-LPSI and 1.5 years for the PPG-LGSI); estimated PPG-LPSI and PPG-LGSI accuracy; and estimated variance of the predicted error (VPE). In each selection cycle, the sample size was equal to 500 genotypes, each with four repetitions and four traits. The selection intensity was 10% ($k_I = 1.755$).

The averages of the estimated PPG-LPSI selection response not including the interval length were 15.14, 14.87, and 13.30, whereas when the interval length was included, the average selection responses were 3.79, 3.72, and 3.33, for one, two, and three predetermined restrictions respectively (Table 6.5). The averages of the estimated PPG-LGSI selection responses not including the interval length for one, two, and three predetermined restrictions were 14.48, 13.47, and 11.26 respectively, and when the interval length was included, the selection responses were 9.65, 8.98, and 7.51 respectively (Table 6.5). These results indicate that when the interval length was included in the estimation of the PPG-LPSI and PPG-LGSI selection responses, PPG-LGSI efficiency was greater than PPG-LPSI efficiency, and vice versa, when the interval length was not included in the PPG-LPSI and PPG-LGSI selection responses, PPG-LPSI efficiency was higher than PPG-LGSI efficiency.

The averages of the estimated VPE values of the PPG-LPSI for one, two, and three predetermined restrictions were 22.42, 30.56, and 41.17 respectively, whereas the estimated VPE values of the PPG-LGSI (see Sect. 6.2.3 for details) were 59.80, 66.95, and 83.98, respectively, that is, in all selection cycles, the VPE of the PPG-LPSI was lower than that of the PPG-LGSI. This means that for this data set, the PPG-LPSI was a better predictor of the net genetic merit than the PPG-LGSI. These results can be explained by observing that the averages of the estimated PPG-LPSI accuracies were 0.88, 0.86, and 0.77, whereas the estimated PPG-LGSI accuracies were 0.65, 0.68, and 0.57 for each predetermined restriction, that is, the estimated PPG-LGSI accuracies were lower than the estimated PPG-LPSI accuracies for this data set.

Table 6.5 Estimated predetermined proportional gain linear phenotypic and genomic selection index (PPG-LPSI and PPG-LGSI respectively) selection responses for 1, 2 and 3 predetermined restrictions for five simulated selection cycles including and not including the interval between selection cycles (4 years for the PPG-LPSI and 1.5 years for the PPG-LGSI); estimated PPG-LPSI and PPG-LGSI accuracy and estimated variance of the predicted error (VPE)

	Estimated PPG-LPSI selection response						Estimated PPG-LPSI accuracy			Estimated PPG-LPSI VPE		
	Not including interval length			Including interval length[a]								
Cycle	1	2	3	1	2	3	1	2	3	1	2	3
1	17.81	16.72	15.23	4.45	4.18	3.81	0.91	0.85	0.77	22.53	50.44	57.14
2	15.69	15.59	14.39	3.92	3.90	3.60	0.88	0.88	0.81	22.66	27.88	37.03
3	14.22	14.16	13.18	3.56	3.54	3.30	0.87	0.86	0.80	21.95	26.14	33.55
4	14.34	14.33	11.56	3.59	3.58	2.89	0.86	0.86	0.70	22.84	25.84	45.46
5	13.64	13.56	12.16	3.41	3.39	3.04	0.86	0.85	0.76	22.13	22.49	32.69
Average	15.14	14.87	13.30	3.79	3.72	3.33	0.88	0.86	0.77	22.42	30.56	41.17
	Estimated PPG-LGSI selection response						Estimated PPG-LGSI accuracy			Estimated PPG-LGSI VPE		
	Not interval length			Including interval length								
Cycle	1	2	3	1	2	3	1	2	3	1	2	3
1	21.22	17.83	16.66	14.15	11.89	11.10	0.65	0.91	0.85	74.75	22.25	35.43
2	13.90	13.52	10.65	9.27	9.01	7.10	0.72	0.70	0.55	58.66	62.06	84.62
3	13.59	13.15	10.70	9.06	8.77	7.13	0.70	0.67	0.55	63.49	67.33	86.30
4	12.30	11.84	9.36	8.20	7.89	6.24	0.61	0.59	0.46	83.26	86.91	103.93
5	11.38	11.03	8.96	7.59	7.35	5.97	0.56	0.54	0.44	93.62	96.21	109.63
Average	14.48	13.47	11.26	9.65	8.98	7.51	0.65	0.68	0.57	74.75	66.95	83.98

[a]The estimated LPSI selection response was divided by 4 and the estimated LGS selection response was divided by 1.5

Table 6.6 Estimated PPG-LPSI heritability ($\widehat{h}_P^2$), values of $W_P = \frac{\widehat{\rho}_{HI_G}}{\widehat{h}_P} L_P$ ($L_P = 4$), and the ratio of the estimated PPG-LPSI accuracy ($\widehat{\rho}_{HI_P}$) to the estimated PPG-LGSI accuracy ($\widehat{\rho}_{HI_{PG}}$): $\widehat{\lambda}_P = \widehat{\rho}_{HI_P}/\widehat{\rho}_{HI_{PG}}$, and values of $\widehat{p} = 100(\widehat{\lambda}_p - 1)$ for 1, 2 and 3 predetermined restrictions for five simulated selection cycles

	PPG-LPSI heritability			Values of W_P			Estimated ratio values ($\widehat{p}$)		
Cycle	1	2	3	1	2	3	1	2	3
1	0.84	0.77	0.83	4.71	4.13	3.72	−18.62	−6.71	−10.20
2	0.80	0.78	0.83	3.22	3.17	2.42	18.30	20.54	32.04
3	0.77	0.76	0.8	3.18	3.09	2.45	19.89	21.59	31.42
4	0.76	0.75	0.78	2.80	2.71	2.10	29.16	31.84	33.75
5	0.75	0.75	0.79	2.57	2.49	1.97	35.26	36.55	42.35
Average	0.72	0.71	0.76	3.29	3.12	2.53	16.80	20.76	25.87

Table 6.6 presents the estimated predetermined PPG-LPSI heritability ($\widehat{h}_P^2$) values, $W_P = \frac{\widehat{\rho}_{HI_G}}{\widehat{h}_P} L_P$ ($L_P = 4$) values, and ratio of the estimated PPG-LPSI accuracy ($\widehat{\rho}_{HI_P}$) to the estimated PPG-LGSI accuracy ($\widehat{\rho}_{HI_{PG}}$), i.e., $\widehat{\lambda}_P = \widehat{\rho}_{HI_P}/\widehat{\rho}_{HI_{PG}}$, and, finally, values of $\widehat{p} = 100(\widehat{\lambda}_P - 1)$ for one, two, and three null restrictions for five simulated selection cycles.

The averages of the W_P values for one, two, and three null restrictions were 3.29, 3.12, and 2.53, respectively, whereas the PPG-LGSI interval length was 1.5 ($L_G = 1.5$). This means that the estimated Technow inequality, $L_G < \frac{\widehat{\rho}_{HI_G}}{\widehat{h}_P} L_P$ (see Chap. 5, Eq. 5.18) was true. Thus, PPG-LGSI efficiency in terms of time was greater than PPG-LPSI efficiency for this data set. These results coincide with those obtained earlier in this chapter, when we compared PPG-LGSI efficiency versus PPG-LPSI efficiency in terms of interval length. However, the average values of $\widehat{p} = 100(\widehat{\lambda}_P - 1)$ (see Chap. 5, Eq. 5.15) were, in percentage terms, 16.80%, 20.76%, and 25.85% for each restriction. These latter results indicate that for this data set, the PPG-LPSI was a better predictor of the net genetic merit than the PPG-LGSI. This is because the estimated PPG-LPSI accuracies were higher than the estimated PPG-LPSI accuracies for this data set. We found similar results when we compared the PPG-LPSI VPE versus PPG-LGSI VPE (Table 6.5).

6.3 The Combined Restricted Linear Genomic Selection Index

The combined restricted linear genomic selection index (CRLGSI) is based on the RLPSI (Chap. 3) and combined linear genomic selection index (CLGSI, Chap. 5) theory. In the RLPSI, the breeder's objective is to improve only $(t - r)$ of t $(r < t)$

traits, leaving r of them fixed; the same is true for the CRLGSI, but in the latter case, it is necessary to impose $2r$ restrictions, i.e., we need to fix r traits and their associated r GEBVs to obtain results similar to those obtained with the RLPSI. This is the main difference between the CRLGSI and the RLPSI.

It can be shown that $Cov(\mathrm{I_C}, \mathbf{a}_C) = \mathbf{\Psi}_C\boldsymbol{\beta}_C$ is the covariance between the breeding value vector ($\boldsymbol{a}'_C = [\mathbf{g}' \quad \boldsymbol{\gamma}']$) and the CLGSI, $I_C = \boldsymbol{\beta}'_C\mathbf{t}_C$ (see Chap. 5 for details), where $\mathbf{t}'_C = [\mathbf{y}' \quad \boldsymbol{\gamma}']$. In the CRLGSI, we want some covariances between the linear combinations of $\mathbf{a}_C$ ($\mathbf{U}'_C\mathbf{a}_C$) and CLGSI to be zero, i.e., $Cov(\mathrm{I_C}, \mathbf{U}'_C\mathbf{a}_C) = \mathbf{U}'_C\mathbf{\Psi}_C\boldsymbol{\beta}_C = \mathbf{0}$, where $\mathbf{U}'_C$ is a matrix $2(t-1) \times 2t$ of 1s and 0s (1 indicates that the trait and its associated GEBV are restricted, and 0 that the trait and its GEBV have no restrictions) and $\mathbf{\Psi}_C = \begin{bmatrix} \mathbf{C} & \mathbf{\Gamma} \\ \mathbf{\Gamma} & \mathbf{\Gamma} \end{bmatrix}$ is a block covariance matrix of $\mathbf{a}'_C = [\mathbf{g}' \quad \boldsymbol{\gamma}']$ where $\mathbf{C}$ and $\mathbf{\Gamma}$ are the covariance matrices of breeding ($\mathbf{g}$) and genomic ($\boldsymbol{\gamma}$) values respectively. This problem can be solved by minimizing the mean squared difference between the CLGSI and H ($E[(H - \mathrm{I_C})^2]$) under the restriction $\mathbf{U}'_C\mathbf{\Psi}_C\boldsymbol{\beta}_C = \mathbf{0}$ similar to the RLGSI in Sect. 6.1.

6.3.1 The Maximized CRLGSI Parameters

Let $\mathbf{T}_C = \begin{bmatrix} \mathbf{P} & \mathbf{\Gamma} \\ \mathbf{\Gamma} & \mathbf{\Gamma} \end{bmatrix}$ be the block covariance matrix of $\mathbf{t}'_C = [\mathbf{y}' \quad \boldsymbol{\gamma}']$ where $\mathbf{P}$ and $\mathbf{\Gamma}$ are the covariance matrices of phenotypic ($\mathbf{y}$) and genomic ($\boldsymbol{\gamma}$) values respectively. Based on the Eq. (6.1) result, it can be shown that the CRLGSI vector of coefficients that minimizes $E[(H - \mathrm{I_C})^2]$ under the restriction $\mathbf{U}'_C\mathbf{\Psi}_C\boldsymbol{\beta}_C = \mathbf{0}$ is

$$\boldsymbol{\beta}_{CR} = \mathbf{K}_C\boldsymbol{\beta}_C, \tag{6.13}$$

where $\mathbf{K}_C = [\mathbf{I} - \mathbf{Q}_C]$, $\mathbf{Q}_C = \mathbf{T}_C^{-1}\mathbf{\Phi}_C\left(\mathbf{\Phi}'_C\mathbf{T}_C^{-1}\mathbf{\Phi}_C\right)^{-1}\mathbf{\Phi}'_C$, $\mathbf{\Phi}_C = \mathbf{U}'_C\mathbf{\Psi}_C$, and $\boldsymbol{\beta}_C = \mathbf{T}_C^{-1}\mathbf{\Psi}_C\mathbf{a}_C$ (the vector of coefficients of the CLGSI, see Chap. 5 for details); $\mathbf{T}_C^{-1}$ is the inverse of matrix $\mathbf{T}_C$, and $\mathbf{I}$ is an identity matrix $2t \times 2t$. When no restrictions are imposed on any of the traits, $\mathbf{U}'_C$ is a null matrix and $\boldsymbol{\beta}_{CR} = \boldsymbol{\beta}_C$ (the vector of coefficients of the CLGSI). That is, the CRLGSI is more general than the CLGSI. Similar to the RLPSI and the RLGSI, matrices $\mathbf{K}_C$ and $\mathbf{Q}_C$ are idempotent ($\mathbf{K}_C = \mathbf{K}_C^2$ and $\mathbf{Q}_C = \mathbf{Q}_C^2$) and orthogonal ($\mathbf{K}_C\mathbf{Q}_C = \mathbf{Q}_C\mathbf{K}_C = \mathbf{0}$), that is, $\mathbf{K}_C$ and $\mathbf{Q}_C$ are projectors. Thus, we can assume that the CRLGSI has similar properties to those described for the RLPSI (see Chap. 3 for details) when matrices $\mathbf{\Psi}_C = \begin{bmatrix} \mathbf{C} & \mathbf{\Gamma} \\ \mathbf{\Gamma} & \mathbf{\Gamma} \end{bmatrix}$ and $\mathbf{T}_C = \begin{bmatrix} \mathbf{P} & \mathbf{\Gamma} \\ \mathbf{\Gamma} & \mathbf{\Gamma} \end{bmatrix}$ are known.

The maximized selection response and the optimized expected genetic gain per trait of the CRLGSI can be written as

$$R_{CR} = \frac{k_I}{L_I}\sqrt{\boldsymbol{\beta}'_{CR}\mathbf{T}_C\boldsymbol{\beta}_{CR}} \tag{6.14}$$

and

$$\mathbf{E}_{CR} = \frac{k_I}{L_I}\frac{\boldsymbol{\Psi}\boldsymbol{\beta}_{CR}}{\sqrt{\boldsymbol{\beta}'_{CR}\mathbf{T}_C\boldsymbol{\beta}_{CR}}}, \tag{6.15}$$

respectively. Although in the RLGSI and the PPG-LGSI the interval between selection cycles is denoted as L_G, in the CRLGSI it is denoted as L_I. This is because the RLPSI and the CRLGSI should have the same interval between selection cycles.

6.3.2 Numerical Examples

To illustrate the CRLGSI theoretical results, we use a real training maize (*Zea mays*) F_2 population with 248 genotypes (each with two repetitions), 233 molecular markers, and three traits: GY (ton ha^{-1}), EHT (cm), and PHT (cm). Matrices $\mathbf{P}$ and $\mathbf{C}$ were estimated based on Eqs. (2.22) to (2.24) described in Chap. 2. The estimated matrices were $\widehat{\mathbf{P}} = \begin{bmatrix} 0.45 & 1.33 & 2.33 \\ 1.33 & 65.07 & 83.71 \\ 2.33 & 83.71 & 165.99 \end{bmatrix}$ and $\widehat{\mathbf{C}} = \begin{bmatrix} 0.07 & 0.61 & 1.06 \\ 0.61 & 17.93 & 22.75 \\ 1.06 & 22.75 & 44.53 \end{bmatrix}$. In a similar manner, we estimated matrix $\boldsymbol{\Gamma}$ using Eqs. (5.21) to (5.23) described in Chap. 5. The estimated matrix was $\widehat{\boldsymbol{\Gamma}} = \begin{bmatrix} 0.07 & 0.65 & 1.05 \\ 0.65 & 10.62 & 14.25 \\ 1.05 & 14.25 & 26.37 \end{bmatrix}$.

To estimate the CRLGSI and its associated parameters (selection response, expected genetic gain per trait, etc.), we need to obtain matrices $\widehat{\mathbf{T}}_C = \begin{bmatrix} \widehat{\mathbf{P}} & \widehat{\boldsymbol{\Gamma}} \\ \widehat{\boldsymbol{\Gamma}} & \widehat{\boldsymbol{\Gamma}} \end{bmatrix}$ and $\widehat{\boldsymbol{\Psi}}_C = \begin{bmatrix} \widehat{\mathbf{C}} & \widehat{\boldsymbol{\Gamma}} \\ \widehat{\boldsymbol{\Gamma}} & \widehat{\boldsymbol{\Gamma}} \end{bmatrix}$ using phenotypic and genomic information and the estimated CRLGSI vector of coefficients $\widehat{\boldsymbol{\beta}}_{CR} = \widehat{\mathbf{K}}_C\widehat{\boldsymbol{\beta}}_C$, where $\widehat{\mathbf{K}}_C = [\mathbf{I} - \widehat{\mathbf{Q}}_C]$, $\widehat{\mathbf{Q}}_C = \widehat{\mathbf{T}}_C^{-1}\widehat{\boldsymbol{\Phi}}_C(\widehat{\boldsymbol{\Phi}}'_C\widehat{\mathbf{T}}_C^{-1}\widehat{\boldsymbol{\Phi}}_C)^{-1}\widehat{\boldsymbol{\Phi}}'_C$, $\widehat{\boldsymbol{\Phi}}_C = \mathbf{U}'_C\widehat{\boldsymbol{\Psi}}_C$, and $\widehat{\boldsymbol{\beta}}_C = \widehat{\mathbf{T}}_C^{-1}\widehat{\boldsymbol{\Psi}}_C\mathbf{a}_C$.

We have indicated that the main difference between the RLGSI and the CRLGSI is matrix $\mathbf{U}'_C$, on which we now need to impose two restrictions: one for the trait and another for its associated GEBV. Consider the (*Zea mays*) F_2 population described earlier and suppose that we restrict trait GY; then, matrix $\mathbf{U}'_C$ should be constructed as $\mathbf{U}'_{C_1} = \begin{bmatrix} 1 & 0 & 0 & 0 & 0 & 0 \\ 0 & 0 & 0 & 1 & 0 & 0 \end{bmatrix}$. If we restrict traits GY and EHT, matrix $\mathbf{U}'_C$ should be constructed as $\mathbf{U}'_{C_2} = \begin{bmatrix} 1 & 0 & 0 & 0 & 0 & 0 \\ 0 & 1 & 0 & 0 & 0 & 0 \\ 0 & 0 & 0 & 1 & 0 & 0 \\ 0 & 0 & 0 & 0 & 1 & 0 \end{bmatrix}$, etc. The procedure for obtaining matrices $\widehat{\mathbf{K}}_C = [\mathbf{I} - \widehat{\mathbf{Q}}_C]$, $\widehat{\mathbf{Q}}_C = \widehat{\mathbf{T}}_C^{-1}\widehat{\mathbf{\Phi}}_C(\widehat{\mathbf{\Phi}}'_C\widehat{\mathbf{T}}_C^{-1}\widehat{\mathbf{\Phi}}_C)^{-1}\widehat{\mathbf{\Phi}}'_C$, and $\widehat{\mathbf{\Phi}}_C = \mathbf{U}'_C\widehat{\mathbf{\Psi}}_C$ is similar to that described in Chap. 3.

Let $\mathbf{w}' = [5 \quad -0.1 \quad -0.1 \quad 0 \quad 0 \quad 0]$ be the vector of economic weights and assume that we restrict trait GY; in this case, according to the estimated matrices $\widehat{\mathbf{P}}$, $\widehat{\mathbf{C}}$, and $\widehat{\mathbf{\Gamma}}$ described earlier, the estimated CRLGSI vector of coefficients was $\widehat{\boldsymbol{\beta}}'_{RG} = [0.076 \quad -0.004 \quad -0.018 \quad 2.353 \quad -0.096 \quad -0.082]$, whence the estimated CRLGSI can be written as

$$\widehat{I}_{CR} = 0.076\text{GY} - 0.004\text{EHT} - 0.018\text{PHT} + 2.353\text{GEBV}_{\text{GY}} - 0.096\text{GEBV}_{\text{EHT}} - 0.082\text{GEBV}_{\text{PHT}}$$

where GEBV_{GY}, GEBV_{EHT}, and GEBV_{PHT} are the GEBVs associated with traits GY, EHT, and PHT respectively. The same procedure is valid for two or more restrictions.

Figure 6.3 presents the frequency distribution of the estimated CRLGSI values for one (Fig. 6.3a) and two null restrictions (Fig. 6.3b) using matrices $\mathbf{U}'_{C_1}$ and $\mathbf{U}'_{C_2}$, and the real data set of the F_2 population. For both restrictions, the frequency distribution of the estimated CRLGSI values approaches normal distribution.

Suppose a selection intensity of 10% ($k_I = 1.755$), matrix $\mathbf{U}'_{C_1} = \begin{bmatrix} 1 & 0 & 0 & 0 & 0 & 0 \\ 0 & 0 & 0 & 1 & 0 & 0 \end{bmatrix}$ and that the vector of economic weights is $\mathbf{w}' = [5 \quad -0.1 \quad -0.1 \quad 0 \quad 0 \quad 0]$; then, according to the estimated matrices $\widehat{\mathbf{P}}$, $\widehat{\mathbf{C}}$, and $\widehat{\mathbf{\Gamma}}$ described earlier, the estimated CRLGSI selection response and the estimated CRLGSI expected genetic gain per trait were $\widehat{R}_{CR} = k_I\sqrt{\widehat{\boldsymbol{\beta}}'_{CR}\widehat{\mathbf{T}}_C\widehat{\boldsymbol{\beta}}_{CR}} = 0.96$ and $\widehat{\mathbf{E}}'_{CR} = k_I\frac{\widehat{\boldsymbol{\beta}}'_{CR}\widehat{\mathbf{\Psi}}}{\sqrt{\widehat{\boldsymbol{\beta}}'_{CR}\widehat{\mathbf{T}}_C\widehat{\boldsymbol{\beta}}_{CR}}} = [0 \quad -3.53 \quad -6.03 \quad 0 \quad -2.93 \quad -4.87]$ respectively, whereas the estimated CRLGSI accuracy was $\widehat{\rho}_{HI_{CR}} = \frac{\widehat{\sigma}_{I_{CR}}}{\widehat{\sigma}_H} = 0.51$ (see Chaps. 3 and 5 for details).

Now, we use the simulated data described in Chap. 2, Sect. 2.8.1 to compare CRLGSI efficiency versus RLGSI efficiency. The criteria for this comparison are the Technow inequality (Eq. 5.18, Chap. 5) and the ratio of the estimated CRLGSI accuracy ($\widehat{\rho}_{HI_{CR}}$) to the estimated RLGSI accuracy ($\widehat{\rho}_{HI_R}$) expressed as percentages (Eq. 5.17, Chap. 5), i.e., $\widehat{p} = 100(\widehat{\lambda}_{CR} - 1)$, where $\widehat{\lambda}_P = \widehat{\rho}_{HI_{CR}}/\widehat{\rho}_{HI_R}$, for one, two, and three null restrictions for five simulated selection cycles.

Table 6.7 presents the estimated CRLGSI heritability ($\widehat{h}^2_C$), the estimated RLGSI accuracy ($\widehat{\rho}_{HI_R}$), the values of $W_C = \frac{\widehat{\rho}_{HI_R}}{\widehat{h}_I}L_I$ ($L_I = 4$), and the values of

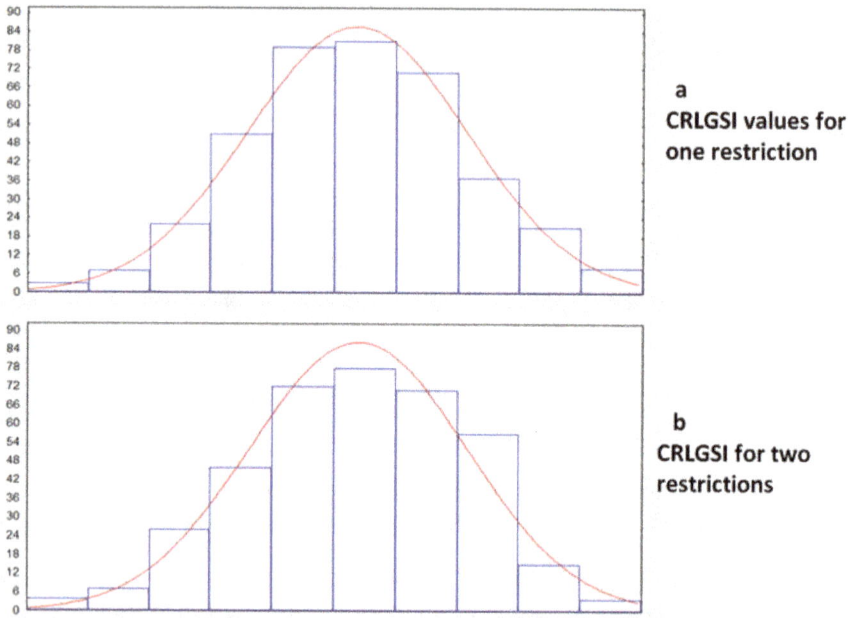

Fig. 6.3 Distribution of 244 estimated combined restricted linear genomic selection index (CRLGSI) values with one (**a**) and two (**b**) null restrictions respectively obtained in a real training population for one selection cycle in one environment

$\widehat{p} = 100(\widehat{\lambda}_{CR} - 1)$, where $\widehat{\lambda}_{CR} = \widehat{\rho}_{HI_{CR}}/\widehat{\rho}_{HI_R}$ and $\widehat{\rho}_{HI_{CR}}$ is the estimated CRLGSI accuracy, for one, two, and three null restrictions for five simulated selection cycles. The averages of the $W_C = \frac{\widehat{\rho}_{HI_R}}{\widehat{h}_C} L_I$ values for one, two, and three null restrictions were 1.26, 0.92, and 0.59 respectively, whereas the RLGSI interval length was 1.5 ($L_G = 1.5$). This means that the estimated Technow inequality ($L_G < \frac{\widehat{\rho}_{HI_G}}{\widehat{h}_I} L_I$) was not true. Thus, for this data set, RLGSI efficiency in terms of time is not greater than CRLGSI efficiency. The inequality $L_G < \frac{\widehat{\rho}_{HI_G}}{\widehat{h}_I} L_I$ was not true because the estimated RLGSI accuracy was very low, whereas CRLGSI heritability was high. Thus, note that the averages of the estimated RLGSI accuracy for one, two, and three null restrictions were 0.25, 0.19, and 0.14 respectively, whereas the averages of the estimated CRLGSI heritability values were 0.72, 0.75, and 0.89 respectively. Thus, according to these results, when the estimated RLGSI accuracy is very low and the estimated CRLGSI heritability is high, RLGSI efficiency will be lower than CRLGSI efficiency in terms of time.

Table 6.7 Estimated combined restricted linear genomic selection index (CRLGSI) heritability ($\widehat{h}_I^2$), estimated RLGSI accuracy ($\widehat{\rho}_{HI_R}$), values of $W_C = \frac{\widehat{\rho}_{HI_R}}{\widehat{h}_I} L_I$ ($L_I = 4$), and values of $\widehat{p} = 100\left(\widehat{\lambda}_{CR} - 1\right)$, where $\widehat{\lambda}_{CR} = \widehat{\rho}_{HI_{CR}} / \widehat{\rho}_{HI_R}$ and $\widehat{\rho}_{HI_{CR}}$ is the estimated CRLGSI accuracy, for 1, 2, and 3 null restrictions for five simulated selection cycles

	CRLGSI heritability			RLGSI accuracy			Values of W_C			Values of $\widehat{p}$		
Cycle	1	2	3	1	2	3	1	2	3	1	2	3
1	0.39	0.41	0.90	0.33	0.28	0.24	2.11	1.75	1.01	15.15	50.00	8.33
2	0.82	0.84	0.89	0.26	0.18	0.13	1.15	0.79	0.55	46.15	72.22	61.54
3	0.80	0.84	0.88	0.24	0.17	0.11	1.07	0.74	0.47	66.67	82.35	81.82
4	0.82	0.85	0.89	0.22	0.15	0.11	0.97	0.65	0.47	68.18	93.33	72.73
5	0.77	0.82	0.91	0.22	0.15	0.11	1.00	0.66	0.46	72.73	93.33	81.82
Average	0.72	0.75	0.89	0.25	0.19	0.14	1.26	0.92	0.59	53.78	78.25	61.25

The last three columns of Table 6.7, from left to right, present the average of the values of $\widehat{p} = 100(\widehat{\lambda}_{CR} - 1)$, for one, two, and three null restrictions of five simulated selection cycles. According to these results, CRLGSI efficiency was 53.78%, 78.25%, and 61.25% higher than RLGSI efficiency. Thus, for this data set, the CRLGSI was a better predictor of the net genetic merit than the RLGSI.

6.4 The Combined Predetermined Proportional Gains Linear Genomic Selection Index

In the PPG-LPSI described in Chap. 3, the vector of the PPG (predetermined proportional gains) was $\mathbf{d}' = [d_1 \quad d_2 \quad \ldots \quad d_r]$. However, because the combined predetermined proportional gains LGSI (CPPG-LGSI) uses phenotypic and GEBV information jointly to predict the net genetic merit, the vector of the PPG ($\mathbf{d}_C$) should be twice the standard vector $\mathbf{d}'$, that is, $\mathbf{d}'_C = [d_1 \quad d_2 \quad \cdots \quad d_r \quad d_{r+1} \quad d_{r+2} \quad \cdots \quad d_{2r}]$, where we would expect that if d_1 is the PPG imposed on trait 1, then d_{r+1} should be the PPG imposed on the GEBV associated with trait 1, etc. In addition, in the CPPG-LGSI, we have three possible options for determining (for each trait and GEBV) the PPG, e.g., for trait 1, $d_1 = d_{r+1}$, $d_1 > d_{r+1}$, or $d_1 < d_{r+1}$. This is the main difference between the standard PPG-LPSI described in Chap. 3 and the CPPG-LGSI.

6.4.1 The Maximized CPPG-LGSI Parameters

It can be shown that the vector of coefficients of the CPPG-LGSI can be written as

$$\boldsymbol{\beta}_{CP} = \boldsymbol{\beta}_{CR} + \theta_{CP}\boldsymbol{\delta}_{CP}, \tag{6.16}$$

where

$$\theta_{\mathrm{CP}} = \frac{\boldsymbol{\beta}'_C\boldsymbol{\Phi}_C\left(\boldsymbol{\Phi}'_C\widehat{\mathbf{T}}_C^{-1}\boldsymbol{\Phi}_C\right)^{-1}\mathbf{d}_C}{\mathbf{d}'_C\left(\boldsymbol{\Phi}'_C\widehat{\mathbf{T}}_C^{-1}\boldsymbol{\Phi}_C\right)^{-1}\mathbf{d}_C} \tag{6.17}$$

is a proportionality constant. In addition, in Eq. (6.16), $\boldsymbol{\beta}_{CR} = \mathbf{K}_C\boldsymbol{\beta}_C$ is the vector of coefficients of the CRLGSI (Eq. 6.13), $\boldsymbol{\delta}_{CP} = \mathbf{T}_C^{-1}\boldsymbol{\Phi}_C\left(\boldsymbol{\Phi}'_C\widehat{\mathbf{T}}_C^{-1}\boldsymbol{\Phi}_C\right)^{-1}\mathbf{d}_C$, $\boldsymbol{\Phi}'_C = \mathbf{U}'_C\boldsymbol{\Psi}_C$, and $\boldsymbol{\beta}_C = \mathbf{T}_C^{-1}\boldsymbol{\Psi}_C\mathbf{a}_C$ (the vector of coefficients of the CLGSI). When $\theta_{\mathrm{CP}} = 0$, $\boldsymbol{\beta}_{CP} = \boldsymbol{\beta}_{CR}$, and if $\theta = 0$ and $\mathbf{U}'_C$ is the null matrix, then $\boldsymbol{\beta}_{CR} = \boldsymbol{\beta}_C$.

Thus, the CPPG-LGSI is more general than the CRLGSI and the CLGSI, and includes the latter two indices as particular cases. In addition, it can be shown that the CPPG-LGSI has the same properties as the PPG-LPSI described in Chap. 3.

The maximized selection response and the expected genetic gain per trait of the CPPG-LGSI can be written as

$$R_{CP} = \frac{k_I}{L_I}\sqrt{\boldsymbol{\beta}'_{CP}\mathbf{T}_C\boldsymbol{\beta}_{CP}} \tag{6.18}$$

and

$$\mathbf{E}_{CP} = \frac{k_I}{L_I}\frac{\boldsymbol{\Psi}\boldsymbol{\beta}_{CP}}{\sqrt{\boldsymbol{\beta}'_{CP}\mathbf{T}_C\boldsymbol{\beta}_{CP}}}, \tag{6.19}$$

respectively. Although in the RLGSI and the PPG-LGSI the interval between selection cycles is denoted as L_G, in the CPPG-LGSI it is denoted as L_I. This is because the RLPSI and the CPPG-LGSI should have the same interval between selection cycles because they use phenotypic information to predict the net genetic merit.

6.4.2 Numerical Examples

Similar to the CRLGSI, to illustrate the CPPG-LGSI results we use the real training maize (*Zea mays*) F_2 population with 248 genotypes, 233 molecular markers, and three traits—GY (ton ha^{-1}), EHT (cm), and PHT (cm)—where $\widehat{\mathbf{P}} = \begin{bmatrix} 0.45 & 1.33 & 2.33 \\ 1.33 & 65.07 & 83.71 \\ 2.33 & 83.71 & 165.99 \end{bmatrix}$, $\widehat{\mathbf{C}} = \begin{bmatrix} 0.07 & 0.61 & 1.06 \\ 0.61 & 17.93 & 22.75 \\ 1.06 & 22.75 & 44.53 \end{bmatrix}$, and $\widehat{\boldsymbol{\Gamma}} = \begin{bmatrix} 0.07 & 0.65 & 1.05 \\ 0.65 & 10.62 & 14.25 \\ 1.05 & 14.25 & 26.37 \end{bmatrix}$ were the estimated matrices of $\mathbf{P}$, $\mathbf{C}$, and $\boldsymbol{\Gamma}$ respectively.

We can obtain the estimated CPPG-LGSI vector of coefficients as $\widehat{\boldsymbol{\beta}}_{CP} = \widehat{\boldsymbol{\beta}}_{CR} + \widehat{\theta}_{CP}\widehat{\boldsymbol{\delta}}_{CP}$ (Eq. 6.16). Suppose that we restrict trait GY and its associated GEBV with matrix $\mathbf{U}'_{C_1} = \begin{bmatrix} 1 & 0 & 0 & 0 & 0 & 0 \\ 0 & 0 & 0 & 1 & 0 & 0 \end{bmatrix}$ and the vector of predetermined restriction $\mathbf{d}'_C = [7 \quad 3.5]$. In Sect. 6.3.2, we showed that the estimated CRLGSI vector of coefficients was $\widehat{\boldsymbol{\beta}}'_{CR} = [0.076 \quad -0.004 \quad -0.018 \quad 2.353 \quad -0.096 \quad -0.082]$; then, we only need to calculate $\widehat{\theta}_{CP}$ and $\widehat{\boldsymbol{\delta}}_{CP}$ to obtain the vector of coefficients $\widehat{\boldsymbol{\beta}}_{CP}$.

Let $\mathbf{w}' = [5 \quad -0.1 \quad -0.1 \quad 0 \quad 0 \quad 0]$ be the vector of economic weights. It can be shown that $\widehat{\theta}_{CP} = 0.00030$ is the estimated value of the proportionality constant and $\boldsymbol{\delta}'_{CP} = [0.56 \quad -77.28 \quad 40.89 \quad 49.44 \quad 77.28 \quad -40.89]$. Thus, the estimated CPPG-LGSI vector of coefficients was $\widehat{\boldsymbol{\beta}}'_{CR} = [0.76 \quad -0.030 \quad -0.004 \quad 2.369 \quad -0.070 \quad -0.096]$, whence the estimated CPPG-LGSI can be written as

$$\widehat{I}_{CP} = 0.076\text{GY} - 0.03\text{EHT} - 0.004\text{PHT} + 2.369\text{GEBV}_{\text{GY}} - 0.070\text{GEBV}_{\text{EHT}} - 0.096\text{GEBV}_{\text{PHT}},$$

where GEBV_{GY}, GEBV_{EHT}, and GEBV_{PHT} are the GEBVs associated with traits GY, EHT, and PHT respectively. The same procedure is valid for two or more restrictions. Note that because $\widehat{\theta}_{CP} = 0.0003$ is very small, the estimated CPPG-LGSI and CRLGSI values were very similar.

Figure 6.4 presents the frequency distribution of the estimated CPPG-LGSI values for one (Fig. 6.4a) and two predetermined restrictions (Fig. 6.4b) using matrices $\mathbf{U}'_{C_1}$ and $\mathbf{U}'_{C_2} = \begin{bmatrix} 1 & 0 & 0 & 0 & 0 & 0 \\ 0 & 1 & 0 & 0 & 0 & 0 \\ 0 & 0 & 0 & 1 & 0 & 0 \\ 0 & 0 & 0 & 0 & 1 & 0 \end{bmatrix}$, the vectors of the PPG

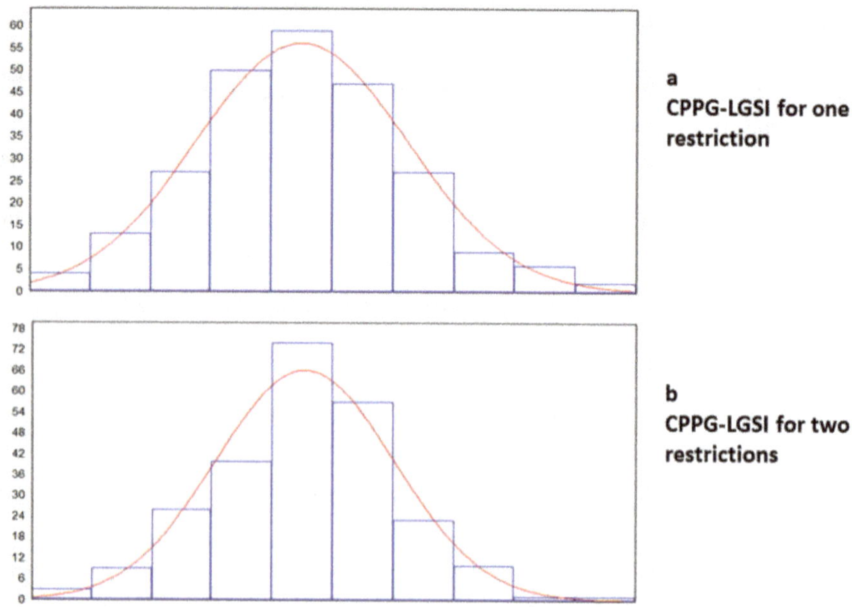

Fig. 6.4 Distribution of 244 estimated combined predetermined proportional gain linear genomic selection index (CPPG-LGSI) values with one (**a**) and two (**b**) predetermined restrictions, $d = 7$ and $\mathbf{d}' = [7 \quad -3]$ respectively, obtained in a real training population for one selection cycle in one environment

$\mathbf{d}'_{C1} = [7 \quad 3.5]$ and $\mathbf{d}'_{C2} = [7 \quad -3 \quad 3.5 \quad -1.5]$, and the real data set F_2. For both restrictions, the frequency distribution of the estimated CPPG-LGSI values approaches normal distribution.

Suppose a selection intensity of 10% ($k_I = 1.755$) and that we restrict trait GY and its associated GEBV. The estimated CPPG-LGSI selection response and expected genetic gain per trait were $\widehat{R}_{CP} = k_I\sqrt{\widehat{\boldsymbol{\beta}}'_{CP}\widehat{\mathbf{T}}_C\widehat{\boldsymbol{\beta}}_{CP}} = 0.98$ and $\widehat{\mathbf{E}}'_{CP} = k_I\frac{\widehat{\boldsymbol{\beta}}'_{CP}\widehat{\boldsymbol{\Psi}}}{\sqrt{\widehat{\boldsymbol{\beta}}'_{CP}\widehat{\mathbf{T}}\widehat{\boldsymbol{\beta}}_{CP}}}$ $= [0.007 \quad -3.647 \quad -5.760 \quad 0.004 \quad -2.829 \quad -4.711]$ respectively, whereas the estimated CPPG-LGSI accuracy was $\widehat{\rho}_{HI_{CP}} = \frac{\widehat{\sigma}_{I_{CP}}}{\widehat{\sigma}_H} = 0.52$. Once again, because $\widehat{\theta}_{\text{CP}} = 0.0003$, the latter results are very similar to the CRLGSI results.

Now, we use the simulated data described in Chap. 2, Sect. 2.8.1, to compare CPPG-LGSI efficiency versus PPG-LGSI efficiency. The criteria for this comparison are the Technow inequality (Chap. 5, Eq. 5.18) and the ratio of CPPG-LGSI accuracy ($\rho_{HI_{CP}}$) to PPG-LGSI accuracy (ρ_{HI_P}) expressed as percentages (Chap. 5, Eq. 5.17), $\widehat{p} = 100\left(\widehat{\lambda}_{CP} - 1\right)$, where $\widehat{\lambda}_{CP} = \widehat{\rho}_{HI_{CP}}/\widehat{\rho}_{HI_P}$ for one, two, and three null restrictions in five simulated selection cycles.

Table 6.8 presents the estimated CPPG-LGSI heritability ($\widehat{h}_I^2$), the estimated PPG-LGSI accuracy ($\widehat{\rho}_{HI_{CP}}$), values of $W_{CP} = \frac{\widehat{\rho}_{HI_G}}{\widehat{h}_I}L_I$ ($L_I = 4$) and $\widehat{p} = 100\left(\widehat{\lambda}_{CP} - 1\right)$, where $\widehat{\lambda}_P = \widehat{\rho}_{HI_{CP}}/\widehat{\rho}_{HI_P}$ and $\widehat{\rho}_{HI_P}$ is the estimated CPPG-LGSI accuracy, for one, two, and three null restrictions in five simulated selection cycles. The averages of the estimated W_{CP} values for one, two, and three predetermined restrictions were 3.60, 3.31, and 2.50 respectively, whereas the PPG-LGSI interval length was 1.5 ($L_G = 1.5$). This means that the estimated Technow inequality, $L_G < \frac{\widehat{\rho}_{HI_G}}{\widehat{h}_I}L_I$, was true. Thus, for this data set, PPG-LGSI efficiency is greater than CPPG-LGSI efficiency in terms of time.

The last three columns of Table 6.8, from left to right, present the values of $\widehat{p} = 100\left(\widehat{\lambda}_{CP} - 1\right)$, for one, two, and three null restrictions in five simulated selection cycles. The average values of $\widehat{p} = 100\left(\widehat{\lambda}_{CP} - 1\right)$ for each of the three restrictions, in percentage terms, were 37.19%, 32.82%, and 37.08% respectively. This means that the CPPG-LGSI efficiency was greater than PPG-LGSI efficiency at predicting the net genetic merit.

Table 6.8 Estimated combined predetermined proportional gain linear genomic selection index (CPPG-LGSI) heritability ($\widehat{h}_I^2$), estimated PPG-LGSI accuracy ($\widehat{\rho}_{HI_{CP}}$), values of $W_{CP} = \frac{\widehat{\rho}_{HI_G}}{\widehat{h}_I} L_I$ ($L_I = 4$), and $\widehat{p} = 100\left(\widehat{\lambda}_{CP} - 1\right)$, where $\widehat{\lambda}_P = \widehat{\rho}_{HI_{CP}} / \rho_{HI_P}$ and $\widehat{\rho}_{HI_P}$ is the estimated CPPG-LGSI accuracy, for one, two, and three null restrictions for five simulated selection cycles

	CPPG-LGSI heritability			PPG-LGSI accuracy			Values of W_{CP}			Values of $\widehat{p}$		
Cycle	1	2	3	1	2	3	1	2	3	1	2	3
1	0.41	0.41	0.85	0.65	0.91	0.85	6.25	5.68	3.69	24.62	−3.30	−12.94
2	0.75	0.86	0.84	0.72	0.70	0.55	3.33	3.02	2.40	25.00	24.29	38.18
3	0.78	0.85	0.81	0.70	0.67	0.55	3.17	2.91	2.44	30.00	32.84	40.00
4	0.78	0.84	0.82	0.61	0.59	0.46	2.76	2.57	2.03	49.18	49.15	56.52
5	0.80	0.83	0.82	0.56	0.54	0.44	2.50	2.37	1.94	57.14	61.11	63.64
Average	0.70	0.76	0.83	0.65	0.68	0.57	3.60	3.31	2.50	37.19	32.82	37.08

References

Beyene Y, Semagn K, Mugo S, Tarekegne A, Babu R et al (2015) Genetic gains in grain yield through genomic selection in eight bi-parental maize populations under drought stress. Crop Sci 55:154–163

Technow F, Bürger A, Melchinger AE (2013) Genomic prediction of northern corn leaf blight resistance in maize with combined or separate training sets for heterotic groups. G3 (Bethesda) 3:197–203

7

ESIM, RESIM and PPG-ESIM

Abstract Based on the canonical correlation, on the singular value decomposition (SVD), and on the linear phenotypic selection indices theory, we describe the eigen selection index method (ESIM), the restricted ESIM (RESIM), and the predetermined proportional gain ESIM (PPG-ESIM), which use only phenotypic information to predict the net genetic merit. The ESIM is an unrestricted linear selection index, but the RESIM and PPG-ESIM are linear selection indices that allow null and predetermined restrictions respectively to be imposed on the expected genetic gains of some traits, whereas the rest remain without any restrictions. The aims of the three indices are to predict the unobservable net genetic merit values of the candidates for selection, maximize the selection response, and the accuracy, and provide the breeder with an objective rule for evaluating and selecting several traits simultaneously. Their main characteristics are: they do not require the economic weights to be known, the first multi-trait heritability eigenvector is used as its vector of coefficients; and because of the properties associated with eigen analysis, it is possible to use the theory of similar matrices to change the direction and proportion of the expected genetic gain values without affecting the accuracy. We describe the foregoing three indices and validate their theoretical results using real and simulated data.

7.1 The Linear Phenotypic Eigen Selection Index Method

The conditions described in Chap. 2 for the linear phenotypic selection index (LPSI) are necessary and sufficient for constructing the linear phenotypic eigen selection index method (ESIM). The ESIM index can be written as $I = \mathbf{b}'\mathbf{y}$, where $\mathbf{b}' = [b_1 \; b_2 \; \cdots \; b_t]$ is the unknown index vector of coefficients, t is the number of traits, and $\mathbf{y}' = [y_1 \quad y_2 \quad \cdots \quad y_t]$ is a known vector of trait phenotypic values. The objectives of ESIM are:

1. To predict the net genetic merit $H = \mathbf{w}'\mathbf{g}$, where $\mathbf{g}' = [g_1 \quad g_2 \quad \ldots \quad g_t]$ is the unknown vector of true breeding values for an individual and $\mathbf{w}' = [w_1 \quad w_2 \quad \ldots \quad w_t]$ is a vector of unknown economic weights.

2. To maximize the ESIM selection response and the accuracy.
3. To select individuals with the highest H values in each selection cycle as parents of the next generation.
4. To provide the breeder with an objective rule for evaluating and selecting several traits simultaneously.

Although in the context of the LPSI $\mathbf{w}$ is a known and fixed vector of economic weights, in the ESIM $\mathbf{w}$ is fixed, but unknown and its values must be estimated in each selection cycle. This latter assumption is the fundamental difference between the ESIM and the LPSI and implies that the ESIM is more general than the LPSI. Thus, when $\mathbf{w}$ is known, the LPSI and ESIM give the same results.

7.1.1 The ESIM Parameters

The theoretical ESIM selection response can be written as

$$R_I = k_I \sigma_H \rho_{HI}, \tag{7.1}$$

where k_I is the standardized selection differential (or selection intensity), $\sigma_H = \sqrt{\mathbf{w}'\mathbf{C}\mathbf{w}}$ is the standard deviation of H, $\rho_{HI} = \frac{\mathbf{w}'\mathbf{C}\mathbf{b}}{\sqrt{\mathbf{w}'\mathbf{C}\mathbf{w}}\sqrt{\mathbf{b}'\mathbf{P}\mathbf{b}}}$ is the correlation, and $\mathbf{w}'\mathbf{C}\mathbf{b} = \sigma_{HI}$ the covariance between H and I respectively, $\sigma_I = \sqrt{\mathbf{b}'\mathbf{P}\mathbf{b}}$ is the standard deviation of I, $\mathbf{C}$ is the covariance matrix of the true breeding values ($\mathbf{g}$), and $\mathbf{P}$ is the covariance matrix of the trait phenotypic values ($\mathbf{y}$).

In the ESIM, it is assumed that k_I and σ_H are fixed, and that $\mathbf{C}$ and $\mathbf{P}$ are known; thus, to maximize Eq. (7.1), it is necessary to maximize $\rho_{HI}^2 = \dfrac{(\mathbf{w}'\mathbf{C}\mathbf{b})^2}{(\mathbf{w}'\mathbf{C}\mathbf{w})(\mathbf{b}'\mathbf{P}\mathbf{b})}$ with respect to vectors $\mathbf{b}$ and $\mathbf{w}$ under the restrictions $\sigma_H^2 = \mathbf{w}'\mathbf{C}\mathbf{w}, \sigma_I^2 = \mathbf{b}'\mathbf{P}\mathbf{b}$, and $0 < \sigma_H^2$, $\sigma_I^2 < \infty$, where $\sigma_H^2 = \mathbf{w}'\mathbf{C}\mathbf{w}$ is the variance of $H = \mathbf{w}'\mathbf{g}$ and $\sigma_I^2 = \mathbf{b}'\mathbf{P}\mathbf{b}$ is the variance of $I = \mathbf{b}'\mathbf{y}$. That is, it is necessary to maximize the function

$$f(\mathbf{b}, \mathbf{w}, \mu, \phi) = (\mathbf{w}'\mathbf{C}\mathbf{b})^2 - \mu(\mathbf{b}'\mathbf{P}\mathbf{b} - \sigma_I^2) - \phi(\mathbf{w}'\mathbf{C}\mathbf{w} - \sigma_H^2) \tag{7.2}$$

with respect to $\mathbf{b}$, $\mathbf{w}$, μ, and ϕ, where μ and ϕ are Lagrange multipliers. The derivative results of Eq. (7.2) with respect to $\mathbf{b}$, $\mathbf{w}$, μ, and ϕ are:

$$(\mathbf{w}'\mathbf{C}\mathbf{b})\mathbf{C}\mathbf{w} - \mu\mathbf{P}\mathbf{b} = \mathbf{0}, \tag{7.3}$$

$$(\mathbf{w}'\mathbf{C}\mathbf{b})\mathbf{C}\mathbf{b} - \phi\mathbf{C}\mathbf{w} = \mathbf{0}, \tag{7.4}$$

$$\mathbf{b}'\mathbf{P}\mathbf{b} = \sigma_I^2 \text{ and } \mathbf{w}'\mathbf{C}\mathbf{w} = \sigma_H^2, \tag{7.5}$$

respectively, where Eq. (7.5) denotes the restrictions imposed for maximizing ρ_{HI}^2. It can be shown that $\mathbf{w}'\mathbf{C}\mathbf{b} = \sqrt{\mu\sigma_I^2} = \sqrt{\phi\sigma_H^2} = \theta^{1/2}$; then, Eqs. (7.3) and (7.4) can be written as

$$\theta^{1/2}\mathbf{Cw} - \frac{\theta}{\sigma_I^2}\mathbf{Pb} = \mathbf{0} \tag{7.6}$$

and

$$\theta^{1/2}\mathbf{Cb} - \frac{\theta}{\sigma_H^2}\mathbf{Cw} = \mathbf{0}, \tag{7.7}$$

respectively. Equation (7.6) is equal to $\mathbf{Cw} = \frac{\theta^{1/2}}{\sigma_I^2}\mathbf{Pb}$; then, vector $\mathbf{w}$ can be written as

$$\mathbf{w}_E = \frac{\theta^{1/2}}{\sigma_I^2}\mathbf{C}^{-1}\mathbf{Pb}. \tag{7.8}$$

By the result of Eq. (7.8), the net genetic merit in the ESIM context is $H_E = \mathbf{w}_E'\mathbf{g}$ and the correlation between H_E and I is $\rho_{H_E I} = \frac{\mathbf{w}_E'\mathbf{Cb}}{\sqrt{\mathbf{w}_E'\mathbf{Cw}_E}\sqrt{\mathbf{b}'\mathbf{Pb}}} = \frac{\sqrt{\mathbf{b}'\mathbf{Pb}}}{\sqrt{\mathbf{b}'\mathbf{PC}^{-1}\mathbf{Pb}}}$. Now, it is necessary to find the vector $\mathbf{b}$ that maximizes $\rho_{H_E I}$, which should be the ESIM index vector of coefficients. Substituting $\mathbf{w}$ with $\mathbf{w}_E$ in Eq. (7.7), we get

$$\mathbf{Cb} - \frac{(\mathbf{w}_E'\mathbf{Cb})^2}{\sigma_I^2\sigma_{H_E}^2}\mathbf{Pb} = \mathbf{0}, \tag{7.9}$$

where $\frac{(\mathbf{w}_E'\mathbf{Cb})^2}{\sigma_I^2\sigma_{H_E}^2} = \rho_{H_E I}^2$ is the square of the correlation between ESIM and $H_E = \mathbf{w}_E'\mathbf{g}$. Let $\rho_{H_E I}^2 = \lambda_E^2$, then Eq. (7.9) can be written as

$$\left(\mathbf{P}^{-1}\mathbf{C} - \lambda_E^2\mathbf{I}\right)\mathbf{b}_E = \mathbf{0}, \tag{7.10}$$

and the optimized ESIM index is $I_E = \mathbf{b}_E'\mathbf{y}$. Note that in Eq. (7.10) $\mathbf{P}^{-1}\mathbf{C}$ is the multi-trait heritability. By Eqs. (7.8) and (7.10), the maximized correlation between $H_E = \mathbf{w}_E'\mathbf{g}$ and $I_E = \mathbf{b}_E'\mathbf{y}$ (or ESIM accuracy) can be written as

$$\rho_{H_E I_E} = \frac{\sigma_{I_E}}{\sigma_{H_E}}, \tag{7.11}$$

where $\sigma_{I_E} = \sqrt{\mathbf{b}_E'\mathbf{Pb}_E}$ is the standard deviation of the variance of $I_E = \mathbf{b}_E'\mathbf{y}$, and $\sigma_{H_E} = \sqrt{\mathbf{b}_E'\mathbf{PC}^{-1}\mathbf{Pb}_E}$ is the standard deviation of the variance of $H_E = \mathbf{w}_E'\mathbf{g}$. Hereafter, we write Eq. (7.11) as $\rho_E = \rho_{H_E I_E}$ or $\lambda_E = \rho_{H_E I_E}$ to simplify the notation.

An additional restriction on Eq. (7.10) is $\mathbf{b}'\mathbf{b} = 1$, because $\rho_{H_E I_E}$ is invariant to the scale change and because if $\mathbf{b}_E$ is an eigenvector of the multi-trait heritability matrix $\mathbf{P}^{-1}\mathbf{C}$, vector $\alpha\mathbf{b}_E$ is also an eigenvector of $\mathbf{P}^{-1}\mathbf{C}$ for all real values of α (Mardia et al. 1982). This means that in the ESIM the magnitude of an eigenvector is unimportant;

only the direction matters (Watkins 2002). Equation (7.10) can also be written as $\mathbf{Cb}_E = \lambda_E^2 \mathbf{Pb}_E$, which is called the *generalized eigenvalue problem* (Watkins 2002). In the latter case, $\mathbf{b}_E$ is called a *generalized eigenvector* and λ_E^2 a *generalized eigenvalue*. The generalized eigenvalues may not exist; that is, they may be infinite. However, if $\mathbf{P}$ is positive definite and has the same size as $\mathbf{C}$, all eigenvalues of $\mathbf{P}^{-1}\mathbf{C}$ exist and are finite (Gentle 2007). Matrix $\mathbf{P}$ is symmetric and positive definite and its eigenvalues are different with a probability of 1 if the number of genotypes is higher than the number of traits (Okamoto 1973).

If the heritability of the ESIM is $h_I^2 = \frac{\mathbf{b}'\mathbf{Cb}}{\mathbf{b}'\mathbf{Pb}}$, then another way of writing Eq. (7.1) is

$$R_I = k_I \sigma_I h_I^2 = k_I \frac{\mathbf{b}'\mathbf{Cb}}{\sqrt{\mathbf{b}'\mathbf{Pb}}}, \tag{7.12}$$

which is similar to the univariate breeder's equation (see Chap. 2, Eq. 2.4). All the parameters of Eq. (7.12) were defined earlier.

The derivative of the ratio $\frac{\mathbf{b}'\mathbf{Cb}}{\sqrt{\mathbf{b}'\mathbf{Pb}}}$ (Eq. 7.12) with respect to $\mathbf{b}$ can be written as $2(\mathbf{b}'\mathbf{Pb})^{1/2}\mathbf{Cb} - (\mathbf{b}'\mathbf{Pb})^{-1/2}(\mathbf{b}'\mathbf{Cb})\mathbf{Pb} = \mathbf{0}$, and, except by a proportionality constant, the result is

$$\left(\mathbf{P}^{-1}\mathbf{C} - h_{I_E}^2 \mathbf{I}\right)\mathbf{b}_E = \mathbf{0}, \tag{7.13}$$

where $h_{I_E}^2 == \frac{\mathbf{b}_E'\mathbf{Cb}_E}{\mathbf{b}_E'\mathbf{Pb}_E}$ is the maximized ESIM heritability. Let $\lambda_E^2 = \rho_E^2 = h_{I_E}^2$, then Eq. (7.13) is equal to Eq. (7.10) and can be written as $\mathbf{b}_E'\mathbf{Cb}_E = \lambda_E^2 \mathbf{b}_E'\mathbf{Pb}_E$, whence the maximized ρ_E^2 in terms of $h_{I_E}^2$ is

$$\rho_E^2 = \frac{\mathbf{b}_E'\mathbf{Cb}_E}{\mathbf{b}_E'\mathbf{Pb}_E}, \tag{7.14}$$

which should give a equivalent result to that of Eq. (7.11).

By Eq. (7.11) and $\sigma_{H_E} = \sqrt{\mathbf{b}_E'\mathbf{PC}^{-1}\mathbf{Pb}_E}$, the maximized ESIM selection response and expected genetic gain per trait can be written as

$$R_E = k_I \sqrt{\mathbf{b}_E'\mathbf{Pb}_E} \tag{7.15}$$

and

$$\mathbf{E}_E = k_I \frac{\mathbf{Cb}_E}{\sqrt{\mathbf{b}_E'\mathbf{Pb}_E}}, \tag{7.16}$$

respectively. Equations (7.15) and (7.16) do not require the economic weights to be known. In the original derivation of the ESIM, Cerón-Rojas et al. (2008) imposed the restrictions $\sigma_{H_E}^2 = 1$ and $\sigma_{I_E}^2 = 1$. Under these restrictions, $\lambda_E = \mathbf{w}_E'\mathbf{C}\mathbf{b}_E$ and Eq. (7.15) can be written as $R_E = k_I\lambda_E$. When $\sigma_{H_E}^2 \neq 1$ Eq. (7.15) is equal to $R_E = k_I\sigma_{H_E}\lambda_E$, where $\sigma_{H_E} = \sqrt{\mathbf{b}_E'\mathbf{P}\mathbf{C}^{-1}\mathbf{P}\mathbf{b}_E}$ and $\lambda_E^2 = \rho_E^2 = h_{I_E}^2$.

Let $\mathbf{T} = \mathbf{P}^{-1}\mathbf{C}$ and $\lambda_E^2 = h_{I_E}^2$; then, Eq. (7.13) can be written as $\mathbf{T}\mathbf{I}\mathbf{b}_E = \lambda_E^2\mathbf{I}\mathbf{b}_E$, where $\mathbf{I} = \mathbf{F}^{-1}\mathbf{F}$ is an identity matrix of size $t \times t$ (t= number of traits), and $\mathbf{F} = diag\{f_1 \quad f_1 \quad \cdots \quad f_t\}$ is a diagonal matrix with values equal to any real number, except zero values. Thus, another way of writing Eqs. (7.10) and (7.13) is

$$\left(\mathbf{T}_2 - \lambda_E^2\mathbf{I}\right)\boldsymbol{\beta} = \mathbf{0}, \tag{7.17}$$

where $\mathbf{T}_2 = \mathbf{F}\mathbf{T}\mathbf{F}^{-1}$ and $\boldsymbol{\beta} = \mathbf{F}\mathbf{b}_E$; $\mathbf{T}$ and $\mathbf{T}_2 = \mathbf{F}\mathbf{T}\mathbf{F}^{-1}$ are similar matrices and both have the same eigenvalues but different eigenvectors (Harville 1997). When the $\mathbf{F}$ values are only 1s, vector $\mathbf{b}_E$ is not affected; when the $\mathbf{F}$ values are only -1s, vector $\mathbf{b}_E$ changes its direction, and if the $\mathbf{F}$ values are different from 1 and -1, matrix $\mathbf{F}$ changes the proportional values of $\mathbf{b}_E$. In practice, $\mathbf{b}_E$ is first obtained from Eq. (7.13) and then multiplied by matrix $\mathbf{F}$ to obtain $\boldsymbol{\beta} = \mathbf{F}\mathbf{b}_E$, that is, $\boldsymbol{\beta}$ is a linear transformation of $\mathbf{b}_E$. Matrix $\mathbf{T}_2 = \mathbf{F}\mathbf{T}\mathbf{F}^{-1}$ is called the *similarity transformation*, and matrix $\mathbf{F}$ is called the *transforming matrix* (Watkins 2002). Cerón-Rojas et al. (2006) introduced an alternative procedure for modifying the $\mathbf{b}_E$ signs that is a particular case of Eq. (7.17). Vector $\boldsymbol{\beta} = \mathbf{F}\mathbf{b}_E$ can substitute $\mathbf{b}_E$ in Eqs. (7.15) and (7.16); and in this case, the optimized ESIM index should be written as $I_E = \boldsymbol{\beta}'\mathbf{y}$.

7.1.2 Statistical ESIM Properties

The ratio of the index accuracies and the variance of the predicted error (VPE) are good criteria for comparing the index efficiencies for predicting the net genetic merit (see Chap. 2 for details). In Eq. (7.11), we obtained the accuracy of the ESIM; now, we derive the VPE of the ESIM.

The variance of $I_E = \mathbf{b}_E'\mathbf{y}$ ($\sigma_{I_E}^2$) and the covariance between $H_E = \mathbf{w}_E'\mathbf{g}$ and $I_E = \mathbf{b}_E'\mathbf{y}(\sigma_{H_EI_E})$ are the same, that is,

$$\sigma_{I_E}^2 = \mathbf{b}_E'\mathbf{P}\mathbf{b}_E \text{ and } \sigma_{H_EI_E} = \mathbf{w}_E'\mathbf{C}\mathbf{b}_E = \mathbf{b}_E'\mathbf{P}\mathbf{C}^{-1}\mathbf{C}\mathbf{b}_E = \mathbf{b}_E'\mathbf{P}\mathbf{b}_E, \tag{7.18}$$

respectively; that is, $\sigma_{I_E}^2 = \sigma_{H_EI_E}$. By Eq. (7.18), the VPE of the ESIM can be written as

$$E\left[(H_E - I_E)^2\right] = \sigma_{H_E}^2 + \sigma_{I_E}^2 - 2\sigma_{H_EI_E} = \sigma_{H_E}^2 - \sigma_{I_E}^2 = \left(1 - \rho_E^2\right)\sigma_{H_E}^2. \tag{7.19}$$

The relative effectiveness of $I_E = \mathbf{b}_E'\mathbf{y}$ in predicting $H_E = \mathbf{w}_E'\mathbf{g}$ is the ratio of $(1-\rho_E^2)\sigma_{H_E}^2$ over $\sigma_{H_E}^2$, i.e., $1-\rho_E^2$; thus, the greater ρ_E^2 is, the more effective $I_E = \mathbf{b}_E'\mathbf{y}$ is at predicting $H_E = \mathbf{w}_E'\mathbf{g}$. The mean squared effect of I_E on H_E, or the total variance of H_E explained by I_E is

$$\sigma_{I_E}^2 = \rho_E^2\sigma_{H_E}^2, \tag{7.20}$$

and the relative mean squared effect can be measured by ρ_E^2 (Anderson 2003). If in Eq. (7.20) $\rho_E^2 = 1$, $\sigma_{I_E}^2 = \sigma_{H_E}^2$, and if $\rho_E^2 = 0$, $\sigma_{I_E}^2 = 0$. That is, the variance of H_E explained by I_E is proportional to ρ_E^2, and when ρ_E^2 is close to 1, $\sigma_{I_E}^2$ is close to $\sigma_{H_E}^2$, and if ρ_E^2 is close to 0, $\sigma_{I_E}^2$ is close to 0. All these results are valid for any index associated with the ESIM, such as the restricted ESIM (RESIM) and the predetermined proportional gains ESIM (PPG-ESIM), which are described in the following sections of this chapter.

7.1.3 The ESIM and the Canonical Correlation Theory

Canonical correlation theory describes the associations between two sets of variables (Hotelling 1935, 1936) and searches for linear combinations, called *canonical variables*, of each of two sets of variables having maximal correlation. The vector of coefficient of these linear combinations is called the *canonical vector* and the correlations between the canonical variables is called the *canonical correlation* (Wilms and Croux 2016).

To see how the ESIM and the canonical correlation theory are related, note that vectors $\mathbf{y}$ and $\mathbf{g}$ (Eq. 7.1) can be ordered in a new vector $\mathbf{x}$ as $\mathbf{x}' = [\mathbf{y}' \quad \mathbf{g}']$, whence the covariance matrix of $\mathbf{x}$ is $\begin{bmatrix} \mathbf{P} & \mathbf{C} \\ \mathbf{C} & \mathbf{C} \end{bmatrix}$. One measure of the association between the *j*th linear combination of $\mathbf{y}(I_E = \mathbf{b}_{E_j}'\mathbf{y})$ and the *j*th linear combination of $\mathbf{g}(H_E = \mathbf{w}_{E_j}'\mathbf{g})$ is the *j*th canonical correlation (λ_j) value obtained from equation $\left(\mathbf{P}^{-1}\mathbf{C} - \lambda_j^2\mathbf{I}\right)\mathbf{b}_{Ej} = \mathbf{0}$, where $\mathbf{b}_{Ej}$ is the *j*th *canonical vector* ($j = 1, 2\cdots, t$) of matrix $\mathbf{P}^{-1}\mathbf{C}$, and $\mathbf{w}_{E_j} = \mathbf{C}^{-1}\mathbf{P}\mathbf{b}_{E_j}$. Thus, in the canonical correlation context, $I_E = \mathbf{b}_{E_j}'\mathbf{y}$ and $H_E = \mathbf{w}_{E_j}'\mathbf{g}$ are *canonical variables.*

In the ESIM, the first eigenvector ($\mathbf{b}_{E_1}$) of matrix $\mathbf{P}^{-1}\mathbf{C}$ should be used on $I_E = \mathbf{b}_{E_1}'\mathbf{y}$; the first eigenvalue ($\lambda_1^2$) and $\mathbf{b}_{E_1}$ of $\mathbf{P}^{-1}\mathbf{C}$ should be used on the ESIM selection response and on the ESIM expected genetic gain per trait, because, in this case, the ESIM has maximum accuracy compared with other indices, such as the LPSI. The latter results in this subsection imply that the sampling statistical properties associated with the canonical correlation theory are also valid for the ESIM.

7.1.4 Estimated ESIM Parameters and Their Sampling Properties

The estimated covariance matrix of the true breeding values (**C**) and that of the trait phenotypic values (**P**) are denoted as $\widehat{\mathbf{C}}$ and $\widehat{\mathbf{P}}$ respectively; they can be obtained by restricted maximum likelihood using Eqs. (2.22) to (2.24) described in Chap. 2. With matrices $\widehat{\mathbf{C}}$ and $\widehat{\mathbf{P}}$, we constructed matrix $\widehat{\mathbf{T}} = \widehat{\mathbf{P}}^{-1}\widehat{\mathbf{C}}$ and equation

$$\left(\widehat{\mathbf{T}} - \widehat{\lambda}^2_{Ej}\mathbf{I}\right)\widehat{\mathbf{b}}_{Ej} = \mathbf{0}, \tag{7.21}$$

$j = 1, 2, \cdots, t$, where t is the number of traits in the ESIM index. Note that $\widehat{\lambda}^2_{Ej}$ is positive only if $\widehat{\mathbf{P}}$ is positive definite (all eigenvalues positive) and $\widehat{\mathbf{C}}$ is positive semidefinite (no negative eigenvalues); in addition, as $\widehat{\mathbf{P}}^{-1}\widehat{\mathbf{C}}$ is an asymmetric matrix, the values of $\widehat{\mathbf{b}}_{Ej}$ and $\widehat{\lambda}^2_{Ej}$ should be obtained using the singular value decomposition (SVD) theory (Anderson 2003).

Matrix $\widehat{\mathbf{T}}$ is square and asymmetric of order $t \times t$ and rank $q \leq$ minimum (p, c), where p and c denote the rank of $\widehat{\mathbf{P}}^{-1}$ and $\widehat{\mathbf{C}}$ respectively; the rank of $\widehat{\mathbf{T}}$ is equal to c only if $\widehat{\mathbf{C}}$ is square and nonsingular. Thus, matrix $\widehat{\mathbf{T}}$ has a maximum of q eigenvalues different from zero (Rao 2002). In addition, $\widehat{\mathbf{T}}\widehat{\mathbf{T}}'$ and $\widehat{\mathbf{T}}'\widehat{\mathbf{T}}$ are symmetric matrices, but $\widehat{\mathbf{T}}\widehat{\mathbf{T}}' \neq \widehat{\mathbf{T}}'\widehat{\mathbf{T}}$. Using the SVD theory, matrix $\widehat{\mathbf{T}}$ can be written as

$$\widehat{\mathbf{T}} = \mathbf{V}_1\mathbf{L}^{1/2}\mathbf{V}_2', \tag{7.22}$$

where $\mathbf{V}_1$ ($\mathbf{V}_1'\mathbf{V}_1 = \mathbf{V}_1\mathbf{V}_1' = \mathbf{I}_q$) and $\mathbf{V}_2$ ($\mathbf{V}_2'\mathbf{V}_2 = \mathbf{V}_2\mathbf{V}_2' = \mathbf{I}_q$) are matrices with the eigenvectors of matrices $\widehat{\mathbf{T}}\widehat{\mathbf{T}}'$ and $\widehat{\mathbf{T}}'\widehat{\mathbf{T}}$ respectively; $\mathbf{L}^{1/2}$ is a diagonal matrix with the square root of the eigenvalues ($\widehat{\lambda}^2_{E_1} \geq \widehat{\lambda}^2_{E_2} \geq \cdots \geq \widehat{\lambda}^2_{E_q} > 0$) of either $\widehat{\mathbf{T}}\widehat{\mathbf{T}}'$ or $\widehat{\mathbf{T}}'\widehat{\mathbf{T}}$ (the eigenvalues of $\widehat{\mathbf{T}}\widehat{\mathbf{T}}'$ and $\widehat{\mathbf{T}}'\widehat{\mathbf{T}}$ are the same). The entries $\widehat{\lambda}^2_{E_1} \geq \widehat{\lambda}^2_{E_2} \geq \cdots \geq \widehat{\lambda}^2_{E_q} > 0$ of $\mathbf{L}^{1/2}$ are uniquely determined, and they are called the *singular values* of $\widehat{\mathbf{T}}$. The columns of $\mathbf{V}_1$ are orthonormal vectors called *left singular vectors* of $\widehat{\mathbf{T}}$, and the columns of $\mathbf{V}_2$ are called *right singular vectors* (Watkins 2002).

Estimators $\widehat{\mathbf{b}}_{E_1}$ and $\widehat{\lambda}^2_{E_1}$ of the first eigenvector $\mathbf{b}_{E_1}$ and the first eigenvalue $\lambda^2_{E_1}$ respectively are the first column of matrix $\mathbf{V}_1$ and the first diagonal element of matrix $\mathbf{L}^{1/2}$. Thus, because $\widehat{\mathbf{T}}\widehat{\mathbf{T}}'$ is a symmetric matrix, the maximum likelihood estimators $\widehat{\lambda}^2_{E_1}$ and $\widehat{\mathbf{b}}_{E_1}$ in the ESIM context can be obtained from

$$\left(\widehat{\mathbf{T}}\widehat{\mathbf{T}}' - \widehat{\mu}_j\mathbf{I}\right)\widehat{\mathbf{b}}_{E_j} = \mathbf{0}, \tag{7.23}$$

where $\widehat{\mu}_j = \widehat{\lambda}^4_{E_j}$, $j = 1, 2, \ldots, t$. In the asymptotic context, $\widehat{\lambda}^2_{E_1}$ and $\widehat{\mathbf{b}}_{E_1}$ are consistent and unbiased estimators (Anderson 2003).

The latter results allow the ESIM index ($I_E = \mathbf{b}_E'\mathbf{y}$) as $\widehat{I}_E = \widehat{\mathbf{b}}'_{E_1}\mathbf{y}$ to be estimated. The estimator of the maximized ESIM selection response and expected genetic gain per trait are $\widehat{R}_E = k_I\sqrt{\widehat{\mathbf{b}}'_{E_1}\widehat{\mathbf{P}}\widehat{\mathbf{b}}_{E_1}}$ and $\widehat{\mathbf{E}}_E = k_I\dfrac{\widehat{\mathbf{C}}\widehat{\mathbf{b}}_{E_1}}{\sqrt{\widehat{\mathbf{b}}'_{E_1}\widehat{\mathbf{P}}\widehat{\mathbf{b}}_{E_1}}}$ respectively, whereas the estimator of the maximized ESIM accuracy is $\widehat{\lambda}_{E_1}$, which should be similar to the estimator of the square root of the maximized ESIM heritability.

In the asymptotic context, the estimator of $\mathbf{b}_{Ej}$ ($\widehat{\mathbf{b}}_{Ej}$) has multivariate normal distribution with expectation $E(\widehat{\mathbf{b}}_{Ej}) = \mathbf{b}_{Ej}$ and variance

$$Var(\widehat{\mathbf{b}}_{Ej}) = \frac{1}{2n}\mathbf{b}_{Ej}\mathbf{b}'_{Ej} + \frac{1}{n}\left(1 - \lambda^2_{Ej}\right)\sum_{i \neq j}^{t}\frac{\lambda^2_{Ej} + \lambda^2_{Ei} - 2\lambda^2_{Ei}\lambda^2_{Ej}}{\left(\lambda^2_{Ei} - \lambda^2_{Ej}\right)^2}\mathbf{b}_{Ei}\mathbf{b}'_{Ei}, \tag{7.24}$$

and, for $i \neq j$, the covariance between $\widehat{\mathbf{b}}_{Ei}$ and $\widehat{\mathbf{b}}_{Ej}$ can be written as

$$Cov(\widehat{\mathbf{b}}_{Ei}, \widehat{\mathbf{b}}_{Ej}) = \frac{\left(1 - \lambda^2_{Ej}\right)\left(1 - \lambda^2_{Ei}\right)\left(\lambda^2_{Ei} + \lambda^2_{Ej}\right)}{n\left(\lambda^2_{Ei} - \lambda^2_{Ej}\right)^2}\mathbf{b}_{Ej}\mathbf{b}'_{Ei}, \tag{7.25}$$

where n is the number of individuals or genotypes (Anderson 1999). The variance of $\widehat{\mathbf{b}}_{Ej}$ and the covariance between $\widehat{\mathbf{b}}_{Ei}$ and $\widehat{\mathbf{b}}_{Ej}$ depend not only on n, but also on eigenvalues λ^2_{Ei} and λ^2_{Ej}. Suppose that $\lambda^2_{Ej} > \lambda^2_{Ei}$; then, when λ^2_{Ej} is very close to 1, $Var(\widehat{\mathbf{b}}_{Ej}) \approx \frac{1}{2n}\mathbf{b}_{Ej}\mathbf{b}'_{Ej}$ ("$\approx$" denotes an approximation) and $Cov(\widehat{\mathbf{b}}_{Ei}, \widehat{\mathbf{b}}_{Ej})$ is very close to 0. By the result of Eq. (7.24), the variance of the first eigenvector ($\widehat{\mathbf{b}}_{E1}$) of $\widehat{\mathbf{P}}^{-1}\widehat{\mathbf{C}}$ can be written as $Var(\widehat{\mathbf{b}}_{E1}) = \frac{1}{2n}\mathbf{b}_{E1}\mathbf{b}'_{E1} + \frac{1}{n}\left(1 - \lambda^2_{E1}\right)\sum_{j=2}^{t}\frac{\lambda^2_{E1} + \lambda^2_{Ej} - 2\lambda^2_{E1}\lambda^2_{Ej}}{\left(\lambda^2_{E1} - \lambda^2_{Ej}\right)^2}\mathbf{b}_{Ej}\mathbf{b}'_{Ej}$. If the first eigenvalue λ^2_{E1} of $\mathbf{P}^{-1}\mathbf{C}$ is very close to 1 ($\lambda^2_{E1} \approx 1$), $Var(\widehat{\mathbf{b}}_{E1}) = \frac{1}{2n}\mathbf{b}_{E1}\mathbf{b}'_{E1}$ and $Cov(\widehat{\mathbf{b}}_{E1}, \widehat{\mathbf{b}}_{Ej}) \approx 0$.

In the asymptotic context, the jth estimator ($\widehat{\lambda}_{Ej}$) of the canonical correlations has normal distribution with expectation $E(\widehat{\lambda}_{Ej}) \approx \lambda_{Ej}$ and variance

$$Var(\widehat{\lambda}_{Ej}) \approx \frac{\left(1 - \lambda^2_{Ej}\right)^2}{n}, \tag{7.26}$$

whereas the jth estimator of the square of the canonical correlations $\widehat{\lambda^2_{Ej}}$ has normal distribution with expectation $E(\widehat{\lambda^2_{Ej}}) \approx \lambda^2_{Ej}$ and variance

$$Var(\widehat{\lambda^2_j}) \approx \frac{4\lambda^2_{Ej}\left(1 - \lambda^2_{Ej}\right)^2}{n}. \tag{7.27}$$

In addition, for $i \neq j$, the correlation between $\widehat{\lambda}^2_{Ej}$ and $\widehat{\lambda}^2_{Ei}$ is zero, i.e., $Corr\left(\widehat{\lambda}^2_{Ei}, \widehat{\lambda}^2_{Ej}\right) = 0$ (Bilodeau and Brenner 1999; Muirhead 2005).

Equation (7.26) implies that under the restrictions $\sigma^2_H = 1$ and $\sigma^2_I = 1$, the expectation and variance of $\widehat{R}_E = k_I\widehat{\lambda}_{E1}$ are $E\left(\widehat{R}_E\right) \approx k_I\lambda_{E1}$ and $Var\left(\widehat{R}_E\right) \approx \frac{k_I^2\left(1-\lambda^2_{E1}\right)^2}{n}$ respectively. However, obtaining the expectation and variance of $\widehat{R}_E = k_I \widehat{\sigma}_H\widehat{\lambda}_{E1}$ or $\widehat{R}_E = k_I\sqrt{\widehat{\mathbf{b}}'_{E1}\widehat{\mathbf{P}}\widehat{\mathbf{b}}_{E1}}$ is more difficult, because in both equations there are two estimators: $\widehat{\sigma}_H$ and $\widehat{\lambda}_1$ in the first one, and $\widehat{\mathbf{P}}$ and $\widehat{\mathbf{b}}_{E1}$ in the second one.

7.1.5 Numerical Examples

We compare ESIM efficiency versus LPSI efficiency using a real data set from commercial egg poultry lines obtained from Akbar et al. (1984). The estimated phenotypic ($\widehat{\mathbf{P}}$) and genetic ($\widehat{\mathbf{C}}$) covariance matrices among the rate of lay (RL, number of eggs), age at sexual maturity (SM, days) and egg weight (EW, kg), were $\widehat{\mathbf{P}} = \begin{bmatrix} 240.57 & -95.62 & 2.07 \\ -95.62 & 167.20 & 4.58 \\ 2.07 & 4.58 & 22.80 \end{bmatrix}$ and $\widehat{\mathbf{C}} = \begin{bmatrix} 29.86 & -17.90 & -4.13 \\ -17.90 & 18.56 & 1.49 \\ -4.13 & 1.49 & 9.24 \end{bmatrix}$ respectively. The number of genotypes and the vector of economic weights were n= 3330 and $\mathbf{w}' = \begin{bmatrix} 19.54 & -3.56 & 17.01 \end{bmatrix}$ respectively, whereas the selection intensity was 10% ($k_I = 1.755$) for both indices.

The estimated LPSI vector of coefficients was $\widehat{\mathbf{b}}'_S = \mathbf{w}'\widehat{\mathbf{P}}^{-1}\widehat{\mathbf{C}} = \begin{bmatrix} 1.82 & -1.38 & 3.25 \end{bmatrix}$, whereas the estimated selection response, expected genetic gain per trait, accuracy, and heritability of the LPSI were $\widehat{R}_S = 1.755\sqrt{\widehat{\mathbf{b}}'_S\widehat{\mathbf{P}}\widehat{\mathbf{b}}_S} = 74.91$, $\quad \widehat{\mathbf{E}}'_S = 1.755\frac{\widehat{\mathbf{b}}'_S\widehat{\mathbf{C}}}{\sqrt{\widehat{\mathbf{b}}'_S\widehat{\mathbf{P}}\widehat{\mathbf{b}}_S}} = \begin{bmatrix} 2.70 & -2.20 & 0.84 \end{bmatrix}$,

$\widehat{\rho}_S = \frac{\sqrt{\widehat{\mathbf{b}}'_S\widehat{\mathbf{P}}\widehat{\mathbf{b}}_S}}{\sqrt{\mathbf{w}'\widehat{\mathbf{C}}\mathbf{w}}} = 0.362$, and $\widehat{h}^2_S = \frac{\widehat{\mathbf{b}}'_S\widehat{\mathbf{C}}\widehat{\mathbf{b}}_S}{\widehat{\mathbf{b}}'_S\widehat{\mathbf{P}}\widehat{\mathbf{b}}_S} = 0.143$ respectively.

Note that because in the ESIM context $\widehat{\mathbf{b}}'_E\widehat{\mathbf{b}}_E = 1$, the best way of comparing ESIM results versus LPSI results is when the LPSI coefficient vector is normalized, i.e., when the LPSI coefficient vector is equal to $\widehat{\mathbf{b}}_S{}^* = \widehat{\mathbf{b}}_S/\widehat{\mathbf{b}}'_S\widehat{\mathbf{b}}_S$ and then $\widehat{\mathbf{b}}'^*_S\widehat{\mathbf{b}}^*_S = 1$; however, it can be shown that the normalization process only affects the estimated LPSI selection response because in that case, $\widehat{R}_S = 74.91$ is divided by $\widehat{\mathbf{b}}'_S\widehat{\mathbf{b}}_S$. For example, for this data set result, $\widehat{\mathbf{b}}'_S\widehat{\mathbf{b}}_S = 15.76$; then, the estimated LPSI selection response using $\widehat{\mathbf{b}}_S{}^* = \widehat{\mathbf{b}}_S/\widehat{\mathbf{b}}'_S\widehat{\mathbf{b}}_S$ is $\widehat{R}_S = \frac{74.91}{15.74} = 4.75$, whereas the rest of the estimated LPSI parameters are the same. When $0 < \widehat{\mathbf{b}}'_S\widehat{\mathbf{b}}_S < 1$ and $1 < \widehat{R}_S$, the values of $\widehat{R}_S$ increase, but when $1 < \widehat{\mathbf{b}}'_S\widehat{\mathbf{b}}_S$, the values of $\widehat{R}_S$ decrease, as in the example.

The product $\widehat{\mathbf{b}}'_S\widehat{\mathbf{b}}_S$ does not affect $\widehat{\rho}_S$ because it is invariant to scale change. Also, $\widehat{\mathbf{b}}'_S\widehat{\mathbf{b}}_S$ does not affect $\widehat{h}^2_S$ and $\widehat{\mathbf{E}}_S$ because $\widehat{\mathbf{b}}'_S\widehat{\mathbf{b}}_S$ appears in the numerator and denominator of both estimated parameters.

In the ESIM, the sign and proportion of the expected genetic gain values for traits RL, SM, and EW should be in accordance with the breeder's interest. For example, if the breeder's interest is that the expected genetic gain per trait for RL should be positive and negative for SM, the sign and proportion of the values of the first eigenvector should be modified using a linear combination of the estimated first eigenvector $\widehat{\mathbf{b}}_{E_1}$, i.e., $\boldsymbol{\beta} = \mathbf{F}\widehat{\mathbf{b}}_{E_1}$, to achieve expected genetic gain per trait values in RL and SM according to the breeder's interest.

The information needed to obtain the estimated ESIM parameters are matrices $\widehat{\mathbf{T}} = \widehat{\mathbf{P}}^{-1}\widehat{\mathbf{C}} = \begin{bmatrix} 0.1102 & -0.0405 & -0.0280 \\ -0.0390 & 0.0864 & -0.0184 \\ -0.1833 & 0.0517 & 0.4115 \end{bmatrix}$ and $\widehat{\mathbf{T}}\widehat{\mathbf{T}}' = \begin{bmatrix} 0.0146 & -0.0073 & -0.0338 \\ -0.0073 & 0.0093 & 0.0041 \\ -0.0338 & 0.0041 & 0.2056 \end{bmatrix}$.

We need to find the eigenvalues and eigenvectors of equation $(\widehat{\mathbf{T}}\widehat{\mathbf{T}}' - \widehat{\mu}_j\mathbf{I})\widehat{\mathbf{b}}_{E_j} = \mathbf{0}$, where $\widehat{\mu}_j = \widehat{\lambda}^4_{E_j}$, to obtain matrices $\mathbf{V}_1$ and $\mathbf{L}^{1/2}$, which form matrix $\widehat{\mathbf{T}} = \mathbf{V}_1\mathbf{L}^{1/2}\mathbf{V}'_2$.

Matrix $\mathbf{V}_1$ is equal to $\mathbf{V}_1 = \begin{bmatrix} -0.1701 & 0.6818 & 0.7115 \\ 0.0259 & -0.7187 & 0.6948 \\ 0.9851 & 0.1366 & 0.1046 \end{bmatrix}$, whereas the diagonal elements of matrix $\mathbf{L}$ are 0.2115, 0.0155, and 0.0025, that is, matrix $\mathbf{L}^{1/2} = \begin{bmatrix} 0.4599 & 0 & 0 \\ 0 & 0.1244 & 0 \\ 0 & 0 & 0.0498 \end{bmatrix}$. Thus, $\widehat{\mu}_1 = \widehat{\lambda}^4_{E_1} = 0.2115$, $\widehat{\lambda}^2_{E_1} = 0.4599$, and the estimated ESIM accuracy was $\widehat{\lambda}_{E_1} = 0.6782$. The estimated ESIM eigenvector of coefficients is the first column of matrix $\mathbf{V}_1$, i.e., $\widehat{\mathbf{b}}'_{E_1} = [-0.1701 \quad 0.0259 \quad 0.9851]$, and the estimated ESIM index can be constructed as $\widehat{I}_E = -0.1701\text{RL} + 0.0259\text{SM} + 0.9851\text{EW}$.

The estimated ESIM selection response and expected genetic gain per trait were $\widehat{R}_E = 1.755\sqrt{\widehat{\mathbf{b}}'_{E_1}\widehat{\mathbf{P}}\widehat{\mathbf{b}}_{E_1}} = 9.54$ and $\widehat{\mathbf{E}}'_E = 1.755\dfrac{\widehat{\mathbf{b}}'_{E_1}\widehat{\mathbf{C}}}{\sqrt{\widehat{\mathbf{b}}'_{E_1}\widehat{\mathbf{P}}\widehat{\mathbf{b}}_{E_1}}} = [-3.10 \quad 1.61 \quad 3.18]$ respectively. Because the estimated LPSI selection response was $\widehat{R}_S = \dfrac{74.91}{15.74} = 4.75$, the estimated ESIM selection response was higher than the estimated LPSI response. In addition, the estimated LPSI expected genetic gain per trait was $\widehat{\mathbf{E}}'_S = [2.70 \quad -2.20 \quad 0.84]$. Now, suppose that the breeder's interest is to increase RL and decrease SM; then, $\widehat{\mathbf{E}}'_S$ is a good result but $\widehat{\mathbf{E}}'_E$ is wrong.

We can change the sign and proportion of $\widehat{\mathbf{E}}'_E$ by transforming $\widehat{\mathbf{b}}_{E_1}$ into $\boldsymbol{\beta} = \mathbf{F}\widehat{\mathbf{b}}_{E_1}$ using a convenient matrix $\mathbf{F}$ such as $\mathbf{F} = \begin{bmatrix} -9 & 0 & 0 \\ 0 & 1 & 0 \\ 0 & 0 & 1 \end{bmatrix}$. In such a case

$\widehat{\boldsymbol{\beta}}' = \widehat{\mathbf{b}}'_{E_1}\mathbf{F} = [1.531 \quad 0.026 \quad 0.981]$, $\widehat{R}_E = 1.755\sqrt{\widehat{\boldsymbol{\beta}}'\widehat{\mathbf{P}}\widehat{\boldsymbol{\beta}}} = 42.44$, and $\widehat{\mathbf{E}}'_E = 1.755\frac{\widehat{\boldsymbol{\beta}}'\widehat{\mathbf{C}}}{\sqrt{\widehat{\boldsymbol{\beta}}'\widehat{\mathbf{P}}\widehat{\boldsymbol{\beta}}}} = [2.990 \quad -1.85 \quad 0.205]$. However, vector $\widehat{\boldsymbol{\beta}}'$ was not normalized. To normalize $\widehat{\boldsymbol{\beta}}'$ we need to divide it by $\widehat{\boldsymbol{\beta}}'\widehat{\boldsymbol{\beta}} = 3.314$, but $\widehat{\boldsymbol{\beta}}'\widehat{\boldsymbol{\beta}}$ should only affect $\widehat{R}_E = 42.44$, which should be divided by 3.314, that is, $\widehat{R}_E = \frac{42.44}{3.314} = 12.806$. According to the theory of similar matrices (Harville 1997), the estimated maximized ESIM accuracy, $\widehat{\lambda}_{E_1} = 0.6782$, should not be affected by matrix **F**.

We can compare ESIM efficiency versus LPSI efficiency to predict the net genetic merit using the ratio of the estimated ESIM accuracy $\widehat{\lambda}_{E_1} = 0.6782$ to LPSI accuracy $\widehat{\rho}_S = 0.362$, i.e., $\frac{\widehat{\lambda}_{E_1}}{\widehat{\rho}_S} = \frac{0.6782}{0.362} = 1.873$, or in percentage terms, $\widehat{p}_E = 100(1.873 - 1) = 87.3$ (see Chap. 5, Eq. 5.17). According to the latter result, the ESIM is a better predictor of the net genetic merit and its efficiency is 87.3% higher than that of the LPSI for this data set.

Now, we compare ESIM efficiency versus LPSI efficiency using the data set described in Sect. 2.8.1 of Chap. 2. From this data set, we ran five phenotypic selection cycles, each with four traits (T_1, T_2, T_3, and T_4), 500 genotypes, and four replicates for each genotype. The economic weights for T_1, T_2, T_3, and T_4 were $1, -1, 1$, and 1 respectively. In this case, matrix **F** is an identity matrix of size 4×4 for all five selection cycles.

Table 7.1 presents the estimated LPSI, the restricted LPSI (RLPSI), and the predetermined proportional gain LPSI (PPG-LPSI) selection response (the latter two for one, two, and three restrictions) for five simulated selection cycles when their vectors of coefficients are normalized. Table 7.1 also presents the estimated ESIM, the RESIM and the PPG-ESIM selection response for one, two, and three restrictions for five simulated selection cycles. The selection intensity was 10% ($k_I = 1.755$) for all five selection cycles. In this subsection, we compare only LPSI results versus ESIM results. The estimated LPSI selection response when the vector of coefficients was not normalized was described in Chap. 2 (Table 2.4). The averages of the estimated LPSI and ESIM selection responses were 4.70 and 6.31 respectively.

Table 7.2 presents the estimated ESIM expected genetic gain per trait, accuracy ($\widehat{\rho}_E$), and the values $\widehat{p}_E = 100(\widehat{\lambda}_E - 1)$, where $\widehat{\lambda}_E = \widehat{\rho}_E/\widehat{\rho}_S$ is the ratio of $\widehat{\rho}_E$ to the estimated LPSI accuracy ($\widehat{\rho}_S$), expressed as percentages. Table 7.2 also presents the accuracy of the PPG-ESIM and the estimated ratio ($\widehat{p}_{PE}$) of the estimated PPG-ESIM accuracy to the estimated PPG-LPSI accuracy, expressed as percentages, for one, two, and three predetermined restrictions for five simulated selection cycles. In this subsection, we use only the estimated ESIM expected genetic gain per trait and $\widehat{p}_E = 100(\widehat{\lambda}_E - 1)$ to compare ESIM efficiency versus LPSI efficiency.

The estimated LPSI expected genetic gains per trait were presented in Chap. 2, Table 2.4. According to the results shown in Table 2.4, the averages of the estimated

Table 7.1 Estimated linear phenotypic selection index (LPSI), restricted null LPSI (RLPSI), and predetermined proportional gains LPSI (PPG-LPSI) selection responses when their vectors of coefficients are normalized; estimated eigen selection index method (ESIM), restricted null ESIM (RESIM), and predetermined proportional gain ESIM (PPG-ESIM) selection responses for one, two, and three restrictions for five simulated selection cycles

		RLPSI response for one, two, and three null restrictions			PPG-LPSI response for one, two, and three predetermined restrictions		
Cycle	LPSI response	1	2	3	1	2	3
1	4.78	4.79	4.44	5.06	4.78	5.41	3.18
2	4.84	4.51	4.39	5.15	4.84	5.19	3.35
3	4.59	4.51	4.39	5.26	4.59	4.83	3.53
4	4.80	4.15	4.06	4.71	4.80	4.96	2.64
5	4.48	4.19	4.22	4.41	4.48	4.14	2.99
Average	4.70	4.43	4.30	4.92	4.70	4.91	3.14
		RESIM response for one, two, and three null restrictions			PPG-ESIM response for one, two, and three predetermined restrictions		
Cycle	ESIM response	1	2	3	1	2	3
1	8.88	4.78	4.64	4.57	8.88	7.1	7.4
2	6.13	4.86	4.69	4.69	6.13	6.04	7.3
3	5.44	4.96	4.79	4.68	5.44	5.87	6.91
4	4.84	4.30	4.19	4.19	4.84	4.91	5.77
5	6.24	3.79	3.78	3.78	6.24	7.49	6.39
Average	6.31	4.54	4.42	4.38	6.31	6.28	6.75

LPSI expected genetic gain per trait T1, T2, T3, and T4 for five simulated selection cycles were 7.26, −3.52, 2.78, and 1.58, whereas according to the results of Table 7.2, the averages of the estimated ESIM expected genetic gains per trait were 5.67, −2.67, 1.81, and 2.9 respectively. This means that the estimated LPSI expected genetic gain for traits T1, T2, and T3 was higher than the estimated ESIM expected genetic gain for those traits.

The average of the $\widehat{p}_E = 100\left(\widehat{\lambda}_E - 1\right)$ values was 9.76 for all five selection cycles (Table 7.2). The latter result is not in accordance with the LPSI and ESIM expected genetic gain per trait; however, note that the $\widehat{p}_E$ values are associated with the estimated LPSI and ESIM selection responses (Table 7.1), not with the expected genetic gain per trait, because $\widehat{\lambda}_E = \frac{\widehat{\rho}_E}{\widehat{\rho}_S} \approx \frac{\widehat{R}_E}{\widehat{R}_S}$, where $\widehat{R}_E$ and $\widehat{R}_S$ are the estimated ESIM and LPSI selection responses respectively. Thus, the $\widehat{p}_E$ values indicate that the efficiency of the ESIM and that of the LPSI were very similar because the former was only 9.76% higher than the latter for this data set.

The equality $\frac{\widehat{\rho}_E}{\widehat{\rho}_S} = \frac{\widehat{R}_E}{\widehat{R}_S}$ is true only when the denominators of both estimated correlations are the same, as in the linear selection indices described in Chaps. 2–6.

Table 7.2 Estimated eigen selection index method (ESIM) expected genetic gain per trait, accuracy ($\widehat{\rho}_E$), and ratio of $\widehat{\rho}_E$ to the estimated LPSI (data not presented) accuracy ($\widehat{\rho}_S$), expressed in percentage terms, $\widehat{p}_E = 100(\widehat{\lambda}_E - 1)$ (where $\widehat{\lambda}_E = \widehat{\rho}_E/\widehat{\rho}_S$)

Cycle	ESIM expected genetic gain per trait				ESIM accuracy	$\widehat{p}_E$ values (in %)
	T1	T2	T3	T4		
1	7.81	−4.62	3.11	2.21	0.98	8.11
2	5.15	−2.98	2.31	3.48	0.96	9.34
3	4.74	−1.15	0.66	3.79	0.97	10.94
4	3.94	−2.44	0.74	3.34	0.95	10.04
5	6.68	−2.15	2.24	2.05	0.95	10.35
Average	5.67	−2.67	1.81	2.97	0.96	9.76

Cycle	PPG-ESIM accuracies for one, two, and three predetermined restrictions			$\widehat{p}_P$ values (in %) for one, two, and three predetermined restrictions		
	1	2	3	1	2	3
1	0.98	0.96	0.99	9.34	8.90	20.99
2	0.96	0.96	0.98	10.94	12.46	25.20
3	0.97	0.97	1.00	10.04	9.71	41.43
4	0.95	0.94	0.99	10.35	13.98	28.95
5	0.98	0.96	0.99	9.34	8.90	20.99
Average	0.96	0.96	0.99	9.76	11.71	29.03

Estimated PPG-ESIM accuracy ($\widehat{\rho}_P$) and estimated ratio ($\widehat{p}_P$) of the $\widehat{\rho}_P$ to the estimated accuracy of the PPG-LPSI (data not presented), expressed in percentages (%), for one, two, and three predetermined restrictions for five simulated selection cycles

Note that $\widehat{\rho}_S = \frac{\sqrt{\widehat{\mathbf{b}}'_S\widehat{\mathbf{P}}\widehat{\mathbf{b}}_S}}{\sqrt{\mathbf{w}'\widehat{\mathbf{C}}\mathbf{w}}}$ and $\widehat{\rho}_E = \frac{\sqrt{\widehat{\mathbf{b}}'_E\widehat{\mathbf{P}}\widehat{\mathbf{b}}_E}}{\sqrt{\mathbf{w}'_E\widehat{\mathbf{C}}\mathbf{w}_E}}$, whereas $\widehat{R}_S = \sqrt{\widehat{\mathbf{b}}'_S\widehat{\mathbf{P}}\widehat{\mathbf{b}}_S}$ and $\widehat{R}_E = \sqrt{\widehat{\mathbf{b}}'_E\widehat{\mathbf{P}}\widehat{\mathbf{b}}_E}$; this means that if $\sqrt{\mathbf{w}'_E\widehat{\mathbf{C}}\mathbf{w}_E} \neq \sqrt{\mathbf{w}'\widehat{\mathbf{C}}\mathbf{w}}$, $\frac{\widehat{\rho}_E}{\widehat{\rho}_S} \neq \frac{\widehat{R}_E}{\widehat{R}_S}$. For the Akbar et al. (1984) data, $\widehat{R}_E = 9.54$ and $\widehat{R}_S = 4.75$, then $\frac{\widehat{R}_E}{\widehat{R}_S} = 2.0$ but $\frac{\widehat{\lambda}_{E_1}}{\widehat{\rho}_S} = 1.873$; that is, $\frac{\widehat{\rho}_E}{\widehat{\rho}_S} \approx \frac{\widehat{R}_E}{\widehat{R}_S}$, where "≈" indicates an approximation.

Figure 7.1 presents the frequency distribution of 500 estimated ESIM values for cycle 2 (Fig. 7.1a) and cycle 5 (Fig. 7.1b), obtained from one selection cycle for 500 genotypes and four traits simulated in one environment. Figure 7.1a, b indicates that the frequency distribution of the estimated ESIM values approaches normal distribution.

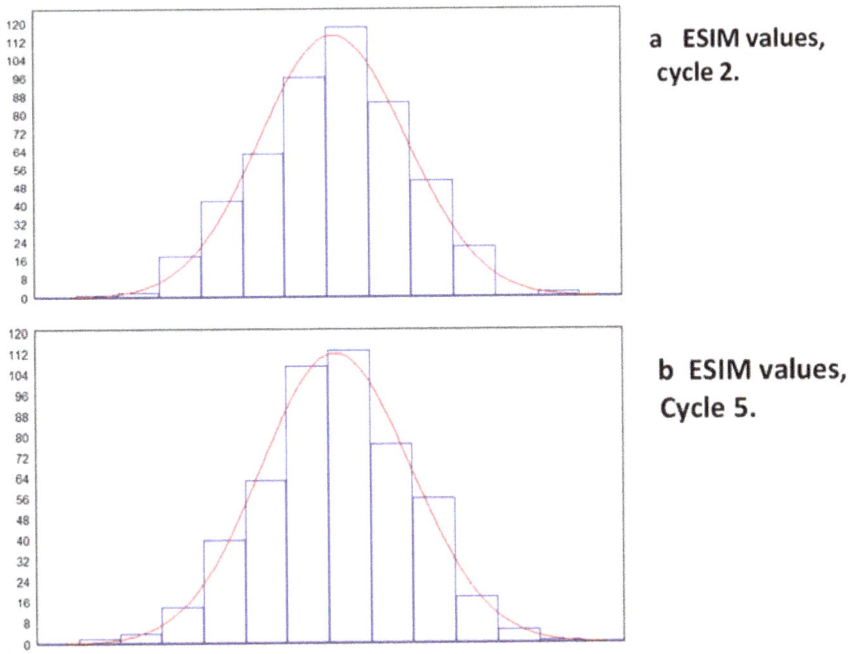

Fig. 7.1 Frequency distribution of 500 estimated eigen selection index method (ESIM) values for (**a**) cycle 2 and (**b**) cycle 5, obtained from one selection cycle for 500 genotypes and four traits simulated in one environment

7.2 The Linear Phenotypic Restricted Eigen Selection Index Method

Similar to the RLPSI (see Chap. 2), the objective of the RESIM is to fix r of t ($r < t$) traits by predicting only the genetic gains of ($t - r$) of them. Let $H = \mathbf{w}'\mathbf{g}$ be the net genetic merit and $I = \mathbf{b}'\mathbf{y}$ the ESIM index. In Chap. 2, we showed that $Cov(I, \mathbf{g}) = \mathbf{Cb}$ is the covariance between the breeding value vector ($\mathbf{g}$) and $I = \mathbf{b}'\mathbf{y}$. Thus, to fix r of t traits, we need r covariances between the linear combinations of $\mathbf{g}$ ($\mathbf{U}'\mathbf{g}$) and $I = \mathbf{b}'\mathbf{y}$ to be zero, i.e., $Cov(I, \mathbf{U}'\mathbf{g}) = \mathbf{U}'\mathbf{Cb} = \mathbf{0}$, where $\mathbf{U}'$ is a matrix with 1s and 0s (1 indicates that the trait is restricted and 0 that the trait has no restrictions). In the RESIM, it is possible to solve this problem by maximizing $\rho_{HI}^2 = \frac{(\mathbf{w}'\mathbf{Cb})^2}{(\mathbf{w}'\mathbf{Cw})(\mathbf{b}'\mathbf{Pb})}$ with respect to vectors $\mathbf{b}$ and $\mathbf{w}$ under the restrictions $\mathbf{U}'\mathbf{Cb} = \mathbf{0}$, $\mathbf{b}'\mathbf{b} = 1$, $\mathbf{w}'\mathbf{Cw} = 1$, and $\mathbf{b}'\mathbf{Pb} = 1$, where $\mathbf{w}'\mathbf{Cw}$ is the variance of $H = \mathbf{w}'\mathbf{g}$ and $\mathbf{b}'\mathbf{Pb}$ is the variance of $I = \mathbf{b}'\mathbf{y}$. Also, the RESIM problem can be solved by maximizing $\frac{\mathbf{b}'\mathbf{Cb}}{\sqrt{\mathbf{b}'\mathbf{Pb}}}$ (Eq. 7.12) with respect to vectors $\mathbf{b}$ only under the restrictions $\mathbf{U}'\mathbf{Cb} = \mathbf{0}$ and $\mathbf{b}'\mathbf{b} = 1$, as we did to obtain Eq. (7.13). Both approaches give the same result, but it is easier to work with the second approach than with the first one.

7.2.1 The RESIM Parameters

To obtain the RESIM vector of coefficients that maximizes the RESIM selection response and the expected genetic gain per trait, we need to maximize the function

$$f(\mathbf{b}, \mathbf{v}') = \frac{\mathbf{b}'\mathbf{C}\mathbf{b}}{\sqrt{\mathbf{b}'\mathbf{P}\mathbf{b}}} - \mathbf{v}'\mathbf{U}'\mathbf{C}\mathbf{b} \tag{7.28a}$$

with respect to $\mathbf{b}$ and $\mathbf{v}'$, where $\mathbf{v}' = \begin{bmatrix} v_1 & v_2 & \cdots & v_{r-1} \end{bmatrix}$ is a vector of Lagrange multipliers. The derivatives of Eq. (7.28a) with respect to $\mathbf{b}$ and $\mathbf{v}'$ can be written as

$$2(\mathbf{b}'\mathbf{P}\mathbf{b})^{1/2}\mathbf{C}\mathbf{b} - (\mathbf{b}'\mathbf{P}\mathbf{b})^{-1/2}(\mathbf{b}'\mathbf{C}\mathbf{b})\mathbf{P}\mathbf{b} - \mathbf{C}\mathbf{U}\mathbf{v} = \mathbf{0} \tag{7.28b}$$

and

$$\mathbf{U}'\mathbf{C}\mathbf{b} = \mathbf{0}, \tag{7.29}$$

respectively, where Eq. (7.29) denotes the restriction imposed for maximizing Eq. (7.28a). Using algebraic methods on Eq. (7.28b) similar to those used to obtain Eqs. (7.10) and (7.13), we get

$$\left(\mathbf{K}\mathbf{P}^{-1}\mathbf{C} - h^2_{I_R}\mathbf{I}_t\right)\mathbf{b}_R = \mathbf{0}, \tag{7.30}$$

where $\mathbf{K} = [\mathbf{I}_t - \mathbf{Q}_R]$, $\mathbf{I}_t$ is an identity matrix of size $t \times t$, $\mathbf{Q}_R = \mathbf{P}^{-1}\mathbf{C}\mathbf{U}(\mathbf{U}'\mathbf{C}\mathbf{P}^{-1}\mathbf{C}\mathbf{U})^{-1}\mathbf{U}'\mathbf{C}$, and $h^2_{I_R} = \frac{\mathbf{b}'_R\mathbf{C}\mathbf{b}_R}{\mathbf{b}'_R\mathbf{P}\mathbf{b}_R}$ is the maximized RESIM heritability obtained under the restriction $\mathbf{U}'\mathbf{C}\mathbf{b} = \mathbf{0}$; $h^2_{I_R}$ is also the square of the maximized correlation between the net genetic merit and $I_R = \mathbf{b}'_R\mathbf{y}$, that is, $h^2_{I_R} = \lambda^2_R$. This means that Eq. (7.30) can be written as

$$\left(\mathbf{K}\mathbf{P}^{-1}\mathbf{C} - \lambda^2_R\mathbf{I}_t\right)\mathbf{b}_R = \mathbf{0}. \tag{7.31}$$

Thus, the optimized RESIM index is $I = \mathbf{b}'_R\mathbf{y}$. The only difference between Eqs. (7.31) and (7.13) is matrix $\mathbf{K}$. Equation (7.31) was obtained by Cerón-Rojas et al. (2008) by maximizing ρ^2_{HI} (Eq. 7.1) with respect to vectors $\mathbf{b}$ and $\mathbf{w}$ under the restriction $\mathbf{U}'\mathbf{C}\mathbf{b} = \mathbf{0}$, $\mathbf{b}'\mathbf{b} = 1$, $\mathbf{w}'\mathbf{C}\mathbf{w} = 1$ and $\mathbf{b}'\mathbf{P}\mathbf{b} = 1$ in a similar manner to the canonical correlation theory. The RESIM expected genetic gain per trait uses the first eigenvector ($\mathbf{b}_R$) of matrix $\mathbf{K}\mathbf{P}^{-1}\mathbf{C}$, whereas the RESIM selection response uses $\mathbf{b}_R$ and the first eigenvalue (λ^2_R) of matrix $\mathbf{K}\mathbf{P}^{-1}\mathbf{C}$. When $\mathbf{U}'$ is a null matrix, $\mathbf{b}_R = \mathbf{b}_E$ (the vector of the ESIM coefficients); thus, the RESIM is more general than the ESIM and includes the ESIM as a particular case.

In the RESIM context, vector $\mathbf{w}$ can be obtained (Cerón-Rojas et al. 2008) as

$$\mathbf{w}_R = \mathbf{C}^{-1}[\lambda_R \mathbf{P}\mathbf{b}_R + \mathbf{\Psi}\mathbf{v}], \tag{7.32}$$

where λ_R and $\mathbf{b}_R$ are the square roots of the first eigenvalue (λ_R^2) and the first eigenvector of matrix $\mathbf{K}\mathbf{P}^{-1}\mathbf{C}$ respectively; $\mathbf{\Psi} = \mathbf{C}\mathbf{U}$ and $\mathbf{v} = \lambda_R^{-1}\left(\mathbf{\Psi}'\mathbf{P}^{-1}\mathbf{\Psi}\right)^{-1}\mathbf{\Psi}'\mathbf{P}^{-1}\mathbf{C}\mathbf{b}_R$. Let $H_R = \mathbf{w}_R'\mathbf{g}$ be the net genetic merit in the RESIM context; then, because the correlation between $I_R = \mathbf{b}_R'\mathbf{y}$ and $H_R = \mathbf{w}_R'\mathbf{g}$ is not affected by scale change, λ_R and λ_R^{-1} can be considered proportional constants and then $\mathbf{\Psi}\mathbf{v}$ can be written as $\mathbf{\Psi}\mathbf{v} = \mathbf{\Psi}\left(\mathbf{\Psi}'\mathbf{P}^{-1}\mathbf{\Psi}\right)^{-1}\mathbf{\Psi}'\mathbf{P}^{-1}\mathbf{C}\mathbf{b}_R = \mathbf{Q}_R'\mathbf{C}\mathbf{b}_R$, where $\mathbf{Q}_R'$ is the transpose of matrix $\mathbf{Q}_R$ described in Eq. (7.30). Thus, another way of writing Eq. (7.32) is

$$\mathbf{w}_R = \mathbf{C}^{-1}\left[\mathbf{P} + \mathbf{Q}_R'\mathbf{C}\right]\mathbf{b}_R. \tag{7.33}$$

By Eq. (7.33) and the restriction $\mathbf{b}'\mathbf{\Psi} = \mathbf{0}$, the covariance between $I_R = \mathbf{b}_R'\mathbf{y}$ and $H_R = \mathbf{w}_R'\mathbf{g}$ ($\sigma_{H_R I_R}$) can be written as

$$\sigma_{H_R I_R} = \mathbf{w}_R'\mathbf{C}\mathbf{b}_R = \mathbf{b}_R'\mathbf{P}\mathbf{b}_R + \mathbf{b}_R'\mathbf{Q}_R'\mathbf{C}\mathbf{b}_R = \mathbf{b}_R'\mathbf{P}\mathbf{b}_R, \tag{7.34}$$

where $\mathbf{b}_R'\mathbf{Q}_R'\mathbf{C}\mathbf{b}_R = 0$ according to the restriction $\mathbf{b}'\mathbf{\Psi} = \mathbf{0}$. Equation (7.34) indicates that the covariance between I_R and H_R ($\sigma_{H_R I_R}$) is equal to the variance of I_R ($\sigma_{I_R}^2 = \mathbf{b}_R'\mathbf{P}\mathbf{b}_R$).

The maximized correlation between I_R and H_R (or RESIM accuracy) can be written as

$$\rho_{H_R I_R} = \frac{\sqrt{\mathbf{b}_R'\mathbf{P}\mathbf{b}_R}}{\sqrt{\mathbf{w}_R'\mathbf{C}\mathbf{w}_R}}, \tag{7.35}$$

where $\mathbf{w}_R'\mathbf{C}\mathbf{w}_R = \sigma_{H_R}^2$ is the variance of H_R, $\mathbf{w}_R = \mathbf{C}^{-1}\left[\mathbf{P} + \mathbf{Q}_R'\mathbf{C}\right]\mathbf{b}_R$, $\mathbf{Q}_R' = \mathbf{\Psi}\left(\mathbf{\Psi}'\mathbf{P}^{-1}\mathbf{\Psi}\right)^{-1}\mathbf{\Psi}'\mathbf{P}^{-1}$, and $\mathbf{\Psi} = \mathbf{C}\mathbf{U}$. When $\mathbf{U}'$ is a null matrix, $\mathbf{w}_R'\mathbf{C}\mathbf{w}_R = \mathbf{b}_E'\mathbf{P}\mathbf{C}^{-1}\mathbf{P}\mathbf{b}_E = \mathbf{w}_E'\mathbf{C}\mathbf{w}_E$, the variance of H_E, and $\sigma_{I_R}^2 = \mathbf{b}_R'\mathbf{P}\mathbf{b}_R = \mathbf{b}_E'\mathbf{P}\mathbf{b}_E = \sigma_{I_E}^2$, the variance of I_E. Hereafter, to simplify the notation, we write Eq. (7.35) as ρ_R or λ_R.

The maximized selection response (R_R) and expected genetic gain per trait ($\mathbf{E}_R$) of the RESIM can be written as

$$R_R = k_I\sqrt{\mathbf{b}_R'\mathbf{P}\mathbf{b}_R} \tag{7.36}$$

and

$$\mathbf{E}_R = k_I\frac{\mathbf{C}\mathbf{b}_R}{\sqrt{\mathbf{b}_R'\mathbf{P}\mathbf{b}_R}}, \tag{7.37}$$

respectively, where $\sqrt{\mathbf{b}_R'\mathbf{P}\mathbf{b}_R} = \sigma_{I_R}$ is the standard deviation of the variance of $I_R = \mathbf{b}_R'\mathbf{y}$. If vector $\mathbf{b}_R$ is transformed as $\boldsymbol{\beta}_R = \mathbf{F}\mathbf{b}_R$, where matrix $\mathbf{F}$ was defined earlier, vector $\mathbf{b}_R$ should be changed by $\boldsymbol{\beta}_R$ in Eqs. (7.36) and (7.37), and in $I_R = \mathbf{b}_R'\mathbf{y}$.

Equation (7.36) can also be written as $R_R = k_I\sigma_{H_R}\lambda_R$, where $\sigma_{H_R} = \sqrt{\mathbf{b}_R'\mathbf{P}\mathbf{C}^{-1}\mathbf{P}\mathbf{b}_R + \mathbf{b}_R'\mathbf{P}\mathbf{C}^{-1}\mathbf{Q}_R'\mathbf{C}\mathbf{b}_R}$ is the standard deviation of the variance of H_R, and $\lambda_R = \rho_{H_R I_R}$ is the first canonical correlation between $H_R = \mathbf{w}_R'\mathbf{g}$ and $I_R = \mathbf{b}_R'\mathbf{y}$. When $\sigma_{H_R} = 1$, λ_R is the covariance between $H_R = \mathbf{w}_R'\mathbf{g}$ and $I_R = \mathbf{b}_R'\mathbf{y}$, and then Eq. (7.36) can be written as $R_R = k_I\lambda_R$. This last result was presented by Cerón-Rojas et al. (2008) in their original paper.

The ratio of the index accuracies and the VPE are also valid in the RESIM context. In Eq. (7.34) we showed that the covariance between $I_R = \mathbf{b}_R'\mathbf{y}$ and $H_R = \mathbf{w}_R'\mathbf{g}$ ($\sigma_{H_R I_R}$) is equal to the variance of $I_R = \mathbf{b}_R'\mathbf{y}$ ($\sigma_{I_R}^2$). This means that the VPE of the RESIM can be written as

$$E\left[(H_R - I_R)^2\right] = \sigma_{H_R}^2 + \sigma_{I_R}^2 - 2\sigma_{H_R I_R} = \sigma_{H_R}^2 - \sigma_{I_R}^2 = \left(1 - \rho_R^2\right)\sigma_{H_R}^2. \tag{7.38}$$

Statistical properties associated with the ESIM and described in Sect. 7.1.2 are also valid for the RESIM.

7.2.2 Estimating the RESIM Parameters

We can estimate the RESIM parameters in a similar manner to the ESIM parameters in Sect. 7.1.4. With matrices $\widehat{\mathbf{C}}$ and $\widehat{\mathbf{P}}$, we constructed matrix $\widehat{\mathbf{S}}_R = \widehat{\mathbf{K}}\widehat{\mathbf{P}}^{-1}\widehat{\mathbf{C}}$ and equation

$$\left(\widehat{\mathbf{S}}_R\widehat{\mathbf{S}}_R' - \widehat{\mu}_{Rj}\mathbf{I}_t\right)\widehat{\mathbf{b}}_{R_j} = \mathbf{0}, \tag{7.39}$$

where $\widehat{\mu}_{Rj} = \widehat{\lambda}_{R_j}^4$, $j = 1, 2, \ldots, t$. The estimated RESIM index ($I_R = \mathbf{b}_R'\mathbf{y}$) is $\widehat{I}_R = \widehat{\mathbf{b}}'_{R_1}\mathbf{y}$ and the estimator of the maximized RESIM selection response and its expected genetic gain per trait can be denoted as $\widehat{R}_R = k_I\sqrt{\widehat{\mathbf{b}}'_{R_1}\widehat{\mathbf{P}}\widehat{\mathbf{b}}_{R_1}}$ and $\widehat{\mathbf{E}}_R = k_I\dfrac{\widehat{\mathbf{C}}\widehat{\mathbf{b}}_{R_1}}{\sqrt{\widehat{\mathbf{b}}'_{R_1}\widehat{\mathbf{P}}\widehat{\mathbf{b}}_{R_1}}}$ respectively, whereas the estimator of the maximized RESIM accuracy is $\widehat{\lambda}_{R_1}$.

7.2.3 Numerical Examples

We compare the RLPSI results with those of the RESIM using the Akbar et al. (1984) data described in Sect. 7.1.5. We restrict the trait RL (number of eggs) in both indices. In Chap. 3, Sect. 3.1.3, we indicated how to construct matrix $\mathbf{U}'$ and, in Sect.

3.1.4 of the same chapter, we described how to obtain matrix $\widehat{\mathbf{K}} = \left[\mathbf{I}_t - \widehat{\mathbf{Q}}\right]$ for one and two restrictions. Matrix $\widehat{\mathbf{K}}$ is the same for the RLPSI and the RESIM. Thus, in this subsection we omit the steps needed to construct matrices $\mathbf{U}'$ and $\widehat{\mathbf{K}}$.

First, we estimate the RLPSI parameters. Assume a selection intensity of 10% ($k_I = 1.755$) and a vector of economic weights $\mathbf{w}' = [\,19.54 \quad -3.56 \quad 17.01\,]$. The estimated RLPSI vector of coefficients for one restriction was $\widehat{\mathbf{b}}' = [\,0.29 \quad -0.84 \quad 5.78\,]$, and the estimated selection response, expected genetic gain per trait, accuracy, and heritability of the RLPSI were $\widehat{R} = 1.755\sqrt{\widehat{\mathbf{b}}'\widehat{\mathbf{P}}\widehat{\mathbf{b}}} = 53.01$, $\widehat{\mathbf{E}}' = 1.755\dfrac{\widehat{\mathbf{b}}'\widehat{\mathbf{C}}}{\sqrt{\widehat{\mathbf{b}}'\widehat{\mathbf{P}}\widehat{\mathbf{b}}}} = [\,0 \quad -0.71 \quad 2.96\,]$, $\widehat{\rho} = \dfrac{\sqrt{\widehat{\mathbf{b}}'\widehat{\mathbf{P}}\widehat{\mathbf{b}}}}{\sqrt{\mathbf{w}'\widehat{\mathbf{C}}\mathbf{w}}} = 0.26$, and $\widehat{h}^2 = \dfrac{\widehat{\mathbf{b}}'\widehat{\mathbf{C}}\widehat{\mathbf{b}}}{\widehat{\mathbf{b}}'\widehat{\mathbf{P}}\widehat{\mathbf{b}}} = 0.33$ respectively. In this case, $\widehat{\mathbf{b}}'\widehat{\mathbf{b}} = 34.25$; then, the estimated RLPSI selection response using the normalized RLPSI vector of coefficients was $\widehat{R} = \dfrac{53.01}{34.25} = 1.55$, and the rest of the estimated RLPSI parameters were the same.

In the RESIM, matrix $\mathbf{F}$ was an identity matrix of size 3×3; that is, we did not use matrix $\mathbf{F}$ to transform the RESIM vector of coefficients. In Sect. 7.1.5 we obtained matrix $\widehat{\mathbf{P}}^{-1}\widehat{\mathbf{C}} = \begin{bmatrix} 0.1102 & -0.0405 & -0.0280 \\ -0.0390 & 0.0864 & -0.0184 \\ -0.1833 & 0.0517 & 0.4115 \end{bmatrix}$, and we have indicated that matrix $\widehat{\mathbf{K}}$ is the same for the RLPSI and the RESIM. In the RESIM, we need matrix $\widehat{\mathbf{S}}_R = \widehat{\mathbf{K}}\widehat{\mathbf{P}}^{-1}\widehat{\mathbf{C}}$ to solve equation $\left(\widehat{\mathbf{S}}_R\widehat{\mathbf{S}}'_R - \widehat{\mu}_{Rj}\mathbf{I}_t\right)\widehat{\mathbf{b}}_{R_j} = \mathbf{0}$, where $\widehat{\mu}_{Rj} = \widehat{\lambda}^4_{R_j}$, whence we shall obtain the eigenvalues and eigenvectors that form matrices $\mathbf{L}_R^{1/2}$, $\mathbf{V}_{R1}$, and $\widehat{\mathbf{S}}_R = \mathbf{V}_{R1}\mathbf{L}_R^{1/2}\mathbf{V}'_{R2}$.

For one null restriction, matrix $\widehat{\mathbf{S}}_R = \widehat{\mathbf{K}}\widehat{\mathbf{P}}^{-1}\widehat{\mathbf{C}} = \begin{bmatrix} 0 & 0.0285 & 0.0232 \\ 0 & 0.0620 & -0.0365 \\ 0 & -0.0630 & 0.3263 \end{bmatrix}$.

This means that $\widehat{\mathbf{S}}_R$ reflects the trait restrictions imposed on the covariance between the RESIM and the vector of genotypic values; thus, if r traits are restricted, r columns of $\widehat{\mathbf{S}}_R$ are equal to zero. Matrix $\widehat{\mathbf{S}}_R\widehat{\mathbf{S}}'_R = \begin{bmatrix} 0.0013 & 0.0009 & 0.0058 \\ 0.0009 & 0.0052 & -0.0158 \\ 0.0058 & -0.0158 & 0.1104 \end{bmatrix}$ and $\mathbf{V}_{R_1} = \begin{bmatrix} 0.0500 & 0.5216 & -0.8517 \\ -0.1446 & 0.8476 & 0.5106 \\ 0.9882 & 0.0976 & 0.1178 \end{bmatrix}$, whereas the $\widehat{\mu}_{Rj} = \widehat{\lambda}^4_{R_j}$ values were 0.1130, 0.0039, and 0.0, whence $\mathbf{L}_R^{1/2} = \begin{bmatrix} 0.3362 & 0 & 0 \\ 0 & 0.0626 & 0 \\ 0 & 0 & 0.0 \end{bmatrix}$. Thus, $\widehat{\mu}_{R_1} = \widehat{\lambda}^4_{R_1} = 0.1130$, $\widehat{\lambda}^2_{R_1} = 0.3362$, and the estimated RESIM accuracy was $\widehat{\lambda}_{E_1} = 0.5798$. The estimated RESIM eigenvector, index, the selection response, and expected genetic gain per trait were $\widehat{\mathbf{b}}'_{R_1} = [0.0500 \;\; -0.1446 \;\; 0.9882]$, $\widehat{I}_R = 0.0500\text{RL} - 0.1446\text{SM} + 0.9882\text{EW}$,

$\widehat{R}_R = 1.755\sqrt{\widehat{\mathbf{b}}'_{R_1}\widehat{\mathbf{P}}\widehat{\mathbf{b}}_{R_1}} = 9.06$, and $\qquad \widehat{\mathbf{E}}'_R = 1.755\frac{\widehat{\mathbf{b}}'_{R_1}\widehat{\mathbf{C}}}{\sqrt{\widehat{\mathbf{b}}'_{R_1}\widehat{\mathbf{P}}\widehat{\mathbf{b}}_{R_1}}} = [0\ \ -0.72\ \ 2.96]$

respectively.

The estimated RLPSI selection response was $\widehat{R} = \frac{53.01}{34.25} = 1.55$; thus, the estimated RESIM selection response was higher than the estimated RLPSI response. In addition, the estimated RLPSI expected genetic gain per trait was $\widehat{\mathbf{E}}' = [0\ \ -0.71\ \ 2.96]$, which is the same as the estimated RESIM expected genetic gain per trait.

We can compare RESIM efficiency versus RLPSI efficiency to predict the net genetic merit using the ratio of the estimated RESIM accuracy $\widehat{\lambda}_{E_1} = 0.5798$ to the RLPSI accuracy $\widehat{\rho} = 0.26$, i.e., $\frac{\widehat{\lambda}_{R_1}}{\widehat{\rho}_S} = \frac{0.5798}{0.26} = 2.23$, or in percentage terms, $\widehat{p}_E = 100(2.23 - 1) = 123$ (see Chap. 5, Eq. 5.17). That is, the RESIM is a better predictor of the net genetic merit and its efficiency was 123% higher than the RLPSI efficiency for this data set.

Now, we compare RESIM efficiency versus RLPSI efficiency using the simulated data set described in Sect. 2.8.1 of Chap. 2 for five phenotypic selection cycles, each with four traits (T_1, T_2, T_3, and T_4), 500 genotypes, and four replicates for each genotype. The economic weights for T_1, T_2, T_3, and T_4 were 1, −1, 1, and 1 respectively. For this data set, matrix **F** was equal to an identity matrix of size 4 × 4 for all five selection cycles.

The first and second parts of columns 3, 4, and 5 of Table 7.1 present the estimated RLPSI and RESIM selection responses respectively for one, two, and three null restrictions for five simulated selection cycles, where the selection intensity was 10% ($k_I = 1.755$) for all five selection cycles. The averages of the estimated RLPSI selection response for each null restriction were 4.43, 4.30, and 4.92, whereas the averages of the estimated RESIM selection response were 4.54, 4.42, and 4.38 respectively. These results indicate that the estimated RLPSI selection response was greater than the estimated RESIM selection response only for three null restrictions.

The first part of Table 7.3 presents the estimated RESIM expected genetic gain per trait for one, two, and three restrictions for five simulated selection cycles. The estimated RLPSI expected genetic gains per trait for one, two, and three restrictions are given in Chap. 3 (Table 3.3). According to the results shown in Table 3.3 (Chap. 3), the averages of the estimated RLPSI expected genetic gains per trait for five simulated selection cycles were −2.52, 2.25, and 2.26 for one restriction; 2.84 and 2.65 for two restrictions; and 3.90 for three restrictions. According to the results shown in Table 7.3, the averages of the estimated RESIM expected genetic gains per trait for five simulated selection cycles were −0.43, −0.75, and 3.90 for one restriction; −0.59 and 3.89 for two restrictions; and 3.90 for three restrictions. This means that the RESIM and RLPSI were the same only for three restrictions, whereas for one and two restrictions, the average of the estimated RESIM expected

Table 7.3 Estimated RESIM and PPG-ESIM expected genetic gain per trait for one, two, and three restrictions for five simulated selection cycles

	Estimated RESIM expected genetic gain per trait											
	One null restriction				Two null restrictions				Three null restrictions			
Cycle	T1	T2	T3	T4	T1	T2	T3	T4	T1	T2	T3	T4
1	0	−0.86	−1.22	4.14	0	0	−0.96	4.12	0	0	0	4.13
2	0	−1.38	−0.004	4.31	0	0	−0.07	4.27	0	0	0	4.27
3	0	1.36	−1.74	4.07	0	0	−1.39	4.09	0	0	0	4.16
4	0	−1.13	−0.34	3.73	0	0	−0.08	3.72	0	0	0	3.72
5	0	−0.14	−0.43	3.22	0	0	−0.43	3.22	0	0	0	3.24
Average	0	−0.43	−0.75	3.90	0	0	−0.59	3.89	0	0	0	3.90
	Estimated PPG-ESIM expected genetic gain per trait											
	One predetermined restriction				Two predetermined restrictions				Three predetermined restrictions			
Cycle	T1	T2	T3	T4	T1	T2	T3	T4	T1	T2	T3	T4
1	7.81	−4.62	3.11	2.21	7.09	−3.04	3.12	2.76	6.62	−2.84	4.73	0.83
2	5.15	−2.98	2.31	3.48	5.41	−2.32	2.41	3.48	6.14	−2.63	4.39	0.92
3	4.74	−1.15	0.66	3.79	5.45	−2.34	1.24	3.26	5.52	−2.37	3.94	1.35
4	3.94	−2.44	0.74	3.34	4.57	−1.96	1.17	3.24	5.03	−2.15	3.59	0.30
5	6.68	−2.15	2.24	2.05	6.93	−2.97	2.25	1.4	5.25	−2.25	3.75	0.72
Average	5.67	−2.67	1.81	2.97	5.89	−2.52	2.04	2.83	5.71	−2.45	4.08	0.82

The selection intensity was 10% ($k_I = 1.755$) and the vectors of the PPG for each predetermined restriction were $\mathbf{d}_1' = 7$, $\mathbf{d}_2' = [7 \quad -3]$ and $\mathbf{d}_3' = [7 \quad -3 \quad 5]$ respectively

genetic gains per trait was higher than that of the estimated RLPSI expected genetic gains per trait only for trait 4.

Figure 7.2 presents the estimated accuracy of the RLPSI and the RESIM for one, two, and three null restrictions for five simulated selection cycles. In all five selection cycles, the estimated RESIM accuracy was greater than the RLPSI accuracy. This means that the RESIM is a better predictor of the net genetic merit than the RLPSI. Additional results associated with the frequency distribution of the estimated RESIM values are presented in Fig. 7.3. Figure 7.3a presents the frequency distribution of the estimated RESIM values with one null restriction for cycle 2, whereas Fig. 7.3b presents the frequency distribution of the estimated RESIM values with two null restrictions for cycle 5; both figures indicate that the estimated RESIM values approach normal distribution.

Finally, in Chap. 10 we present the results of comparing the ESIM with the LPSI and the RESIM with the RLPSI for many selection cycles. Such results are similar to those obtained in this chapter.

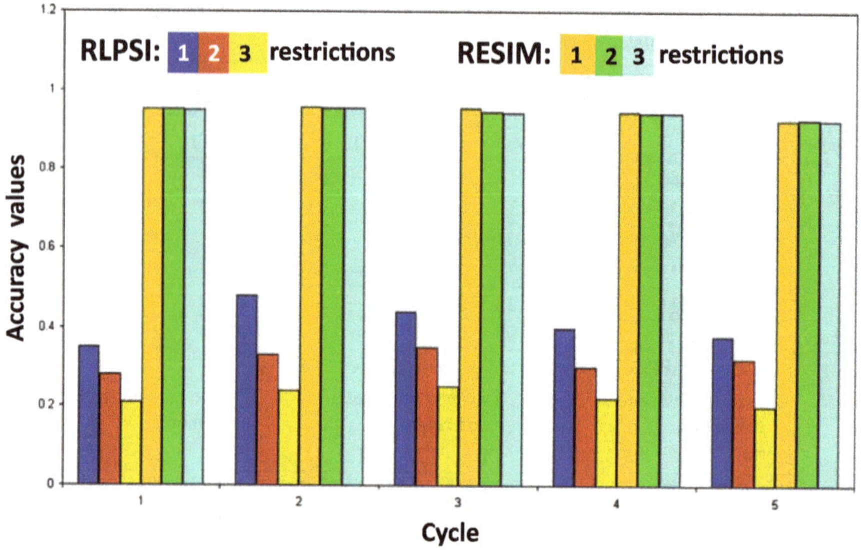

Fig. 7.2 Estimated correlation values between the restricted linear phenotypic selection index (RLPSI) and the net genetic merit ($H = \mathbf{w}'\mathbf{g}$); estimated correlation values between the restricted eigen selection index method (RESIM) and H for one, two and three null restrictions for four traits and 500 genotypes in one environment simulated for five selection cycles

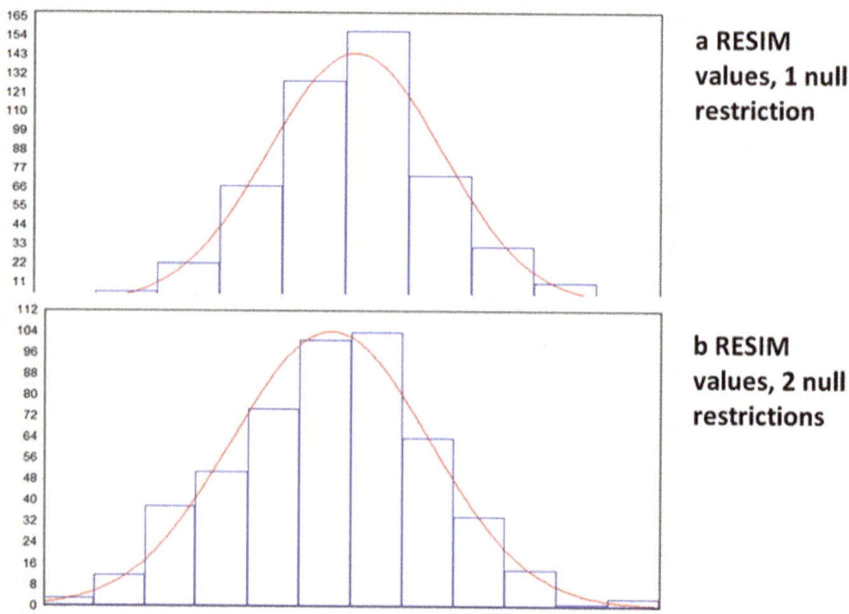

Fig. 7.3 Frequency distribution of 500 estimated RESIM values for (**a**) cycle 2 and (**b**) cycle 5, obtained from one selection cycle for 500 genotypes and four traits simulated in one environment

7.3 The Linear Phenotypic Predetermined Proportional Gain Eigen Selection Index Method

In a similar manner to the PPG-LPSI (see Chap. 3), in the PPG-ESIM the breeder pre-sets optimal levels (predetermined proportional gains) on certain traits before the selection is carried out. Let $\mathbf{d}' = [d_1 \quad d_2 \quad \cdots \quad d_r]$ be the vector of the PPGs (predetermined proportional gains) imposed by the breeder on r traits and assume that μ_q is the population mean of the qth trait before selection. The objective of the PPG-ESIM is to change μ_q to $\mu_q + d_q$, where d_q is a predetermined change in μ_q (in the RESIM, $d_q = 0$, $q = 1, 2, \cdots, r$, where r is the number of PPGs). That is, the PPG-ESIM attempts to make some traits change their expected genetic gain values based on a predetermined level, whereas the rest of the traits remain without restrictions.

The simplest way to solve the foregoing problem is by maximizing the PPG-ESIM heritability under the restriction $\mathbf{D}'\mathbf{U}'\mathbf{C}\mathbf{b} = \mathbf{0}$, where

$\mathbf{D}' = \begin{bmatrix} d_r & 0 & \cdots & 0 & -d_1 \\ 0 & d_r & \cdots & 0 & -d_2 \\ \vdots & \vdots & \ddots & \vdots & \vdots \\ 0 & 0 & \cdots & d_r & -d_{r-1} \end{bmatrix}$ (see Chap. 3 for details) is a matrix $(r-1) \times r$,

r is the number of PPGs, d_q ($q = 1, 2 \ldots, r$) is the qth element of vector $\mathbf{d}$, $\mathbf{U}'$ is the RLPSI matrix of restrictions of 1s and 0s, and $\mathbf{C}$ is the covariance matrix of genotypic values. Matrix $\mathbf{D}'$ is a Mallard (1972) matrix of PPGs used to impose predetermined restrictions.

The Mallard (1972) matrix of predetermined restrictions can be written as $\mathbf{M}' = \mathbf{D}'\boldsymbol{\Psi}'$, where $\boldsymbol{\Psi}' = \mathbf{U}'\mathbf{C}$ and $\mathbf{U}'$ is the Kempthorne and Nordskog (1959) matrix of restrictions of 1s and 0s (1 indicates that the trait is restricted, i.e., $d_q = 0$, and 0 that the trait has no restrictions).

To find the PPG-ESIM vector of coefficients that maximizes the PPG-ESIM selection response and expected genetic gain per trait, we can maximize $\rho_{HI}^2 =$

$\frac{(\mathbf{w}'\mathbf{C}\mathbf{b})^2}{(\mathbf{w}'\mathbf{C}\mathbf{w})(\mathbf{b}'\mathbf{P}\mathbf{b})}$ with respect to vectors $\mathbf{b}$ and $\mathbf{w}$ under the restrictions $\mathbf{M}'\mathbf{b} = \mathbf{0}$,

$\mathbf{b}'\mathbf{b} = 1$, $\mathbf{w}'\mathbf{C}\mathbf{w} = 1$, and $\mathbf{b}'\mathbf{P}\mathbf{b} = 1$, where $\mathbf{w}'\mathbf{C}\mathbf{w}$ is the variance of $H = \mathbf{w}'\mathbf{g}$ and $\mathbf{b}'\mathbf{P}\mathbf{b}$ is the variance of $I = \mathbf{b}'\mathbf{y}$, as did Cerón-Rojas et al. (2016) according to the canonical correlation theory, or we can solve this problem by maximizing $\frac{\mathbf{b}'\mathbf{C}\mathbf{b}}{\sqrt{\mathbf{b}'\mathbf{P}\mathbf{b}}}$ (Eq. 7.12) only with respect to vectors $\mathbf{b}$ under the restriction $\mathbf{M}'\mathbf{b} = \mathbf{0}$ and $\mathbf{b}'\mathbf{b} = 1$, as we did to obtain the RESIM vector of coefficients. Both approaches give the same result, but we use the latter approach because it is easier to work with.

7.3.1 The PPG-ESIM Parameters

To obtain the PPG-ESIM vector of coefficients, we need to maximize the function

$$f(\mathbf{b},\mathbf{v}') = \frac{\mathbf{b}'\mathbf{C}\mathbf{b}}{\sqrt{\mathbf{b}'\mathbf{P}\mathbf{b}}} - \mathbf{v}'\mathbf{M}'\mathbf{b} \tag{7.40}$$

with respect to vectors $\mathbf{b}$ and $\mathbf{v}'$, where $\mathbf{v}' = \begin{bmatrix} v_1 & v_2 & \cdots & v_{r-1} \end{bmatrix}$ is a vector of Lagrange multipliers. The derivatives of Eq. (7.40) with respect to $\mathbf{b}$ and $\mathbf{v}'$ were:

$$2(\mathbf{b}'\mathbf{P}\mathbf{b})^{1/2}\mathbf{C}\mathbf{b} - (\mathbf{b}'\mathbf{P}\mathbf{b})^{-1/2}(\mathbf{b}'\mathbf{C}\mathbf{b})\mathbf{P}\mathbf{b} - \mathbf{M}\mathbf{v} = \mathbf{0} \tag{7.41}$$

and

$$\mathbf{M}'\mathbf{b} = \mathbf{0}, \tag{7.42}$$

respectively, where Eq. (7.42) denotes the restriction imposed for maximizing Eq. (7.40). By using algebraic methods on Eq. (7.41) similar to those used to obtain Eq. (7.10) we get

$$\left(\mathbf{K}_P\mathbf{P}^{-1}\mathbf{C} - \lambda_P^2\mathbf{I}_t\right)\mathbf{b}_P = \mathbf{0}, \tag{7.43}$$

where $\mathbf{K}_P = [\mathbf{I}_t - \mathbf{Q}_P]$, $\mathbf{Q}_P = \mathbf{P}^{-1}\boldsymbol{\Psi}\mathbf{D}(\mathbf{D}'\boldsymbol{\Psi}'\mathbf{P}^{-1}\boldsymbol{\Psi}\mathbf{D})^{-1}\mathbf{D}'\boldsymbol{\Psi}'$, $\boldsymbol{\Psi}' = \mathbf{U}'\mathbf{C}$, $\mathbf{I}_t$ is an identity matrix $t \times t, \lambda_P^2 = h_{I_P}^2$, and $\mathbf{b}_P$ are the first eigenvalue and the first eigenvector of matrix $\mathbf{K}_P\mathbf{P}^{-1}\mathbf{C}$ respectively. Note that $h_{I_P}^2$ is PPG-ESIM heritability and λ_P is the maximum correlation between $I_P = \mathbf{b}_P'\mathbf{y}$ and $H = \mathbf{w}'\mathbf{g}$. When $\mathbf{D}' = \mathbf{U}'$, $\mathbf{b}_P = \mathbf{b}_R$ (the vector of coefficients of the RESIM), and when $\mathbf{U}'$ is a null matrix, $\mathbf{b}_P = \mathbf{b}_E$ (the vector of coefficients of the ESIM). That is, the PPG-ESIM is more general than the RESIM and the ESIM and includes the latter two indices as particular cases. Matrices $\mathbf{K}_P = [\mathbf{I}_t - \mathbf{Q}_P]$ and $\mathbf{Q}_P = \mathbf{P}^{-1}\boldsymbol{\Psi}\mathbf{D}(\mathbf{D}'\boldsymbol{\Psi}'\mathbf{P}^{-1}\boldsymbol{\Psi}\mathbf{D})^{-1}\mathbf{D}'\boldsymbol{\Psi}'$ are the same as those obtained in the PPG-LPSI (see Chap. 3). Also, vector $\mathbf{b}_P$ can be transformed as $\boldsymbol{\beta}_P = \mathbf{F}\mathbf{b}_P$; matrix $\mathbf{F}$ was defined earlier.

Let $\mathbf{S}_P = \boldsymbol{\Psi}'\mathbf{P}^{-1}\boldsymbol{\Psi}$; then, under the assumption $\mathbf{D}'\mathbf{d} = \mathbf{0}$, it is possible to show that $\mathbf{D}(\mathbf{D}'\mathbf{S}_P\mathbf{D})^{-1}\mathbf{D}' = \mathbf{S}_P^{-1} - \mathbf{S}_P^{-1}\mathbf{d}\left(\mathbf{d}'\mathbf{S}_P^{-1}\mathbf{d}\right)^{-1}\mathbf{d}'\mathbf{S}_P^{-1}$ (see Chap. 3), whence by substituting $\mathbf{S}_P^{-1} - \mathbf{S}_P^{-1}\mathbf{d}\left(\mathbf{d}'\mathbf{S}_P^{-1}\mathbf{d}\right)^{-1}\mathbf{d}'\mathbf{S}_P^{-1}$ for $\mathbf{D}(\mathbf{D}'\mathbf{S}_P\mathbf{D})^{-1}\mathbf{D}'$ in matrix $\mathbf{Q}_P = \mathbf{P}^{-1}\boldsymbol{\Psi}\mathbf{D}(\mathbf{D}'\boldsymbol{\Psi}'\mathbf{P}^{-1}\boldsymbol{\Psi}\mathbf{D})^{-1}\mathbf{D}'\boldsymbol{\Psi}'$, matrix $\mathbf{K}_P\mathbf{P}^{-1}\mathbf{C}$ can be written as

$$\mathbf{K}_P\mathbf{P}^{-1}\mathbf{C} = \left[\mathbf{I}_t - \mathbf{P}^{-1}\boldsymbol{\Psi}\mathbf{S}^{-1}\boldsymbol{\Psi}'\right]\mathbf{P}^{-1}\mathbf{C} + \mathbf{A}_P, \tag{7.44}$$

where $\boldsymbol{\Psi}' = \mathbf{U}'\mathbf{C}$, $\mathbf{A}_P = \boldsymbol{\delta}\boldsymbol{\alpha}'$, $\boldsymbol{\delta} = \mathbf{P}^{-1}\boldsymbol{\Psi}(\boldsymbol{\Psi}'\mathbf{P}^{-1}\boldsymbol{\Psi})^{-1}\mathbf{d}$, and $\boldsymbol{\alpha}' = \frac{\mathbf{d}'\mathbf{S}^{-1}\boldsymbol{\Psi}'\mathbf{P}^{-1}\mathbf{C}}{\mathbf{d}'\mathbf{S}^{-1}\mathbf{d}}$. When $\mathbf{A}_P$ is a null matrix, $\mathbf{K}_P\mathbf{P}^{-1}\mathbf{C} = \mathbf{K}\mathbf{P}^{-1}\mathbf{C}$ (matrix of the RESIM), and if $\mathbf{U}'$ is a null matrix, $\mathbf{K}_P\mathbf{P}^{-1}\mathbf{C} = \mathbf{P}^{-1}\mathbf{C}$ (matrix of the ESIM), this means that Eq. (7.44) is a mathematical equivalent form of matrix $\mathbf{K}_P\mathbf{P}^{-1}\mathbf{C}$ and that Eq. (7.44) does not require matrix $\mathbf{D}'$. The easiest way to obtain $\mathbf{b}_P$ and λ_P is to use matrix $[\mathbf{I}_t - \mathbf{P}^{-1}\boldsymbol{\Psi}\mathbf{S}^{-1}\boldsymbol{\Psi}']\mathbf{P}^{-1}\mathbf{C} + \mathbf{A}_P$ in Eq. (7.43) instead of matrix $\mathbf{K}_P\mathbf{P}^{-1}\mathbf{C}$.

In the PPG-ESIM context, vector $\mathbf{w}$ can be obtained as

$$\mathbf{w}_P = \mathbf{C}^{-1}[\lambda_P \mathbf{P}\mathbf{b}_P + \mathbf{M}\mathbf{v}_P], \tag{7.45}$$

whence $H = \mathbf{w}'\mathbf{g}$ can be written as $H_P = \mathbf{w}_P'\mathbf{g}$. In Eq. (7.45), λ_P is the maximum correlation between $I_P = \mathbf{b}_P'\mathbf{y}$ and $H_P = \mathbf{w}_P'\mathbf{g}$, $\mathbf{b}_P$ is the first eigenvector of matrix $\mathbf{K}_P\mathbf{P}^{-1}\mathbf{C}$, $\mathbf{v}_P = \lambda_P^{-1}(\mathbf{M}'\mathbf{P}^{-1}\mathbf{M})^{-1}\mathbf{M}'\mathbf{P}^{-1}\mathbf{C}\mathbf{b}_P$, $\mathbf{M}' = \mathbf{D}'\mathbf{\Psi}'$, and $\mathbf{\Psi}' = \mathbf{U}'\mathbf{C}$. In a similar manner to the RESIM context, we can assume that λ_P and λ_P^{-1} are proportionality constants and it can be shown that the covariance between $I_P = \mathbf{b}_P'\mathbf{y}$ and $H_P = \mathbf{w}_P'\mathbf{g}$ ($\sigma_{H_P I_P}$) is equal to the variance of $I_P = \mathbf{b}_P'\mathbf{y}$ ($\sigma_{I_P}^2 = \mathbf{b}_p'\mathbf{P}\mathbf{b}_P$), that is, $\sigma_{H_P I_P} = \mathbf{w}_P'\mathbf{C}\mathbf{b}_P = \mathbf{b}_p'\mathbf{P}\mathbf{b}_P$.

The accuracy of the PPG-ESIM can also be written as

$$\rho_{H_P I_P} = \frac{\sqrt{\mathbf{b}_P'\mathbf{P}\mathbf{b}_P}}{\sqrt{\mathbf{w}_P'\mathbf{C}\mathbf{w}_P}}, \tag{7.46}$$

where $\sigma_{H_P}^2 = \mathbf{w}_P'\mathbf{C}\mathbf{w}_P = \mathbf{b}_P'\mathbf{P}\mathbf{C}^{-1}\mathbf{P}\mathbf{b}_P + \mathbf{b}_P'\mathbf{P}\mathbf{C}^{-1}\mathbf{Q}_P'\mathbf{C}\mathbf{b}_P$ is the variance of H_P. When $\mathbf{D}' = \mathbf{U}'$, $\mathbf{w}_P'\mathbf{C}\mathbf{w}_P = \mathbf{w}_R'\mathbf{C}\mathbf{w}_R$ (the variance of H_R), and when $\mathbf{U}'$ is a null matrix, $\mathbf{w}_P'\mathbf{C}\mathbf{w}_P = \mathbf{w}_E'\mathbf{C}\mathbf{w}_E$ (the variance of H_E). Hereafter, to simplify the notation, we write Eq. (7.46) as ρ_P or λ_P.

Let $\boldsymbol{\beta}_P = \mathbf{F}\mathbf{b}_P$ be the PPG-ESIM transformed vector of coefficients by matrix $\mathbf{F}$. By Eqs. (7.1) and (7.46), the maximized selection response (R_P) and expected genetic gain per trait ($\mathbf{E}_P$) of the PPG-ESIM can be written as

$$R_P = k_I\sqrt{\boldsymbol{\beta}_P'\mathbf{P}\boldsymbol{\beta}_P} \tag{7.47}$$

and

$$\mathbf{E}_P = k_I\frac{\mathbf{C}\boldsymbol{\beta}_P}{\sqrt{\boldsymbol{\beta}_P'\mathbf{P}\boldsymbol{\beta}_P}}, \tag{7.48}$$

respectively, where $\sqrt{\boldsymbol{\beta}_P'\mathbf{P}\boldsymbol{\beta}_P} = \sigma_{I_P}$ is the standard deviation of the variance of $I_P = \boldsymbol{\beta}_P'\mathbf{y}$. Equations (7.47) and (7.48) do not require economic weights. When $\mathbf{F}$ is an identity matrix, $\boldsymbol{\beta}_P = \mathbf{b}_P$, $I_P = \mathbf{b}_P'\mathbf{y}$, $R_P = k_I\sqrt{\mathbf{b}_P'\mathbf{P}\mathbf{b}_P}$, and $\mathbf{E}_P = k_I\frac{\mathbf{C}\mathbf{b}_P}{\sqrt{\mathbf{b}_P'\mathbf{P}\mathbf{b}_P}}$.

Equation (7.47) can also be written as $R_P = k_I\sigma_{H_P}\lambda_P$, where $\sigma_{H_P} = \sqrt{\mathbf{b}_P'\mathbf{P}\mathbf{C}^{-1}\mathbf{P}\mathbf{b}_P + \mathbf{b}_P'\mathbf{P}\mathbf{C}^{-1}\mathbf{Q}_P'\mathbf{C}\mathbf{b}_P}$ is the standard deviation of the variance of H_P, and λ_P is the canonical correlation between H_P and $I_P = \boldsymbol{\beta}_P'\mathbf{y}$. When $\sigma_{H_P} = 1$, Eq. (7.47) can be written as $R_P = k_I\lambda_P$, where λ_P is the covariance between $I_P = \mathbf{b}_P'\mathbf{y}$ and $H = \mathbf{w}_P'\mathbf{g}$.

The prediction efficiency of the PPG-ESIM can be obtained in a similar manner to the ESIM and RESIM. The accuracy of the PPG-ESIM (Eq. 7.46) can be used to construct the ratio of index accuracies. The PPG-ESIM mean square error or the VPE can be obtained as

$$E\left[(H_P - I_P)^2\right] = \sigma^2_{H_P} + \sigma^2_{I_P} - 2\sigma_{H_P I_P} = \sigma^2_{H_P} - \sigma^2_{I_P} = \left(1 - \rho^2_P\right)\sigma^2_{H_P}. \quad (7.49)$$

Additional properties associated with the ESIM are also valid for the PPG-ESIM.

7.3.2 Estimating PPG-ESIM Parameters

The procedure used to estimate PPG-ESIM parameters is the same as that described for RESIM. Let $\widehat{\mathbf{C}}$ and $\widehat{\mathbf{P}}$ be the estimated matrices of $\mathbf{C}$ and $\mathbf{P}$. In the PPG-ESIM context, we use matrix $\widehat{\mathbf{S}} = \widehat{\mathbf{K}}_P\widehat{\mathbf{P}}^{-1}\widehat{\mathbf{C}}$ to obtain the estimated eigenvalues and eigenvectors of equation

$$\left(\widehat{\mathbf{S}} - \widehat{\lambda}^2_{Pj}\mathbf{I}_t\right)\widehat{\mathbf{b}}_{Pj} = \mathbf{0}, \quad (7.50)$$

$j = 1, 2, \cdots, t$, where t is the number of traits in the PPG-ESIM index, $\widehat{\mathbf{K}}_P = \left[\mathbf{I}_t - \widehat{\mathbf{Q}}_P\right]$, $\mathbf{I}_t$ is an identity matrix of size $t \times t$ and $\widehat{\mathbf{Q}}_P = \widehat{\mathbf{P}}^{-1}\widehat{\mathbf{\Psi}}\mathbf{D}\left(\mathbf{D}'\widehat{\mathbf{\Psi}'}\widehat{\mathbf{P}}^{-1}\widehat{\mathbf{\Psi}}\mathbf{D}\right)^{-1}\mathbf{D}'\widehat{\mathbf{\Psi}'}$. As $\widehat{\mathbf{S}}$ is an asymmetric matrix, the values of $\widehat{\mathbf{b}}_{Pj}$ and $\widehat{\lambda}^2_{Pj}$ should be obtained using SVD (singular value decomposition).

According to SVD, we need to solve equation

$$\left(\widehat{\mathbf{S}}\widehat{\mathbf{S}}' - \widehat{\mu}_{Pj}\mathbf{I}_t\right)\widehat{\mathbf{b}}_{P_j} = \mathbf{0}, \quad (7.51)$$

where $\widehat{\mu}_{Pj} = \widehat{\lambda}^4_{P_j}$ (j= 1, 2, . . ., t). By Eq. (7.51), the estimated PPG-ESIM index ($I_P = \mathbf{b}'_P\mathbf{y}$) is $\widehat{I}_P = \widehat{\mathbf{b}}'_{P_1}\mathbf{y}$. The estimator of the maximized PPG-ESIM selection response, and its expected genetic gain per trait, can be denoted as $\widehat{R}_P = k_I\sqrt{\widehat{\mathbf{b}}'_{P_1}\widehat{\mathbf{P}}\widehat{\mathbf{b}}_{P_1}}$ and $\widehat{\mathbf{E}}_P = k_I\dfrac{\widehat{\mathbf{C}}\widehat{\mathbf{b}}_{P_1}}{\sqrt{\widehat{\mathbf{b}}'_{P_1}\widehat{\mathbf{P}}\widehat{\mathbf{b}}_{P_1}}}$ respectively, whereas the estimator of the maximized accuracy of the PPG-ESIM is $\widehat{\lambda}_{P_1}$.

7.3.3 Numerical Examples

We compare the results of the PPG-LPSI and the PPG-ESIM using the Akbar et al. (1984) data described earlier. We restrict traits RL and SM, on both indices using the PPG vector $\mathbf{d}' = [3 \quad -1]$. In Chap. 3, Sect. 3.1.4, we indicated how to construct matrix $\mathbf{U}'$ and, in Sect. 3.2.4 of the same chapter, we described how to obtain matrix $\widehat{\mathbf{K}}_P$ for one and two restrictions. Matrix $\widehat{\mathbf{K}}_P$ is the same for the PPG-LPSI and the PPG-ESIM. Thus, we omit the steps for constructing matrices $\mathbf{U}'$ and $\widehat{\mathbf{K}}_P$.

Assume a selection intensity of 10% (k_I = 1.755) and that the vector of economic weights is $\mathbf{w}' = [19.54 \quad -3.56 \quad 17.01]$. The estimated PPG-LPSI vector of coefficients for two predetermined restrictions was $\widehat{\mathbf{b}}' = [1.70 \quad 1.04 \quad 2.93]$, and its estimated selection response, expected genetic gain per trait, accuracy, and heritability were $\widehat{R} = 1.755\sqrt{\widehat{\mathbf{b}}'\widehat{\mathbf{P}}\widehat{\mathbf{b}}} = 49.02$, $\widehat{\mathbf{E}}' = 1.755\frac{\widehat{\mathbf{b}}'\widehat{\mathbf{C}}}{\sqrt{\widehat{\mathbf{b}}'\widehat{\mathbf{P}}\widehat{\mathbf{b}}}} = [1.25 \quad -0.42 \quad 1.36]$, $\widehat{\rho} = \frac{\sqrt{\widehat{\mathbf{b}}'\widehat{\mathbf{P}}\widehat{\mathbf{b}}}}{\sqrt{\mathbf{w}'\widehat{\mathbf{C}}\mathbf{w}}} = 0.24$, and $\widehat{h}^2 = \frac{\widehat{\mathbf{b}}'\widehat{\mathbf{C}}\widehat{\mathbf{b}}}{\widehat{\mathbf{b}}'\widehat{\mathbf{P}}\widehat{\mathbf{b}}} = 0.12$ respectively. In this case, $\widehat{\mathbf{b}}'\widehat{\mathbf{b}} = 12.57$; then, the estimated PPG-LPSI selection response using the normalized PPG-LPSI vector of coefficients was $\widehat{R} = \frac{49.02}{12.57} = 3.90$, whereas the rest of the estimated PPG-LPSI parameters were the same.

In the PPG-ESIM, we need matrix $\widehat{\mathbf{S}} = \widehat{\mathbf{K}}_P\widehat{\mathbf{P}}^{-1}\widehat{\mathbf{C}}$ to obtain the eigenvalues and eigenvectors of $(\widehat{\mathbf{S}}\widehat{\mathbf{S}}' - \widehat{\mu}_{Pj}\mathbf{I}_t)\widehat{\mathbf{b}}_{P_j} = \mathbf{0}$ that make up matrices $\mathbf{L}_P^{1/2}$, $\mathbf{V}_{P_1}$, and $\widehat{\mathbf{S}} = \mathbf{V}_{P_1}\mathbf{L}_P^{1/2}\mathbf{V}'_{P_2}$, where $\widehat{\mu}_{Pj} = \widehat{\lambda}^4_{P_j}$. It can be shown that $\widehat{\mathbf{S}} = \widehat{\mathbf{K}}_P\widehat{\mathbf{P}}^{-1}\widehat{\mathbf{C}} = \begin{bmatrix} 0.1047 & -0.0349 & -0.0279 \\ 0.0678 & -0.0226 & -0.0213 \\ -0.1970 & 0.0657 & 0.4119 \end{bmatrix}$, $\widehat{\mathbf{S}}\widehat{\mathbf{S}}' = \begin{bmatrix} 0.0130 & 0.0085 & -0.0344 \\ 0.0085 & 0.0056 & -0.0236 \\ -0.0344 & -0.0236 & 0.2118 \end{bmatrix}$, and $\mathbf{V}_{P_1} = \begin{bmatrix} -0.1663 & 0.8292 & 0.5336 \\ -0.1138 & 0.5214 & -0.8457 \\ 0.9795 & 0.2014 & -0.0076 \end{bmatrix}$, whereas the $\widehat{\mu}_{Pj} = \widehat{\lambda}^4_{P_j}$ values were 0.2214, 0.0099, and 0.0, whence $\mathbf{L}_P^{1/2} = \begin{bmatrix} 0.4705 & 0 & 0 \\ 0 & 0.0997 & 0 \\ 0 & 0 & 0.0 \end{bmatrix}$. Thus, $\widehat{\mu}_{P_1} = \widehat{\lambda}^4_{P_1} = 0.2214$, $\widehat{\lambda}^2_{P_1} = 0.4705$, and the estimated maximized PPG-ESIM accuracy was $\widehat{\lambda}_{P_1} = 0.6859$.

We transformed the first eigenvector $\widehat{\mathbf{b}}'_{P_1} = [-0.1663 \quad -0.1138 \quad 0.9795]$ using matrix $\mathbf{F} = \begin{bmatrix} -9 & 0 & 0 \\ 0 & 1 & 0 \\ 0 & 0 & 1 \end{bmatrix}$ to obtain vector $\widehat{\boldsymbol{\beta}}_P = \widehat{\mathbf{b}}'_{P_1}\mathbf{F} = [1.4968 \quad -0.1138 \quad 0.9795]$ and $\widehat{\boldsymbol{\beta}}'_P\widehat{\boldsymbol{\beta}}_P = 3.21$, whence the estimates of the index, the selection response, and expected genetic gain per trait of the PPG-ESIM were $\widehat{I}_P = 1.4968\text{RL} - 0.1138\text{SM} + 0.9795\text{EW}$, $\widehat{R}_P = \frac{1.755\sqrt{\widehat{\boldsymbol{\beta}}'_P\widehat{\mathbf{P}}\widehat{\boldsymbol{\beta}}_P}}{\widehat{\boldsymbol{\beta}}'_P\widehat{\boldsymbol{\beta}}_P} = \frac{43.01}{3.21} = 13.39$, and $\widehat{\mathbf{E}}'_P = 1.755\frac{\widehat{\boldsymbol{\beta}}'_P\widehat{\mathbf{C}}}{\sqrt{\widehat{\boldsymbol{\beta}}'_P\widehat{\mathbf{P}}\widehat{\boldsymbol{\beta}}_P}} = [3.05 \quad -1.96 \quad 0.19]$ respectively. The estimated PPG-LPSI selection response was $\widehat{R} = \frac{49.02}{12.57} = 3.90$, which means that the estimated PPG-ESIM selection response was greater than the estimated PPG-LPSI response.

We compared PPG-ESIM efficiency versus LPSI efficiency to predict the net genetic merit using the ratio of the estimated PPG-ESIM accuracy ($\widehat{\lambda}_{P_1} = 0.6859$) to PPG-LPSI accuracy ($\widehat{\rho} = 0.24$), i.e., $\frac{\widehat{\lambda}_{P_1}}{\widehat{\rho}} = \frac{0.6859}{0.24} = 2.858$ or, in percentage terms, $\widehat{p}_P = 100(2.858 - 1) = 185.80$. Then, the PPG-ESIM was a better predictor of the net genetic merit and its efficiency was 185.80% higher than that of the PPG-LPSI for this data set.

Now, we compare PPG-ESIM efficiency versus PPG-LPSI efficiency using the data set described in Sect. 2.8.1 of Chap. 2 for five phenotypic selection cycles, each with four traits (T_1, T_2, T_3, and T_4), 500 genotypes, and four replicates for each genotype. The economic weights for T_1, T_2, T_3, and T_4 were 1, −1, 1, and 1 respectively. For this data set, matrix **F** was an identity matrix of size 4 × 4 for all five selection cycles.

The first and second parts of columns 6, 7, and 8 in Table 7.1 present the estimated PPG-LPSI and PPG-ESIM selection responses for one, two, and three predetermined restrictions for five simulated selection cycles. The selection intensity was 10% ($k_I = 1.755$) and the vectors of PPG for each predetermined restriction were $\mathbf{d}_1' = 7$, $\mathbf{d}_2' = [7 \quad -3]$, and $\mathbf{d}_3' = [7 \quad -3 \quad 5]$ respectively, for all five selection cycles. The estimated PPG-LPSI selection response when the vector of coefficients was not normalized was presented in Chap. 3 (Table 3.5). The averages of the estimated PPG-LPSI selection response for each predetermined restriction were 4.70, 4.91, and 3.14, whereas the averages of the estimated PPG-ESIM selection response were 6.31, 6.28, and 6.75 respectively. These results indicate that the estimated PPG-ESIM selection response was greater than the estimated PPG-LPSI selection response for all predetermined restrictions.

The second part of Table 7.2 presents the estimated PPG-ESIM accuracy ($\widehat{\rho}_P$) and the ratio of $\widehat{\rho}_P$ to the estimated PPG-LPSI accuracy ($\widehat{\rho}$), expressed in percentage terms, $\widehat{p}_P = 100\left(\widehat{\lambda}_P - 1\right)$, where $\widehat{\lambda}_P = \widehat{\rho}_P / \widehat{\rho}$, for one, two, and three predetermined restrictions for five simulated selection cycles. The estimated PPG-LPSI accuracies were presented in Chap. 3 (Table 3.6). The average estimated PPG-ESIM efficiency for each restriction was 9.76%, 11.71%, and 29.03% greater than the PPG-LPSI efficiency for this data set in all five selection cycles.

The second part of Table 7.3 presents the estimated PPG-ESIM expected genetic gain per trait for one, two, and three predetermined restrictions for five simulated selection cycles. The estimated PPG-LPSI expected genetic gains per trait for one, two, and three predetermined restrictions were presented in Chap. 3, Table 3.5, where it can be seen that the averages of the estimated PPG-LPSI expected genetic gains per trait for five simulated selection cycles were 6.85, −3.25, 2.62 and 1.48 for one restriction; 6.93, −2.97, 2.65 and 1.45 for two restrictions; and 5.20, −2.23, 3.72 and 1.43 for three restrictions, whereas for the same set of restrictions, the averages of the estimated PPG-ESIM expected genetic gain per trait were 5.67, −2.67, 1.81, and 2.97 for one restriction; 5.89, −2.52, 2.04, and 2.83 for two restrictions; and 5.71, −2.45, 4.08, and 0.82 for three restrictions (Table 7.3). Because the vectors of predetermined proportional gains for each predetermined restriction were $\mathbf{d}_1' = 7$, $\mathbf{d}_2' = [7 \quad -3]$, and $\mathbf{d}_3' = [7 \quad -3 \quad 5]$, the averages of the estimated PPG-LPSI expected genetic gains per trait were closer than those of the estimated PPG-ESIM expected genetic gains per trait for one and two predetermined restrictions, whereas for three restrictions, the results of both selection indices were similar.

References

Akbar MK, Lin CY, Gyles NR, Gavora JS, Brown CJ (1984) Some aspects of selection indices with constraints. Poult Sci 63:1899–1905

Anderson TW (1999) Asymptotic theory for canonical correlation analysis. J Multivar Anal 70:1–29

Anderson TW (2003) An introduction to multivariate statistical analysis, 3rd edn. Wiley, New Jersey

Bilodeau M, Brenner D (1999) Theory of multivariate statistics. Springer, New York

Cerón-Rojas JJ, Crossa J, Sahagún-Castellanos J, Castillo-González F, Santacruz-Varela A (2006) A selection index method based on eigen analysis. Crop Sci 46:1711–1721

Cerón-Rojas JJ, Sahagún-Castellanos J, Castillo-González F, Santacruz-Varela A, Crossa J (2008) A restricted selection index method based on eigenanalysis. J Agric Biol Environ Stat 13 (4):421–438

Cerón-Rojas JJ, Crossa J, Toledo FH, Sahagún-Castellanos J (2016) A predetermined proportional gains eigen selection index method. Crop Sci 56:2436–2447

Gentle JE (2007) Matrix algebra theory, computations, and applications in statistics. Springer, New York

Harville DA (1997) Matrix algebra from a statistician's perspective. Springer, New York

Hotelling H (1935) The most predictable criterion. J Educ Psychol 26:139–142

Hotelling H (1936) Relations between two sets of variables. Biometrika 28:321–377

Kempthorne O, Nordskog AW (1959) Restricted selection indices. Biometrics 15:10–19

Mallard J (1972) The theory and computation of selection indices with constraints: a critical synthesis. Biometrics 28:713–735

Mardia KV, Kent JT, Bibby JM (1982) Multivariate analysis. Academic, New York

Muirhead RJ (2005) Aspects of multivariate statistical theory. Wiley, Hoboken

Okamoto M (1973) Distinctness of the eigenvalues of a quadratic form in a multivariate sample. Ann Stat 1(4):763–765

Rao CR (2002) Linear statistical inference and its applications, 2nd edn. Wiley, New York

Watkins DS (2002) Fundamentals of matrix computations, 2nd edn. Wiley, New York

Wilms I, Croux C (2016) Robust sparse canonical correlation analysis. BMC Syst Biol 10:72

8

MESIM, GESIM, GW-ESIM and RGESIM: Parameters and Numerical Examples

Abstract The three main linear phenotypic eigen selection index methods are the eigen selection index method (ESIM), the restricted ESIM (RESIM) and the predetermined proportional gain ESIM (PPG-ESIM). The ESIM is an unrestricted index, but the RESIM and PPG-ESIM allow null and predetermined restrictions respectively to be imposed on the expected genetic gains of some traits, whereas the rest remain without any restrictions. These indices are based on the canonical correlation, on the singular value decomposition, and on the linear phenotypic selection indices theory. We extended the ESIM theory to the molecular-assisted and genomic selection context to develop a molecular ESIM (MESIM), a genomic ESIM (GESIM), and a genome-wide ESIM (GW-ESIM). Also, we extend the RESIM and PPG-ESIM theory to the restricted genomic ESIM (RGESIM), and to the predetermined proportional gain genomic ESIM (PPG-GESIM) respectively. The latter five indices use marker and phenotypic information jointly to predict the net genetic merit of the candidates for selection, but although MESIM uses only statistically significant markers linked to quantitative trait loci, the GW-ESIM uses all genome markers and phenotypic information and the GESIM, RGESIM, and PPG-GESIM use the genomic estimated breeding values and the phenotypic values to predict the net genetic merit. Using real and simulated data, we validated the theoretical results of all five indices.

8.1 The Molecular Eigen Selection Index Method

The molecular eigen selection index method (MESIM) is very similar to the linear molecular selection index (LMSI) described in Chap. 4; thus, it uses the same set of information to predict the net genetic merit of individual candidates for selection, and therefore needs the same set of conditions as those of the LMSI. The only difference between the two indices is how the vector of coefficients is obtained and the assumption associated with the vector of economic weights. Thus, although the LMSI obtains the vector of coefficients according to the linear phenotypic selection index (LPSI) described in Chap. 2 and assumes that the economic weights are known and fixed, the MESIM assumes that the economic weights are unknown and fixed and obtains the vector of coefficients according to the ESIM theory.

8.1.1 The MESIM Parameters

In the MESIM context, the net genetic merit can be written as

$$H = \mathbf{w}_1'\mathbf{g} + \mathbf{w}_2'\mathbf{s} = \begin{bmatrix}\mathbf{w}_1' & \mathbf{w}_2'\end{bmatrix}\begin{bmatrix}\mathbf{g} \\ \mathbf{s}\end{bmatrix} = \mathbf{w}'\mathbf{a}, \tag{8.1}$$

where $\mathbf{g}' = [g_1 \quad \cdots \quad g_t]$ is the vector of true breeding values, t is the number of traits, $\mathbf{w}_1' = [w_1 \quad \cdots \quad w_t]$ is a vector of unknown economic weights associated with $\mathbf{g}$, $\mathbf{w}_2' = [0_1 \quad \cdots \quad 0_t]$ is a null vector associated with the vector of marker score values $\mathbf{s}' = [s_1 \quad s_2 \quad \ldots \quad s_t]$, $\mathbf{w}' = [\mathbf{w}_1' \quad \mathbf{w}_2']$ and $\mathbf{a}' = [\mathbf{g}' \quad \mathbf{s}']$ (Chap. 4 for details). The MESIM index can be written as

$$I = \boldsymbol{\beta}_\mathbf{y}'\mathbf{y} + \boldsymbol{\beta}_s'\mathbf{s} = \begin{bmatrix}\boldsymbol{\beta}_\mathbf{y}' & \boldsymbol{\beta}_s'\end{bmatrix}\begin{bmatrix}\mathbf{y} \\ \mathbf{s}\end{bmatrix} = \boldsymbol{\beta}'\mathbf{t}, \tag{8.2}$$

where $\mathbf{y}' = [y_1 \quad \cdots \quad y_t]$ is the vector of phenotypic values; $\mathbf{s}' = [s_1 \quad s_2 \quad \ldots \quad s_t]$ is the vector of marker scores; $\boldsymbol{\beta}_\mathbf{y}'$ and $\boldsymbol{\beta}_s$ are vectors of phenotypic and marker score weight values respectively, $\boldsymbol{\beta}' = [\boldsymbol{\beta}_\mathbf{y}' \quad \boldsymbol{\beta}_G']$ and $\mathbf{t}' = [\mathbf{y}' \quad \mathbf{s}']$. The objectives of the MESIM are the same as those of the ESIM (see Chap. 7 for details).

Let $Var(H) = \mathbf{w}'\boldsymbol{\Psi}_M\mathbf{w} = \sigma_H^2$ be the variance of H, $Var(I) = \boldsymbol{\beta}'\mathbf{T}_M\boldsymbol{\beta} = \sigma_I^2$ the variance of I, and $Cov(H, I) = \mathbf{w}'\boldsymbol{\Psi}_M\boldsymbol{\beta}$ the covariance between H and I, where $\boldsymbol{\Psi}_M = Var\begin{bmatrix}\mathbf{g} \\ \mathbf{s}\end{bmatrix} = \begin{bmatrix}\mathbf{C} & \mathbf{S}_M \\ \mathbf{S}_M & \mathbf{S}_M\end{bmatrix}$ and $\mathbf{T}_M = Var\begin{bmatrix}\mathbf{y} \\ \mathbf{s}\end{bmatrix} = \begin{bmatrix}\mathbf{P} & \mathbf{S}_M \\ \mathbf{S}_M & \mathbf{S}_M\end{bmatrix}$ are block matrices of size $2t \times 2t$ (t is the number of traits) of covariance matrices where $\mathbf{P}$, $\mathbf{S}_M$, and $\mathbf{C}$ are covariance matrices $t \times t$ of phenotypic ($\mathbf{y}$), marker score ($\mathbf{s}$), and genetic breeding ($\mathbf{g}$) values respectively. Let $\rho_{HI} = \frac{\mathbf{w}'\boldsymbol{\Psi}_M\boldsymbol{\beta}}{\sqrt{\mathbf{w}'\boldsymbol{\Psi}_M\mathbf{w}}\sqrt{\boldsymbol{\beta}'\mathbf{T}_M\boldsymbol{\beta}}}$ and $h_I^2 = \frac{\boldsymbol{\beta}'\boldsymbol{\Psi}_M\boldsymbol{\beta}}{\boldsymbol{\beta}'\mathbf{T}_M\boldsymbol{\beta}}$ be the correlation between H and I, and the heritability of I respectively; then, the MESIM selection response can be written as

$$R = k_I\sigma_H\rho_{HI} \tag{8.3}$$

and

$$R = k_I\sigma_I h_I^2, \tag{8.4}$$

where k_I is the standardized selection differential (or selection intensity) associated with MESIM; $\sigma_H = \sqrt{\mathbf{w}'\boldsymbol{\Psi}_M\mathbf{w}}$ and $\sigma_I = \sqrt{\boldsymbol{\beta}'\mathbf{T}_M\boldsymbol{\beta}}$ are the standard deviations of the variance of H and I respectively. It is assumed that k_I is fixed, and that matrices $\mathbf{T}_M$ and $\mathbf{\Psi}_M$ are known; therefore, we can maximize R by maximizing ρ_{HI} (Eq. 8.3) with respect to vectors $\mathbf{w}$ and $\boldsymbol{\beta}$, or by maximizing h_I^2 (Eq. 8.4) only with respect to vector $\boldsymbol{\beta}$.

Maximizing h_I^2 only with respect to $\boldsymbol{\beta}$ is simpler than maximizing ρ_{HI} with respect to $\mathbf{w}$ and $\boldsymbol{\beta}$; however, in the latter case the maximization process of ρ_{HI} gives more information associated with MESIM parameters than when h_I^2 is maximized only with respect to $\boldsymbol{\beta}$ (see Chap. 7, Eq. 7.13, for details). In this subsection, we maximize ρ_{HI} with respect to vectors $\mathbf{w}$ and $\boldsymbol{\beta}$ similar to the ESIM in Chap. 7, Sect. 7.1.1. Thus, we omit the steps and details of the maximization process of ρ_{HI}.

We maximize $\rho_{HI} = \frac{\mathbf{w}'\mathbf{\Psi}_M\boldsymbol{\beta}}{\sqrt{\mathbf{w}'\mathbf{\Psi}_M\mathbf{w}}\sqrt{\boldsymbol{\beta}'\mathbf{T}_M\boldsymbol{\beta}}}$ with respect to vectors $\mathbf{w}$ and $\boldsymbol{\beta}$ under the restrictions $\sigma_H^2 = \mathbf{w}'\mathbf{\Psi}\mathbf{w}$, $\sigma_I^2 = \boldsymbol{\beta}'\mathbf{T}\boldsymbol{\beta}$, and $0 < \sigma_H^2, \sigma_I^2 < \infty$, where σ_H^2 is the variance of $H = \mathbf{w}'\mathbf{a}$ and σ_I^2 is the variance of $I = \boldsymbol{\beta}'\mathbf{t}$. Thus, it is necessary to maximize the function

$$f(\boldsymbol{\beta}, \mathbf{w}, \mu, \phi) = \mathbf{w}'\mathbf{\Psi}\boldsymbol{\beta} - 0.5\mu\left(\boldsymbol{\beta}'\mathbf{T}\boldsymbol{\beta} - \sigma_I^2\right) - 0.5\phi\left(\mathbf{w}'\mathbf{\Psi}\mathbf{w} - \sigma_H^2\right) \tag{8.5}$$

with respect to $\boldsymbol{\beta}$, $\mathbf{w}$, μ, and ϕ, where μ and ϕ are Lagrange multipliers. The derivatives of Eq. (8.5) with respect to $\boldsymbol{\beta}$, $\mathbf{w}$, μ, and ϕ are:

$$\mathbf{\Psi}\mathbf{w} - \mu\mathbf{T}\boldsymbol{\beta} = \mathbf{0}, \tag{8.6}$$

$$\mathbf{\Psi}\boldsymbol{\beta} - \phi\mathbf{\Psi}\mathbf{w} = \mathbf{0}, \tag{8.7}$$

$$\boldsymbol{\beta}'\mathbf{T}\boldsymbol{\beta} = \sigma_I^2 \quad \text{and} \quad \mathbf{w}'\mathbf{\Psi}\mathbf{w} = \sigma_H^2, \tag{8.8}$$

respectively, where Eq. (8.8) denotes the restrictions imposed for maximizing ρ_{HI}. It can be shown (see Chap. 7) that vector $\mathbf{w}$ can be obtained as

$$\mathbf{w}_M = \mathbf{\Psi}_M^{-1}\mathbf{T}_M\boldsymbol{\beta} \tag{8.9}$$

and the net genetic merit in the MESIM context can be written as $H_M = \mathbf{w}_M'\mathbf{a}$; thus, the correlation between $H_M = \mathbf{w}_M'\mathbf{a}$ and I is $\rho_{H_MI} = \frac{\sqrt{\boldsymbol{\beta}'\mathbf{T}\boldsymbol{\beta}}}{\sqrt{\boldsymbol{\beta}'\mathbf{T}\mathbf{\Psi}^{-1}\mathbf{T}\boldsymbol{\beta}}}$ and the MESIM vector of coefficients ($\boldsymbol{\beta}$) that maximizes ρ_{H_MI} can be obtained from equation

$$\left(\mathbf{T}^{-1}\mathbf{\Psi} - \lambda_M^2\mathbf{I}_{2t}\right)\boldsymbol{\beta}_M = \mathbf{0}, \tag{8.10}$$

where $\mathbf{I}_{2t}$ is an identity matrix of size $2t \times 2t$ (t is the number of traits), and λ_M^2 and $\boldsymbol{\beta}_M$ are the *eigenvalue* and *eigenvector* of matrix $\mathbf{T}_M^{-1}\mathbf{\Psi}_M$. The words *eigenvalue* and *eigenvector* are derived from the German word *eigen*, which means *owned by* or *peculiar to*. Eigenvalues and eigenvectors are sometimes called *characteristic values* and *characteristic vectors*, *proper values* and *proper vectors*, or *latent values* and *latent vectors* (Meyer 2000). The square root of λ_M^2 (λ_M) is the canonical correlation between $H_M = \mathbf{w}_M'\mathbf{a}$ and $I_M = \boldsymbol{\beta}_M'\mathbf{t}$, and the optimized MESIM index can be written as $I_M = \boldsymbol{\beta}_M'\mathbf{t}$. Using a similar procedure to that described in Chap. 7 (Eq. 7.17), it can be show that vector $\boldsymbol{\beta}_M$ can be transformed into $\boldsymbol{\beta}_C = \mathbf{F}\boldsymbol{\beta}_M$, where $\mathbf{F}$ is a diagonal matrix with values equal to any real number, except zero values.

The maximized correlation between $H_M = \mathbf{w}'_M\mathbf{a}$ and $I_M = \boldsymbol{\beta}'_M\mathbf{t}$, or MESIM accuracy, is

$$\rho_{H_M I_M} = \frac{\sqrt{\boldsymbol{\beta}'_M \mathbf{T}_M \boldsymbol{\beta}_M}}{\sqrt{\boldsymbol{\beta}'_M \mathbf{T}_M \boldsymbol{\Psi}_M^{-1} \mathbf{T}_M \boldsymbol{\beta}_M}} = \frac{\sigma_{I_M}}{\sigma_{H_M}}, \tag{8.11}$$

where $\sigma_{I_M} = \sqrt{\boldsymbol{\beta}'_M \mathbf{T}_M \boldsymbol{\beta}_M}$ is the standard deviation of $I_M = \boldsymbol{\beta}'_M\mathbf{t}$, and $\sigma_{H_M} = \sqrt{\boldsymbol{\beta}'_M \mathbf{T}_M \boldsymbol{\Psi}_M^{-1} \mathbf{T}_M \boldsymbol{\beta}_M}$ is the standard deviation of $H_M = \mathbf{w}'_M\mathbf{a}$.

The maximized selection response and expected genetic gain per trait of MESIM are

$$R_M = k_I \sqrt{\boldsymbol{\beta}'_{M_1} \mathbf{T}_M \boldsymbol{\beta}_{M_1}} \tag{8.12}$$

and

$$\mathbf{E}_M = k_I \frac{\boldsymbol{\Psi}_M \boldsymbol{\beta}_{M_1}}{\sqrt{\boldsymbol{\beta}'_{M_1} \mathbf{T}_M \boldsymbol{\beta}_{M_1}}}, \tag{8.13}$$

respectively, where $\boldsymbol{\beta}_{M_1}$ is the first eigenvector of matrix $\mathbf{T}_M^{-1}\boldsymbol{\Psi}_M$. If vector $\boldsymbol{\beta}_{M_1}$ is multiplied by matrix $\mathbf{F}$, we obtain $\boldsymbol{\beta}_{C_1} = \mathbf{F}\boldsymbol{\beta}_{M_1}$; in this case, we can replace $\boldsymbol{\beta}_{M_1}$ with $\boldsymbol{\beta}_{C_1} = \mathbf{F}\boldsymbol{\beta}_{M_1}$ in Eqs. (8.12) and (8.13), and the optimized MESIM index should be written as $I_M = \boldsymbol{\beta}'_{C_1}\mathbf{y}$.

8.1.2 *Estimating MESIM Parameters*

We estimate the MESIM parameters using the same procedure described in Chap. 7 (Sect. 7.1.4) to estimate the ESIM parameters. Let $\widehat{\mathbf{C}}$, $\widehat{\mathbf{P}}$, and $\widehat{\mathbf{S}}_M$ be the estimates of the genotypic, phenotypic, and marker scores covariance matrices, $\widehat{\mathbf{T}}_M = \begin{bmatrix} \widehat{\mathbf{P}} & \widehat{\mathbf{S}}_M \\ \widehat{\mathbf{S}}_M & \widehat{\mathbf{S}}_M \end{bmatrix}$ and $\widehat{\boldsymbol{\Psi}}_M = \begin{bmatrix} \widehat{\mathbf{C}} & \widehat{\mathbf{S}}_M \\ \widehat{\mathbf{S}}_M & \widehat{\mathbf{S}}_M \end{bmatrix}$ the estimated block matrices (Chap. 4) and $\widehat{\mathbf{W}} = \widehat{\mathbf{T}}_M^{-1}\widehat{\boldsymbol{\Psi}}_M$; then, to find the estimators $\widehat{\boldsymbol{\beta}}_{M_1}$ and $\widehat{\lambda}^2_{M_1}$ of the first eigenvector ($\boldsymbol{\beta}_{M_1}$) and the first eigenvalue ($\lambda^2_{M_1}$) respectively, we need to solve the equation

$$(\widehat{\mathbf{W}}\widehat{\mathbf{W}}' - \widehat{\mu}_j \mathbf{I})\widehat{\boldsymbol{\beta}}_{M_j} = \mathbf{0}, \tag{8.14}$$

where $\widehat{\mu}_j = \widehat{\lambda}^4_{M_j}$, $j = 1, 2, \ldots, 2t$. For additional details, see Eqs. (7.22) and (7.23), and Sect. 7.1.5 of Chap. 7. The result of Equation (8.14) allow the MESIM index

($I_M = \boldsymbol{\beta}'_{M_1}\mathbf{t}$) to be estimated as $\widehat{I}_M = \widehat{\boldsymbol{\beta}}'_{M_1}\mathbf{t}$, whereas the estimator of the maximized ESIM selection response and its expected genetic gain per trait can be denoted by

$$\widehat{R}_M = k_I\sqrt{\widehat{\boldsymbol{\beta}}'_{M_1}\widehat{\mathbf{T}}_M\widehat{\boldsymbol{\beta}}_{M_1}} \text{ and } \widehat{\mathbf{E}}_M = k_I\frac{\widehat{\boldsymbol{\Psi}}_M\widehat{\boldsymbol{\beta}}_{M_1}}{\sqrt{\widehat{\boldsymbol{\beta}}'_{M_1}\widehat{\mathbf{T}}_M\widehat{\boldsymbol{\beta}}_{M_1}}}, \tag{8.15}$$

respectively.

8.1.3 Numerical Examples

To validate the MESIM theoretical results, we use a real maize (*Zea mays*) F_2 population with 247 genotypes (each with two repetitions), 195 molecular markers, and two traits—plant height (PHT, cm) and ear height (EHT, cm)—evaluated in one environment. We coded the marker homozygous loci for the allele from the first parental line by 1, whereas the marker homozygous loci for the allele from the second parental line was coded by -1 and the marker heterozygous loci by 0. The estimated phenotypic, genetic, and marker scores covariance matrices were $\widehat{\mathbf{P}} = \begin{bmatrix} 191.81 & 106.89 \\ 106.89 & 167.93 \end{bmatrix}$, $\widehat{\mathbf{C}} = \begin{bmatrix} 83.00 & 57.44 \\ 57.44 & 59.80 \end{bmatrix}$, and $\widehat{\mathbf{S}}_M = \begin{bmatrix} 15.750 & 0.983 \\ 0.983 & 28.083 \end{bmatrix}$ respectively, and the vector of economic weights was $\mathbf{a}' = [\mathbf{w}' \quad \mathbf{0}']$, where $\mathbf{w}' = [-1 \quad -1]$ and $\mathbf{0}' = [0 \quad 0]$. Details of how to estimate the marker scores and their variance were given in Chap. 4.

We compare LMSI versus MESIM efficiency. The estimated LMSI vector of coefficients was $\widehat{\boldsymbol{\beta}}' = \mathbf{a}'\widehat{\boldsymbol{\Psi}}_M\widehat{\mathbf{T}}_M^{-1} = [-0.59 \quad -0.18 \quad -0.41 \quad -0.82]$. Using a 10% selection intensity ($k_I = 1.755$), the estimated LMSI selection response and the expected genetic gain per trait were $\widehat{R} = k_I\sqrt{\widehat{\boldsymbol{\beta}}'\widehat{\mathbf{T}}_M\widehat{\boldsymbol{\beta}}} = 20.41$ and $\widehat{\mathbf{E}}' = k_I\frac{\widehat{\boldsymbol{\beta}}'\widehat{\boldsymbol{\Psi}}_M}{\sqrt{\widehat{\boldsymbol{\beta}}'\widehat{\mathbf{T}}_M\widehat{\boldsymbol{\beta}}}} = [-10.09 \quad -10.31 \quad -2.53 \quad -4.39]$ respectively, whereas the estimated LMSI accuracy was $\widehat{\rho}_{H\widehat{I}} = \frac{\widehat{\sigma}_I}{\widehat{\sigma}_H} = 0.72$.

Vector $\widehat{\boldsymbol{\beta}}'_{M_1} = [0.089 \quad -0.061 \quad -0.536 \quad 0.837]$ was the original estimated MESIM vector of coefficients. Using matrix $\mathbf{F} = \begin{bmatrix} -0.1 & 0 & 0 & 0 \\ 0 & -0.1 & 0 & 0 \\ 0 & 0 & 0.75 & 0 \\ 0 & 0 & 0 & -0.75 \end{bmatrix}$, vector $\widehat{\boldsymbol{\beta}}'_{M_1}$ was transformed as $\widehat{\boldsymbol{\beta}}'_{C_1} = \widehat{\boldsymbol{\beta}}'_{M_1}\mathbf{F} = [-0.009 \quad 0.006 \quad -0.402 \quad 0.628]$ and then the estimated MESIM index was $\widehat{I}_M = -0.009\,\text{PHT} + 0.006\,\text{EHT} - 0.402\,S_{\text{PHT}} + 0.628\,S_{\text{EHT}}$, where S_{PHT} and S_{EHT} denote the marker scores associated with PHT and EHT respectively. The estimated MESIM expected genetic gain, selection response, and accuracy were $\widehat{\mathbf{E}}'_M = k_I \frac{\widehat{\boldsymbol{\beta}}'_{C_1}\widehat{\boldsymbol{\Psi}}_M}{\sqrt{\widehat{\boldsymbol{\beta}}'_{C_1}\widehat{\mathbf{T}}_M\widehat{\boldsymbol{\beta}}_{C_1}}} = [-3.438 \quad -8.516 \quad -3.319 \quad -8.372]$, $\widehat{R}_M = k_I\sqrt{\widehat{\boldsymbol{\beta}}'_{C_1}\widehat{\mathbf{T}}_M\widehat{\boldsymbol{\beta}}_{C_1}} = 6.573$ and $\widehat{\rho}_{H_M\widehat{I}_M} = \frac{\widehat{\sigma}_{I_M}}{\widehat{\sigma}_{H_M}} = 0.99$ respectively.

The inner product of the estimated LMSI and MESIM vector of coefficients were 1.221 and 0.556 respectively, whence the estimated LMSI selection response (20.41) divided by 1.221 was 16.716, and the estimated MESIM selection response (6.573) divided by 0.556 was 11.821. That is, the estimated LMSI selection response was higher than the estimated MESIM selection response for this data set. Similar results were found when we compared the estimated LMSI expected genetic gain per trait with the estimated MESIM expected genetic gain per trait. Finally, Fig. 8.1 presents the frequency distribution of the 247 estimated MESIM values for the real data set described earlier, which approaches normal distribution, as we would expect.

Now with a selection intensity of 10% ($k_I = 1.755$), we compare the LMSI and MESIM efficiency using the simulated data set described in Sect. 2.8.1 of Chap. 2 for four phenotypic selection cycles, each with four traits (T_1, T_2, T_3 and T_4), 500 genotypes, and four replicates of each genotype. The economic weights for T_1, T_2, T_3, and T_4 were 1, −1, 1, and 1 respectively. For this data set, we did not use the linear transformation $\widehat{\boldsymbol{\beta}}_{C_1} = \mathbf{F}\widehat{\boldsymbol{\beta}}_{M_1}$.

The estimated selection responses of the linear marker, combined genomic and genome-wide selection indices (LMSI, CLGSI, and GW-LMSI respectively; see

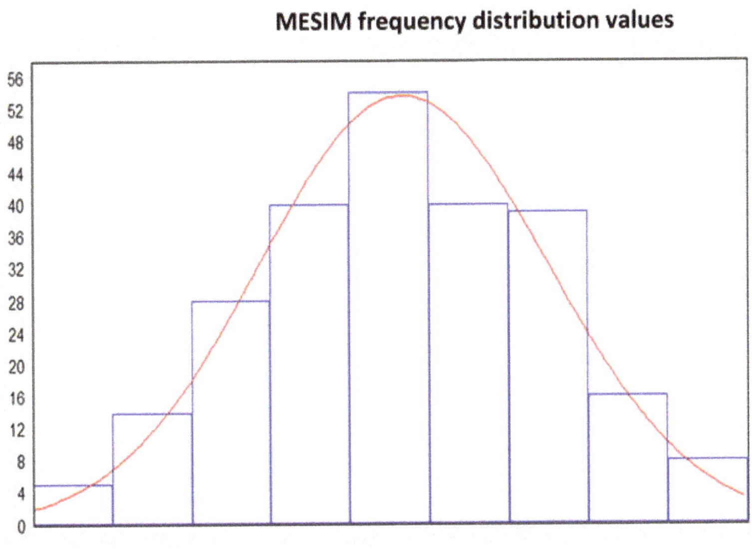

Fig. 8.1 Frequency distribution of 247 estimated molecular eigen selection index method (MESIM) values for one selection cycle in an environment for a real maize (*Zea mays*) F_2 population with 195 molecular markers and two traits, plant height (PHT, cm) and ear height (EHT, cm), and their associated marker scores S_{PHT} and S_{EHT} respectively

Chaps. 4 and 5 for details) for four simulated selection cycles when their vectors of coefficients were normalized, are presented in Table 8.1. Also, in this table the selection responses of the estimated linear molecular, genomic, and genome-wide eigen selection index methods (MESIM, GESIM, and GW-ESIM respectively; details in Sect. 8.2) are shown for four simulated selection cycles. The average of the estimated LMSI selection response was 2.22, whereas the average of the estimated MESIM selection response was 1.69. The estimated LMSI selection response was higher than that of the MESIM.

Table 8.2 presents the estimated LMSI and MESIM expected genetic gains for four traits (T1, T2, T3, and T4) and their associated marker scores (S1, S2, S3, and S4) for four simulated selection cycles. The averages of the estimated LMSI

Table 8.1 Estimated linear molecular, combined genomic, and genome-wide selection index (LMSI, CLGSI and GW-LMSI respectively) selection responses when their vectors of coefficients are normalized for four simulated selection cycles

	Estimated selection response					
Cycle	LMSI	CLGSI	GW-LMSI	MESIM	GESIM	GW-ESIM
1	0.02	1.24	0.93	0.50	3.95	0.73
2	4.94	0.80	0.80	1.21	3.07	1.06
3	3.69	0.34	0.93	3.91	2.05	0.77
4	0.23	0.35	0.83	1.15	1.90	1.14
Average	2.22	0.68	0.87	1.69	2.74	0.93

Estimated linear molecular, genomic, and genome-wide eigen selection index method (MESIM, GESIM, and GW-ESIM respectively) selection responses for four simulated selection cycles. The selection intensity was 10% ($k_I = 1.755$)

Table 8.2 Estimated linear molecular selection index (LMSI) and estimated linear molecular eigen selection index method (MESIM) expected genetic gains for four traits (T1, T2, T3, and T4) and their associated marker scores (S1, S2, S3, and S4) for four simulated selection cycles. The selection intensity was 10% ($k_I = 1.755$)

	Estimated LMSI expected genetic gain							
	Traits				Marker scores			
Cycle	T1	T2	T3	T4	S1	S2	S3	S4
1	24.48	−0.01	0.74	−0.87	4.18	−1.14	0.72	0.79
2	7.14	−3.39	2.62	1.55	3.78	−2.30	1.02	1.37
3	9.17	−3.04	1.87	1.21	6.22	−1.51	1.02	0.26
4	10.16	−1.95	1.17	1.88	8.63	−3.83	0.09	0.13
Average	12.74	−2.10	1.60	0.94	5.70	−2.19	0.71	0.64
	Estimated MESIM expected genetic gain							
	Traits				Marker scores			
Cycle	T1	T2	T3	T4	S1	S2	S3	S4
1	27.48	2.60	−1.03	−2.64	3.85	0.00	−0.04	−0.43
2	8.82	−4.75	0.37	2.11	14.06	4.09	0.38	−2.76
3	9.83	1.74	0.72	0.37	8.03	1.76	0.31	0.34
4	11.47	−1.13	−1.64	1.53	8.66	−3.96	−1.47	0.04
Average	14.40	−0.38	−0.39	0.34	8.65	0.47	−0.21	−0.70

expected genetic gains for the four traits and their associated marker scores were 12.74, −2.10, 1.60, 0.94, 5.70, −2.19, 0.71, and 0.64 respectively, whereas the averages of the estimated MESIM expected genetic gains for the four traits and their associated marker scores were 14.40, −0.38, −0.39, 0.34, 8.65, 0.47, −0.21, and −0.70 respectively. Except for trait T1 and its associated molecular scores, the estimated LMSI expected genetic gains per trait were higher than the estimated MESIM expected genetic gains. Thus, for this data set, LMSI efficiency was greater than MESIM efficiency.

Chapter 11 presents RIndSel, a user-friendly graphical unit interface in JAVA that is useful for estimating the LMSI and ESIM parameters and selecting parents for the next selection cycle.

8.2 The Linear Genomic Eigen Selection Index Method

The linear genomic eigen selection index method (GESIM) is based on the standard CLGSI described in Chap. 5, and uses genomic estimated breeding values (GEBVs) and phenotypic values jointly to predict the net genetic merit. Thus, conditions for constructing a valid GESIM are the same as those for constructing the CLGSI. Also, the MESIM theory described in Sect. 8.1 is directly applied to the GESIM and only minor changes are necessary in GESIM theory. For example, instead of marker scores, the GESIM uses GEBVs to predict the net genetic merit; thus, the details of the estimation process are the same as for the MESIM.

8.2.1 The GESIM Parameters

In the GESIM context, the net genetic merit can be written as

$$H = \mathbf{w}_1'\mathbf{g} + \mathbf{w}_2'\boldsymbol{\gamma} = \begin{bmatrix} \mathbf{w}_1' & \mathbf{w}_2' \end{bmatrix} \begin{bmatrix} \mathbf{g} \\ \boldsymbol{\gamma} \end{bmatrix} = \mathbf{w}'\boldsymbol{\alpha}, \tag{8.16}$$

where $\mathbf{g}' = [g_1 \quad \cdots \quad g_t]$ is the vector of true breeding values, t is the number of traits, $\mathbf{w}_1' = [w_1 \quad \cdots \quad w_t]$ is a vector of unknown economic weights associated with $\mathbf{g}$, $\mathbf{w}_2' = [0_1 \quad \cdots \quad 0_t]$ is a null vector associated with the vector of genomic breeding values $\boldsymbol{\gamma}' = [\gamma_1 \quad \gamma_2 \quad \cdots \quad \gamma_t]$, $\mathbf{w}' = [\mathbf{w}_1' \quad \mathbf{w}_2']$, and $\boldsymbol{\alpha}' = [\mathbf{g}' \quad \boldsymbol{\gamma}']$. The estimator of $\boldsymbol{\gamma}$ is the GEBV (see Chap. 5 for additional details). The GESIM index can be written as

$$I = \boldsymbol{\beta}_y'\mathbf{y} + \boldsymbol{\beta}_\gamma'\boldsymbol{\gamma} = \begin{bmatrix} \boldsymbol{\beta}_y' & \boldsymbol{\beta}_\gamma' \end{bmatrix} \begin{bmatrix} \mathbf{y} \\ \boldsymbol{\gamma} \end{bmatrix} = \boldsymbol{\beta}'\mathbf{f}, \tag{8.17}$$

where $\mathbf{y}' = [y_1 \quad \cdots \quad y_t]$ is the vector of phenotypic values; $\boldsymbol{\beta}_y'$ and $\boldsymbol{\beta}_\gamma$ are vectors of weights of phenotypic and genomic breeding values weights respectively; $\boldsymbol{\beta}' = [\boldsymbol{\beta}_y' \quad \boldsymbol{\beta}_\gamma']$ and $\mathbf{f}' = [\mathbf{y}' \quad \boldsymbol{\gamma}']$.

Let $Var(H) = \mathbf{w}'\mathbf{A}\mathbf{w} = \sigma_H^2$ be the variance of $H = \mathbf{w}'\boldsymbol{\alpha}$, $Var(I) = \boldsymbol{\beta}'\boldsymbol{\Phi}\boldsymbol{\beta} = \sigma_I^2$ the variance of $I = \boldsymbol{\beta}'\mathbf{f}$, and $Cov(H, I) = \mathbf{w}'\mathbf{A}\boldsymbol{\beta} = \sigma_{HI}$ the covariance between H and I, where $\mathbf{A} = Var\begin{bmatrix}\mathbf{g}\\ \boldsymbol{\gamma}\end{bmatrix} = \begin{bmatrix}\mathbf{C} & \boldsymbol{\Gamma}\\ \boldsymbol{\Gamma} & \boldsymbol{\Gamma}\end{bmatrix}$ and $\boldsymbol{\Phi} = Var\begin{bmatrix}\mathbf{y}\\ \boldsymbol{\gamma}\end{bmatrix} = \begin{bmatrix}\mathbf{P} & \boldsymbol{\Gamma}\\ \boldsymbol{\Gamma} & \boldsymbol{\Gamma}\end{bmatrix}$ are block matrices $2t \times 2t$ (t is the number of traits) of covariance matrices and $\mathbf{P}$, $\boldsymbol{\Gamma}$, and $\mathbf{C}$ are covariance matrices of phenotypic ($\mathbf{y}$), genomic ($\boldsymbol{\gamma}$), and genetic ($\mathbf{g}$) values respectively. Then, $\rho_{HI} = \frac{\mathbf{w}'\mathbf{A}\boldsymbol{\beta}}{\sqrt{\mathbf{w}'\mathbf{A}\mathbf{w}}\sqrt{\boldsymbol{\beta}'\boldsymbol{\Phi}\boldsymbol{\beta}}}$ is the correlation between $H = \mathbf{w}'\boldsymbol{\alpha}$ and $I = \boldsymbol{\beta}'\mathbf{f}$ and the GESIM selection response can be written as

$$R = k_I \sigma_H \rho_{HI}, \tag{8.18}$$

where k_I is the standardized selection differential (or selection intensity) associated with the GESIM and $\sigma_H = \sqrt{\mathbf{w}'\mathbf{A}\mathbf{w}}$ is the standard deviation of the variance of H. It is assumed that k_I is fixed, and that matrices $\boldsymbol{\Phi}$ and $\mathbf{A}$ are known; then, we can maximize R by maximizing ρ_{HI} with respect to vectors $\mathbf{w}$ and $\boldsymbol{\beta}$ under the restrictions $\sigma_H^2 = \mathbf{w}'\mathbf{A}\mathbf{w}$, $\sigma_I^2 = \boldsymbol{\beta}'\boldsymbol{\Phi}\boldsymbol{\beta}$, and $0 < \sigma_H^2$, $\sigma_I^2 < \infty$; similar to the MESIM.

It can be shown that the vector $\mathbf{w}$ in the GESIM context is

$$\mathbf{w}_G = \mathbf{A}^{-1}\boldsymbol{\Phi}\boldsymbol{\beta} \tag{8.19}$$

and that the net genetic merit can be written as $H_G = \mathbf{w}_G'\boldsymbol{\alpha}$. The correlation between $H_G = \mathbf{w}_G'\boldsymbol{\alpha}$ and $I = \boldsymbol{\beta}'\mathbf{f}$ is $\rho_{H_G I} = \frac{\sqrt{\boldsymbol{\beta}'\boldsymbol{\Phi}\boldsymbol{\beta}}}{\sqrt{\boldsymbol{\beta}'\boldsymbol{\Phi}\mathbf{A}^{-1}\boldsymbol{\Phi}\boldsymbol{\beta}}}$ and the GESIM index vector of coefficients that maximizes $\rho_{H_G I}$ can be obtained from the equation

$$\left(\boldsymbol{\Phi}^{-1}\mathbf{A} - \lambda_G^2 \mathbf{I}_{2t}\right)\boldsymbol{\beta}_G = \mathbf{0}, \tag{8.20}$$

where $\mathbf{I}_{2t}$ is an identity matrix of size $2t \times 2t$ (t is the number of traits); the optimized GESIM index can be written as $I_G = \boldsymbol{\beta}_G'\mathbf{f}$. By Eqs. (8.19) and (8.20), GESIM accuracy can be written as

$$\rho_{H_G I_G} = \frac{\sigma_{I_G}}{\sigma_{H_G}}, \tag{8.21}$$

where $\sigma_{I_G} = \sqrt{\boldsymbol{\beta}_G'\boldsymbol{\Phi}\boldsymbol{\beta}_G}$ is the standard deviation of $I_G = \boldsymbol{\beta}_G'\mathbf{f}$, and $\sigma_{H_G} = \sqrt{\boldsymbol{\beta}_G'\boldsymbol{\Phi}\mathbf{A}^{-1}\boldsymbol{\Phi}\boldsymbol{\beta}_G}$ is the standard deviation of $H_G = \mathbf{w}_G'\boldsymbol{\alpha}$. In Eq. (8.20), $\lambda_G^2 = \rho_{H_G I_G}^2$ is the square of the canonical correlation between H_G and I_G, and $\boldsymbol{\beta}_G$ is the canonical vector associated with $\lambda_G^2 = \rho_{H_G I_G}^2$.

The maximized GESIM selection response and expected genetic gain per trait are

$$R_G = k_I\sqrt{\boldsymbol{\beta}_G'\boldsymbol{\Phi}\boldsymbol{\beta}_G} \tag{8.22}$$

and

$$\mathbf{E}_G = k_I\frac{\mathbf{A}\boldsymbol{\beta}_G}{\sqrt{\boldsymbol{\beta}_G'\boldsymbol{\Phi}\boldsymbol{\beta}_G}}, \tag{8.23}$$

respectively, where $\boldsymbol{\beta}_G$ is the first eigenvector of matrix $\boldsymbol{\Phi}^{-1}\mathbf{A}$. Vector $\boldsymbol{\beta}_G$ can be transformed as $\boldsymbol{\beta}_{CG} = \mathbf{F}\boldsymbol{\beta}_G$, where $\mathbf{F}$ is a diagonal matrix defined earlier.

8.2.2 Numerical Examples

To compare the CLGSI versus GESIM theoretical results, we use a real maize (*Zea mays*) F_2 population with 244 genotypes (each with two repetitions), 233 molecular markers, and three traits—grain yield (GY, ton ha^{-1}), ear height (EHT, cm), and plant height (PHT, cm). We estimated matrices $\mathbf{P}$ and $\mathbf{C}$ using Eqs. (2.22) to (2.24) described in Chap. 2, whence the estimated matrices were $\widehat{\mathbf{P}} = \begin{bmatrix} 0.45 & 1.33 & 2.33 \\ 1.33 & 65.07 & 83.71 \\ 2.33 & 83.71 & 165.99 \end{bmatrix}$ and $\widehat{\mathbf{C}} = \begin{bmatrix} 0.07 & 0.61 & 1.06 \\ 0.61 & 17.93 & 22.75 \\ 1.06 & 22.75 & 44.53 \end{bmatrix}$. In a similar manner, we estimated matrix $\boldsymbol{\Gamma}$ by applying Eqs. (5.21) to (5.23) described in Chap. 5 using phenotypic and marker information jointly; the estimated matrix was $\widehat{\boldsymbol{\Gamma}} = \begin{bmatrix} 0.07 & 0.65 & 1.05 \\ 0.65 & 10.62 & 14.25 \\ 1.05 & 14.25 & 26.37 \end{bmatrix}$. The selection intensity for making a selection cycle was 10% ($k_I = 1.755$) and the vector of economic weights was $\mathbf{w}' = [5 \quad -0.1 \quad -0.1 \quad 0 \quad 0 \quad 0]$. To obtain the estimated vector of coefficient of CLGSI ($\widehat{\boldsymbol{\beta}} = \widehat{\boldsymbol{\Phi}}^{-1}\widehat{\mathbf{A}}\mathbf{w}$) and GESIM (Eq. 8.20), it is necessary to construct matrices $\widehat{\mathbf{A}} = \begin{bmatrix} \widehat{\mathbf{C}} & \widehat{\boldsymbol{\Gamma}} \\ \widehat{\boldsymbol{\Gamma}} & \widehat{\boldsymbol{\Gamma}} \end{bmatrix}$ and $\widehat{\boldsymbol{\Phi}} = \begin{bmatrix} \widehat{\mathbf{P}} & \widehat{\boldsymbol{\Gamma}} \\ \widehat{\boldsymbol{\Gamma}} & \widehat{\boldsymbol{\Gamma}} \end{bmatrix}$.

The estimated CLGSI vector of coefficients for the traits GY, EHT, and PHT and their associated GEBVs (GEBV_{GY}, GEBV_{EHT}, and GEBV_{PHT} respectively) was $\widehat{\boldsymbol{\beta}}' = [0.08 \quad -0.02 \quad -0.01 \quad 4.92 \quad -0.08 \quad -0.09]$, whereas the estimated CLGSI selection response, accuracy, and expected genetic gain per trait were $\widehat{R} = k_I\sqrt{\widehat{\boldsymbol{\beta}}'\widehat{\boldsymbol{\Phi}}\widehat{\boldsymbol{\beta}}} = 1.54$, $\widehat{\rho}_{HI} = \frac{\widehat{\sigma}_I}{\widehat{\sigma}_H} = 0.814$, and $\widehat{\mathbf{E}}' = k_I\frac{\widehat{\boldsymbol{\beta}}'\widehat{\mathbf{A}}}{\sqrt{\widehat{\boldsymbol{\beta}}'\widehat{\boldsymbol{\Phi}}\widehat{\boldsymbol{\beta}}}} = [0.36 \quad 1.04 \quad 1.70 \quad 0.36 \quad 1.53 \quad 2.38]$ respectively. Finally, $\widehat{I} = 0.08\text{GY} - 0.02\text{EHT} - 0.01\text{PHT} + 4.92\text{GEBV}_{\text{GY}} - 0.08\text{GEBV}_{\text{EHT}} - 0.09\text{GEBV}_{\text{PHT}}$ was the estimated CLGSI.

The estimated GESIM vector of coefficients, selection response, accuracy, and expected genetic gain per trait were $\widehat{\boldsymbol{\beta}}'_{G_1} = [-0.207 \quad 0.029 \quad 0.041 \quad 0.820 \quad 0.337 \quad 0.411]$, $\widehat{R}_G = k_I\sqrt{\widehat{\boldsymbol{\beta}}'_{G_1}\widehat{\boldsymbol{\Phi}}\widehat{\boldsymbol{\beta}}_{G_1}} = 6.288$, $\widehat{\rho}_{\widehat{H}_G\widehat{I}_G} = \dfrac{\sqrt{\widehat{\boldsymbol{\beta}}'_{G_1}\widehat{\boldsymbol{\Phi}}\widehat{\boldsymbol{\beta}}_{G_1}}}{\sqrt{\widehat{\boldsymbol{\beta}}'_{G_1}\widehat{\boldsymbol{\Phi}}\widehat{\mathbf{A}}^{-1}\widehat{\boldsymbol{\Phi}}\widehat{\boldsymbol{\beta}}_{G_1}}} = 0.9056$, and $\widehat{\mathbf{E}}'_G = k_1\dfrac{\widehat{\boldsymbol{\beta}}'_{G_1}\widehat{\mathbf{A}}}{\sqrt{\widehat{\boldsymbol{\beta}}'_{G_1}\widehat{\boldsymbol{\Phi}}\widehat{\boldsymbol{\beta}}_{G_1}}} = [0.369 \quad 5.528 \quad 9.186 \quad 0.370 \quad 5.250 \quad 8.702]$ respectively.

Fig. 8.2 presents the frequency distribution of the 244 estimated GESIM index values for one (Fig. 8.2a) and three traits (Fig. 8.2b) using the real data set described earlier. The frequency distribution of the estimated GESIM index values approaches the normal distribution for both indices.

Now, we compare the estimated CLGSI and GESIM selection response and expected genetic gain per trait using the simulated data set described in Sect. 2.8.1 of Chap. 2 for four phenotypic selection cycles, each with four traits (T_1, T_2, T_3 and T_4), 500 genotypes, and four replicates per genotype. The economic weights of T_1, T_2, T_3, and T_4 were 1, −1, 1, and 1 respectively and the selection intensity for both

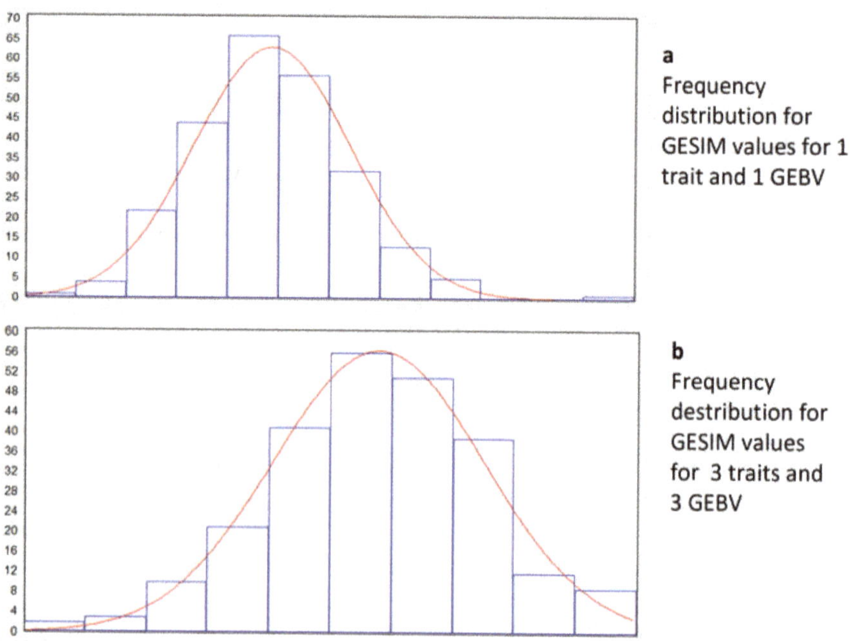

Fig. 8.2 Frequency distribution of the 244 estimated genomic eigen selection index method (GESIM) values for the one-trait case (**a**) and for the three-trait case (**b**) for one selection cycle in an environment for a real maize (*Zea mays*) F_2 population with 233 molecular markers. Note that the frequency distribution of the estimated GESIM index values approaches normal distribution for both indices

Table 8.3 Estimated combined linear genomic selection index (CLGSI) and estimated GESIM expected genetic gains for four traits (T1, T2, T3, and T4) and their associated genomic estimated breeding values (GEBV1, GEBV2, GEBV3, and GEBV4) for four simulated selection cycles. The selection intensity was 10% ($k_I = 1.755$)

	Estimated CLGSI expected genetic gain							
	Traits				Genomic estimated breeding value			
Cycle	T1	T2	T3	T4	GEBV1	GEBV2	GEBV3	GEBV4
1	7.46	−3.69	3.26	1.60	7.28	−4.38	3.72	3.29
2	7.08	−3.45	2.91	1.17	7.08	−3.63	3.66	2.67
3	7.81	−3.51	2.06	0.76	7.30	−3.92	2.35	2.40
4	7.46	−2.76	2.48	0.81	6.84	−2.79	2.79	2.40
Average	7.45	−3.35	2.68	1.09	7.13	−3.68	3.13	2.69
	Estimated GESIM expected genetic gain							
	Traits				Genomic estimated breeding value			
Cycle	T1	T2	T3	T4	GEBV1	GEBV2	GEBV3	GEBV4
1	8.28	−3.51	2.93	0.92	7.77	−4.27	3.52	2.64
2	7.89	−3.09	2.42	0.82	7.40	−3.41	3.29	2.38
3	8.47	−3.26	1.69	0.46	7.55	−3.78	2.11	2.16
4	8.08	−2.46	2.04	0.66	7.15	−2.67	2.53	2.39
Average	8.18	−3.08	2.27	0.71	7.46	−3.53	2.86	2.39

indices was 10% ($k_I = 1.755$). For this data set, matrix **F** was an identity matrix of size 8×8 in all four selection cycles.

For this data set, the averages of the estimated CLGSI and GESIM selection responses were 0.68 and 2.74 (Table 8.1) respectively. The estimated CLGSI selection response was lower than the estimated GESIM selection response. Table 8.3 presents the estimated CLGSI and GESIM expected genetic gain for four traits (T1, T2, T3, and T4) and their associated genomic estimated breeding values (GEBV1, GEBV2, GEBV3, and GEBV4) for four simulated selection cycles. The averages of the estimated CLGSI expected genetic gains for the four traits and their associated GEBVs were 7.45, −3.35, 2.68, 1.09, 7.13, −3.68, 3.13, and 2.69 respectively, whereas the averages of the estimated GESIM expected genetic gains for the four traits and their associated GEBVs were 8.18, −3.08, 2.27, 0.71, 7.46, −3.53, 2.86, and 2.39 respectively. The estimated CLGSI and GESIM expected genetic gains per trait were very similar.

8.3 The Genome-Wide Linear Eigen Selection Index Method

The MESIM requires regressing phenotypic values on marker coded values to predict the marker score values for each individual candidate for selection, and then combining the marker scores with phenotypic information using the MESIM to obtain a final prediction of the net genetic merit. In addition, the GESIM requires fitting of a statistical model to estimate all available marker effects in the training population; these estimates are then used to obtain GEBVs, which are predictors of breeding values. Crossa and Cerón-Rojas (2011) extended the ESIM theory to a genome-wide linear molecular ESIM (GW-ESIM) similar to the GW-LMSI described in Chap. 4. The GW-LMSI and GW-ESIM are very similar and only minor changes are necessary in GW-ESIM; for example, instead of estimating the GW-LMSI vector of coefficients according to the LPSI method (Chap. 2), the GW-ESIM vector of coefficients is estimated according to the singular value decomposition (SVD) described in Chap. 7.

8.3.1 *The GW-ESIM Parameters*

In the GW-ESIM context, the net genetic merit can be written as

$$H = \mathbf{w}_1'\mathbf{g} + \mathbf{w}_2'\mathbf{m} = \begin{bmatrix} \mathbf{w}_1' & \mathbf{w}_2' \end{bmatrix} \begin{bmatrix} \mathbf{g} \\ \mathbf{m} \end{bmatrix} = \mathbf{w}'\mathbf{x}, \tag{8.24}$$

where $\mathbf{g}' = [g_1 \ \cdots \ g_t]$ is the vector of true breeding values, t is the number of traits, $\mathbf{w}_1' = [w_1 \ \cdots \ w_t]$ is the vector of unknown economic weights associated with the breeding values; $\mathbf{w}_2' = [0_1 \ \cdots \ 0_N]$ is a null vector associated with the vector of marker code values $\mathbf{m}' = [m_1 \ \cdots \ m_N]$, where m_j (j = 1, 2, ..., N = number of markers) is the jth marker in the training population; $\mathbf{w}' = [\mathbf{w}_1' \ \mathbf{w}_2']$ and $\mathbf{x} = [\mathbf{g}' \ \mathbf{m}']$. The GW-ESIM ($I$) index combines the phenotypic value and all the marker information of individuals to predict Eq. (8.24) values in each selection cycle and can be written as

$$I = \boldsymbol{\beta}_y'\mathbf{y} + \boldsymbol{\beta}_m'\mathbf{m} = \begin{bmatrix} \boldsymbol{\beta}_y' & \boldsymbol{\beta}_m' \end{bmatrix} \begin{bmatrix} \mathbf{y} \\ \mathbf{m} \end{bmatrix} = \boldsymbol{\beta}'\mathbf{q}, \tag{8.25}$$

where $\boldsymbol{\beta}_y'$ and $\boldsymbol{\beta}_m$ are vectors of phenotypic and marker weights respectively; $\mathbf{y}' = [y_1 \ \cdots \ y_t]$ is the vector of phenotypic values; $\mathbf{m}$ was defined in Eq. (8.24); $\boldsymbol{\beta}' = [\boldsymbol{\beta}_y' \ \boldsymbol{\beta}_m']$ and $\mathbf{q}' = [\mathbf{y}' \ \mathbf{m}']$.

Let $\sigma_I^2 = \boldsymbol{\beta}'\mathbf{Q}\boldsymbol{\beta}$ and $\sigma_H^2 = \mathbf{w}'\mathbf{Z}\mathbf{w}$ be the variance of $I = \boldsymbol{\beta}'\mathbf{q}$ and $H = \mathbf{w}'\mathbf{z}$ respectively, and $\sigma_{HI} = \mathbf{w}'\mathbf{Z}\boldsymbol{\beta}$ the covariance between I and H, where $\mathbf{Q} = Var\begin{bmatrix} \mathbf{y} \\ \mathbf{m} \end{bmatrix} = \begin{bmatrix} \mathbf{P} & \mathbf{G}_M' \\ \mathbf{G}_M & \mathbf{M} \end{bmatrix}$ and $\mathbf{X} = Var\begin{bmatrix} \mathbf{g} \\ \mathbf{m} \end{bmatrix} = \begin{bmatrix} \mathbf{C} & \mathbf{G}_M' \\ \mathbf{G}_M & \mathbf{M} \end{bmatrix}$ are block matrices of size $(t + N) \times (t + N)$ (t is the number of traits and N is the number of markers) where $\mathbf{P} = Var(\mathbf{y})$, $\mathbf{M} = Var(\mathbf{m})$, $\mathbf{C} = Var(\mathbf{g})$, and $\mathbf{G}_M = cov(\mathbf{y}, \mathbf{m}) = cov(\mathbf{g}, \mathbf{m})$ are covariance matrices of phenotypic ($\mathbf{y}$), coded marker ($\mathbf{m}$), and genetic ($\mathbf{g}$) values respectively, whereas $\mathbf{G}_M$ is the covariance matrix between $\mathbf{y}$ and $\mathbf{m}$, and between $\mathbf{g}$ and $\mathbf{m}$ (for details see Chap. 4); $\mathbf{w}$ and $\boldsymbol{\beta}$ were defined earlier. Note that although the size of matrices **P** and **C** are $t \times t$, the sizes of matrices **M** and $\mathbf{G}_M$ are $N \times N$ and $N \times t$ respectively. Thus, if the number of markers is very high, the size of matrices **M** and $\mathbf{G}_M$ could also be very high.

In Chap. 4 we described matrix **M** as

$$\mathbf{M} = \begin{bmatrix} 1 & (1-2\theta_{11}) & \ldots & (1-2\theta_{1N}) \\ (1-2\theta_{21}) & 1 & \ldots & (1-2\theta_{2N}) \\ \vdots & \vdots & \ddots & \vdots \\ (1-2\theta_{N1}) & (1-2\theta_{N2}) & \ldots & 1 \end{bmatrix}, \tag{8.26}$$

where $(1-2\theta_{ij})$ and θ_{ij} ($i, j = 1, 2, \ldots, N =$ number of markers) are the covariance (or correlation) and the recombination frequency between the ith and jth marker respectively, whereas matrix $\mathbf{G}_M$ can be written as

$$\mathbf{G}_M = \begin{bmatrix} (1-2r_{11})\alpha_{11} & (1-2r_{11})\alpha_{12} & \ldots & (1-2r_{1N})\alpha_{1N_Q} \\ (1-2r_{21})\alpha_{21} & (1-2r_{22})\alpha_{22} & \ldots & (1-2r_{2N})\alpha_{2N_Q} \\ \vdots & \vdots & \ddots & \vdots \\ (1-2r_{t1})\alpha_{t1} & (1-2r_{N2})\alpha_{t2} & \ldots & (1-2r_{NN})\alpha_{tN_Q} \end{bmatrix}, \tag{8.27}$$

where $(1-2r_{ik})\alpha_{qk}$ ($i = 1, 2, \ldots, N$, $k = 1, 2, \ldots, N_Q =$ number of quantitative trait loci (QTL), $q = 1, 2, \ldots, t$) is the covariance between the qth trait and the ith marker; r_{ik} is the recombination frequency between the ith and kth QTL, and α_{qk} is the effect of the kth QTL over the qth trait.

Let $\rho_{HI} = \frac{\mathbf{w}'\mathbf{X}\boldsymbol{\beta}}{\sqrt{\mathbf{w}'\mathbf{X}\mathbf{w}}\sqrt{\boldsymbol{\beta}'\mathbf{Q}\boldsymbol{\beta}}}$ be the correlation between $I = \boldsymbol{\beta}'\mathbf{q}$ and $H = \mathbf{w}'\mathbf{x}$; then, the GW-ESIM selection response can be written as

$$R = k_I \sigma_H \rho_{HI}, \tag{8.28}$$

where k_I is the standardized selection differential (or selection intensity) associated with GW-ESIM and $\sigma_H = \sqrt{\mathbf{w}'\mathbf{X}\mathbf{w}}$ is the standard deviation of the variance of H.

Assuming that k_I is fixed, and that matrices **Q** and **X** are known, we can maximize R (Eq. 8.28) by maximizing ρ_{HI} with respect to vectors $\mathbf{w}'$ and $\boldsymbol{\beta}$ under the restrictions $\sigma_H^2 = \mathbf{w}'\mathbf{X}\mathbf{w}$, $\sigma_I^2 = \boldsymbol{\beta}'\mathbf{Q}\boldsymbol{\beta}$, and $0 < \sigma_H^2, \sigma_I^2 < \infty$, similar to the MESIM and GESIM. It can be shown that vector **w** can be written as

$$\mathbf{w}_W = \mathbf{X}^{-1}\mathbf{Q}\boldsymbol{\beta} \tag{8.29}$$

and that $H_W = \mathbf{w}_W'\mathbf{x}$ is the net genetic merit in the GW-ESIM context. The correlation between $H_W = \mathbf{w}_W'\mathbf{x}$ and $I = \boldsymbol{\beta}'\mathbf{q}$ is $\rho_{H_W I} = \frac{\sqrt{\boldsymbol{\beta}'\mathbf{Q}\boldsymbol{\beta}}}{\sqrt{\boldsymbol{\beta}'\mathbf{Q}\mathbf{X}^{-1}\mathbf{Q}\boldsymbol{\beta}}}$ and the GW-ESIM vector of coefficients ($\boldsymbol{\beta}$) that maximizes $\rho_{H_W I}$ can be obtained from equation

$$\left(\mathbf{Q}^{-1}\mathbf{Z} - \lambda_W^2 \mathbf{I}_{(t+N)}\right)\boldsymbol{\beta}_W = \mathbf{0}, \tag{8.30}$$

where $\mathbf{I}_{(t+N)}$ is an identity matrix of size $(t + N) \times (t + N)$ and $I_W = \boldsymbol{\beta}'_W\mathbf{q}$ is the optimized GW-ESIM. The accuracy of the GW-ESIM can be written as

$$\rho_{H_W I_W} = \frac{\sqrt{\boldsymbol{\beta}'_W\mathbf{Q}\boldsymbol{\beta}_W}}{\sqrt{\boldsymbol{\beta}'_W\mathbf{Q}\mathbf{X}^{-1}\mathbf{Q}\boldsymbol{\beta}_W}} = \frac{\sigma_{I_W}}{\sigma_{H_W}}, \tag{8.31}$$

where $\sigma_{I_W} = \sqrt{\boldsymbol{\beta}'_W\mathbf{Q}\boldsymbol{\beta}_W}$ is the standard deviation of $I_W = \boldsymbol{\beta}'_W\mathbf{q}$, and $\sigma_{H_W} = \sqrt{\boldsymbol{\beta}'_W\mathbf{Q}\mathbf{X}^{-1}\mathbf{Q}\boldsymbol{\beta}_W}$ is the standard deviation of $H_W = \mathbf{w}'_W\mathbf{x}$. In Eq. (8.30) $\lambda_W^2 = \rho^2_{H_W I_W}$ is the square of the canonical correlation between H_W and I_W.

The maximized GW-ESIM selection response and expected genetic gain per trait are

$$R_W = k_I\sqrt{\boldsymbol{\beta}'_W\mathbf{Q}\boldsymbol{\beta}_W} \tag{8.32}$$

and

$$\mathbf{E}_W = k_1\frac{\mathbf{X}\boldsymbol{\beta}_W}{\sqrt{\boldsymbol{\beta}'_W\mathbf{Q}\boldsymbol{\beta}_W}}, \tag{8.33}$$

respectively, where $\boldsymbol{\beta}_W$ is the first eigenvector of Eq. (8.30).

8.3.2 Estimating GW-ESIM Parameters

In Chap. 2, Eqs. (2.22) to (2.24), we described the restricted maximum likelihood methods to estimate matrices $\mathbf{C}$ and $\mathbf{P}$, which can be denoted by $\widehat{\mathbf{C}}$ and $\widehat{\mathbf{P}}$. In Chap. 4, we described how to estimate matrices $\mathbf{M}$ and $\mathbf{G}_M$, which can be denoted by $\widehat{\mathbf{M}}$ and $\widehat{\mathbf{G}}_M$. With these estimates, we constructed the block estimated matrices as $\widehat{\mathbf{Q}} = \begin{bmatrix} \widehat{\mathbf{P}} & \widehat{\mathbf{G}}'_M \\ \widehat{\mathbf{G}}_M & \widehat{\mathbf{M}} \end{bmatrix}$ and $\widehat{\mathbf{X}} = \begin{bmatrix} \widehat{\mathbf{C}} & \widehat{\mathbf{G}}'_M \\ \widehat{\mathbf{G}}_M & \widehat{\mathbf{M}} \end{bmatrix}$, whence we obtained the equation

$$\left(\widehat{\mathbf{Q}}^{-}\widehat{\mathbf{X}} - \widehat{\lambda}^2_{Wj}\mathbf{I}\right)\widehat{\boldsymbol{\beta}}_{Wj} = \mathbf{0}, \tag{8.34}$$

$j = 1, 2, \ldots, (t + N)$, where $(t + N)$ is the number of traits and markers in the GW-ESIM index. Similar to the MESIM, we obtained estimators $\widehat{\boldsymbol{\beta}}_{W_1}$ and $\widehat{\lambda}^2_{W_1}$ of the first eigenvector $\boldsymbol{\beta}_{W_1}$ and the first eigenvalue $\widehat{\lambda}^2_{W_1}$ respectively, from equation

$$\left(\widehat{\mathbf{E}}\widehat{\mathbf{E}}' - \widehat{\mu}_j\mathbf{I}\right)\widehat{\boldsymbol{\beta}}_{W_j} = \mathbf{0}, \tag{8.35}$$

where $\widehat{\mathbf{E}} = \widehat{\mathbf{Q}}^{-}\widehat{\mathbf{X}}$ and $\widehat{\mu}_j = \widehat{\lambda}^4_{W_j}$. These results allow the GW-ESIM index selection response and its expected genetic gain per trait to be estimated as $\widehat{I}_W = \widehat{\boldsymbol{\beta}}'_{W_1}\widehat{\mathbf{q}}$, $\widehat{R}_W = k_I\sqrt{\widehat{\boldsymbol{\beta}}'_{W_1}\widehat{\mathbf{Q}}\widehat{\boldsymbol{\beta}}'_{W_1}}$ and $\widehat{\mathbf{E}}_w = k_I\frac{\widehat{\mathbf{X}}\widehat{\boldsymbol{\beta}}'_{W_1}}{\sqrt{\widehat{\boldsymbol{\beta}}'_{W_1}\widehat{\mathbf{Q}}\widehat{\boldsymbol{\beta}}'_{W_1}}}$ respectively, whereas the estimator of GW-ESIM accuracy is $\widehat{\lambda}_{W_1}$.

8.3.3 Numerical Examples

We compare the estimated GW-LMSI and GW-ESIM selection responses using the simulated data set described in Sect. 2.8.1 of Chap. 2, with a selection intensity of 10% ($k_I = 1.755$). Table 8.1 presents the estimated GW-LMSI selection response for four simulated selection cycles when their vectors of coefficients are normalized, whence it can be seen that the average estimated GW-LMSI selection response was 0.87. Table 8.1 also presents the estimated GW-ESIM selection response for four simulated selection cycles; the average of the estimated GW-ESIM selection responses was 0.93. Thus, for this data set, the estimated GW-LMSI and selection responses were very similar.

8.4 The Restricted Linear Genomic Eigen Selection Index Method

The restricted linear genomic eigen selection index method (RGESIM) is based on the restricted linear phenotypic ESIM (RESIM) theory described in Chap. 7. In the RESIM, the breeder's objective is to improve only $(t - r)$ of t $(r < t)$ traits, leaving r of them fixed. The same is true for RGESIM, but in this case, we should impose $2r$ restrictions, i.e., we need to fix r traits and their associated r GEBV to obtain results similar to those obtained with the RESIM (see Chap. 7 for details). This is the main difference between the RGESIM and the RESIM.

It can be shown that $Cov(I, \boldsymbol{\alpha}) = \mathbf{A}\boldsymbol{\beta}$ is the covariance between the breeding value vector ($\boldsymbol{\alpha}' = [\mathbf{g}' \quad \boldsymbol{\gamma}']$) and the GESIM index ($I = \boldsymbol{\beta}'\mathbf{f}$). In the RGESIM, we want some covariances between the linear combinations of $\boldsymbol{\alpha}$ ($\mathbf{U}'_G\boldsymbol{\alpha}$) and $I = \boldsymbol{\beta}'\mathbf{f}$ to be zero, i.e., $Cov(I_G, \mathbf{U}'_G\boldsymbol{\alpha}) = \mathbf{U}'_G\mathbf{A}\boldsymbol{\beta} = \mathbf{0}$, where $\mathbf{U}'_G$ is a matrix $2(t - 1) \times 2t$ of 1s and 0s (1 indicates that the trait and its associated GEBV are restricted, and 0 indicates that the trait and its GEBV have no restrictions). We can solve this problem by maximizing $\frac{\boldsymbol{\beta}'\mathbf{A}\boldsymbol{\beta}}{\sqrt{\boldsymbol{\beta}'\boldsymbol{\Phi}\boldsymbol{\beta}}}$ with respect to vector $\boldsymbol{\beta}$ under the restriction $\mathbf{U}'_G\mathbf{A}\boldsymbol{\beta} = \mathbf{0}$ and $\boldsymbol{\beta}'\boldsymbol{\beta} = 1$ similar to the RESIM, or by maximizing the correlation between $H = \mathbf{w}'\boldsymbol{\alpha}$ and

$I = \boldsymbol{\beta}'\mathbf{f}$, $\rho_{HI} = \frac{\mathbf{w}'\mathbf{A}\boldsymbol{\beta}}{\sqrt{\mathbf{w}'\mathbf{A}\mathbf{w}}\sqrt{\boldsymbol{\beta}'\boldsymbol{\Phi}\boldsymbol{\beta}}}$, with respect to vectors $\mathbf{w}'$ and $\boldsymbol{\beta}$ under the restrictions $\mathbf{U}_G'\mathbf{A}\boldsymbol{\beta} = \mathbf{0}$, $\sigma_H^2 = \mathbf{w}'\mathbf{A}\mathbf{w}$, $\sigma_I^2 = \boldsymbol{\beta}'\boldsymbol{\Phi}\boldsymbol{\beta}$ and $0 < \sigma_H^2, \sigma_I^2 < \infty$, as we did for the GESIM.

8.4.1 The RGESIM Parameters

To obtain the RGESIM vector of coefficients, we maximize the function

$$f(\boldsymbol{\beta}, \mathbf{v}') = \frac{\boldsymbol{\beta}'\mathbf{A}\boldsymbol{\beta}}{\sqrt{\boldsymbol{\beta}'\boldsymbol{\Phi}\boldsymbol{\beta}}} - \mathbf{v}'\mathbf{U}'_G\mathbf{A}\boldsymbol{\beta} \tag{8.36}$$

with respect to $\boldsymbol{\beta}$ and $\mathbf{v}'$, where $\mathbf{v}' = [v_1 \ v_2 \ \cdots \ v_{2(r-1)}]$ is a vector of Lagrange multipliers. The derivatives of function $f(\boldsymbol{\beta}, \mathbf{v}')$ with respect to $\boldsymbol{\beta}$ and $\mathbf{v}'$ can be written as

$$2(\boldsymbol{\beta}'\boldsymbol{\Phi}\boldsymbol{\beta})^{1/2}\mathbf{A}\boldsymbol{\beta} - (\boldsymbol{\beta}'\boldsymbol{\Phi}\boldsymbol{\beta})^{-1/2}(\boldsymbol{\beta}'\mathbf{A}\boldsymbol{\beta})\boldsymbol{\Phi}\boldsymbol{\beta} - \mathbf{A}\mathbf{U}_G\mathbf{v} = \mathbf{0}, \tag{8.37}$$

$$\mathbf{U}_G'\mathbf{A}\boldsymbol{\beta} = \mathbf{0}, \tag{8.38}$$

respectively, where Eq. (8.38) denotes the restriction imposed for maximizing Eq. (8.36). Using algebraic methods on Eq. (8.37), we get

$$\left(\mathbf{K}_{RG}\boldsymbol{\Phi}^{-1}\mathbf{A} - \lambda_{RG}^2\mathbf{I}_{2t}\right)\boldsymbol{\beta}_{RG} = \mathbf{0}, \tag{8.39}$$

where $\lambda_{RG}^2 = h_{I_{RG}}^2$, $h_{I_{RG}}^2$ is the RGESIM heritability obtained under the restriction $\mathbf{U}_G'\mathbf{A}\boldsymbol{\beta} = \mathbf{0}$; $\mathbf{K}_{RG} = [\mathbf{I}_{2t} - \mathbf{Q}_{RG}]$, $\mathbf{I}_{2t}$ is an identity matrix of size $2t \times 2t$, and $\mathbf{Q}_{RG} = \boldsymbol{\Phi}^{-1}\mathbf{A}\mathbf{U}_G\left(\mathbf{U}_G'\mathbf{A}\boldsymbol{\Phi}^{-1}\mathbf{A}\mathbf{U}_G\right)^{-1}\mathbf{U}_G'\mathbf{A}$. When $\mathbf{U}_G'$ is a null matrix, $\boldsymbol{\beta}_{RG}' = \boldsymbol{\beta}_G'$ (the vector of the GESIM coefficients); thus, the RGESIM is more general than the GESIM and includes the GESIM as a particular case. The RGESIM index $I_{GR} = \boldsymbol{\beta}_{RG}'\mathbf{y}$ and its selection response and expected genetic gain per trait use the first eigenvector of matrix $\mathbf{K}_G\boldsymbol{\Phi}^{-1}\mathbf{A}$. It can be shown that the vector of coefficients of $H = \mathbf{w}_{RG}'\boldsymbol{\alpha}$ in the RGESIM can be written as

$$\mathbf{w}_{RG} = \mathrm{A}^{-1}\left[\boldsymbol{\Phi} + \mathbf{Q}_{RG}'\mathbf{A}\right]\boldsymbol{\beta}_{RG}, \tag{8.40}$$

where $\mathbf{Q}_{RG}' = \mathbf{A}\mathbf{U}_G\left(\mathbf{U}_G'\mathbf{A}\boldsymbol{\Phi}^{-1}\mathbf{A}\mathbf{U}_G\right)^{-1}\mathbf{U}_G'\mathbf{A}\boldsymbol{\Phi}^{-1}$.

Note that the restriction $\mathbf{U}_G'\mathbf{A}\boldsymbol{\beta} = \mathbf{0}$ can be written as $\boldsymbol{\beta}'\mathbf{A}\mathbf{U}_G = \mathbf{0}$; this means that $\boldsymbol{\beta}'\mathbf{Q}'_{RG} = \mathbf{0}$ and that the covariance between $H_{RG} = \mathbf{w}_{RG}'\boldsymbol{\alpha}$ and $I_{RG} = \boldsymbol{\beta}_{RG}'\mathbf{f}$ ($\sigma_{H_{RG}I_{RG}}$) can be written as

$$\sigma_{H_{RG}I_{RG}} = \mathbf{w}_{RG}'\mathbf{A}\boldsymbol{\beta}_{RG}' = \boldsymbol{\beta}_{RG}'\boldsymbol{\Phi}\boldsymbol{\beta}_{RG} + \boldsymbol{\beta}_{RG}'\mathbf{Q}_{RG}'\mathbf{C}\boldsymbol{\beta}_{RG} = \boldsymbol{\beta}_{RG}'\boldsymbol{\Phi}\boldsymbol{\beta}_{RG}. \tag{8.41}$$

Equation (8.41) indicates that $\sigma_{H_{RG}I_{RG}}$ is equal to the variance of $I_{RG} = \boldsymbol{\beta}'_{RG}\mathbf{f}$ ($\sigma^2_{I_{RG}} = \boldsymbol{\beta}'_{RG}\boldsymbol{\Phi}\boldsymbol{\beta}_{RG}$); therefore, the maximized correlation between I_{RG} and H_{RG} or RGESIM accuracy can be written as

$$\rho_{H_{RG}I_{RG}} = \frac{\sqrt{\boldsymbol{\beta}'_{RG}\boldsymbol{\Phi}\boldsymbol{\beta}_{RG}}}{\sqrt{\mathbf{w}'_{RG}\mathbf{A}\mathbf{w}_{RG}}}, \tag{8.42}$$

where $\mathbf{w}'_{RG}\mathbf{A}\mathbf{w}_{RG}$ is the variance of H_{RG}. Hereafter, to simplify the notation, we write Eq. (8.42) as λ_{RG}.

The maximized selection response and the expected genetic gain per trait of the RGESIM are

$$R_{RG} = k_I\sqrt{\boldsymbol{\beta}'_{RG}\boldsymbol{\Phi}\boldsymbol{\beta}_{RG}} \tag{8.43}$$

and

$$\mathbf{E}_{RG} = k_I\frac{\mathbf{A}\boldsymbol{\beta}_{RG}}{\sqrt{\boldsymbol{\beta}'_{RG}\boldsymbol{\Phi}\boldsymbol{\beta}_{RG}}}, \tag{8.44}$$

respectively, where $\boldsymbol{\beta}_{RG}$ is the first eigenvector of matrix $\mathbf{K}_{RG}\boldsymbol{\Phi}^{-1}\mathbf{A}$.

8.4.2 Estimating RGESIM Parameters

In Sect. 8.2, we indicated how to estimate matrices $\mathbf{P}$, $\boldsymbol{\Gamma}$, and $\mathbf{C}$ using phenotypic and genomic information, whence we can estimate matrices $\mathbf{A} = \begin{bmatrix} \mathbf{C} & \boldsymbol{\Gamma} \\ \boldsymbol{\Gamma} & \boldsymbol{\Gamma} \end{bmatrix}$ and $\boldsymbol{\Phi} = \begin{bmatrix} \mathbf{P} & \boldsymbol{\Gamma} \\ \boldsymbol{\Gamma} & \boldsymbol{\Gamma} \end{bmatrix}$. Those methods are also valid for the RGESIM. This means that the SVD methods described for estimating MESIM parameters are also valid for estimating RGESIM parameters.

8.4.3 Numerical Examples

With a selection intensity of 10% ($k_I = 1.755$), we compare the CRLGSI (for details see Chap. 6) versus the RGESIM theoretical results using a real maize (*Zea mays*) F_2 population with 244 genotypes (each with two repetitions), 233 molecular markers, and three traits—GY (ton ha^{-1}), EHT (cm), and PHT (cm)—described in Sect. 8.2.2, where $\widehat{\mathbf{P}} = \begin{bmatrix} 0.45 & 1.33 & 2.33 \\ 1.33 & 65.07 & 83.71 \\ 2.33 & 83.71 & 165.99 \end{bmatrix}$, $\widehat{\mathbf{C}} = \begin{bmatrix} 0.07 & 0.61 & 1.06 \\ 0.61 & 17.93 & 22.75 \\ 1.06 & 22.75 & 44.53 \end{bmatrix}$, and $\widehat{\boldsymbol{\Gamma}} = \begin{bmatrix} 0.07 & 0.65 & 1.05 \\ 0.65 & 10.62 & 14.25 \\ 1.05 & 14.25 & 26.37 \end{bmatrix}$ were the estimated matrices of **P**, **C**, and **Γ** respectively.

We have indicated that the main difference between the RLPSI and the CRLGSI is the matrix $\mathbf{U}'_C$, on which we now need to impose two restrictions: one for the trait and another for its associated GEBV. Consider the data set described earlier and suppose that we restrict the trait GY (ton ha^{-1}) and its associated GEBV$_{GY}$; then, matrix $\mathbf{U}'_C$ should be constructed as $\mathbf{U}'_{C1} = \begin{bmatrix} 1 & 0 & 0 & 0 & 0 & 0 \\ 0 & 0 & 0 & 1 & 0 & 0 \end{bmatrix}$. If we restrict traits GY and EHT (cm) and their associated GEBV$_{GY}$ and GEBV$_{EHT}$, matrix $\mathbf{U}'_C$ should be constructed as $\mathbf{U}'_{C2} = \begin{bmatrix} 1 & 0 & 0 & 0 & 0 & 0 \\ 0 & 1 & 0 & 0 & 0 & 0 \\ 0 & 0 & 0 & 1 & 0 & 0 \\ 0 & 0 & 0 & 0 & 1 & 0 \end{bmatrix}$, etc. The procedure for obtaining matrices $\widehat{\mathbf{K}}_{RG} = \left[\mathbf{I}_{2t} - \widehat{\mathbf{Q}}_{RG}\right]$ and $\widehat{\mathbf{Q}}_{RG} = \widehat{\boldsymbol{\Phi}}^{-1}\widehat{\mathbf{A}}\mathbf{U}_G\left(\mathbf{U}'_G\widehat{\mathbf{A}}\widehat{\boldsymbol{\Phi}}^{-1}\widehat{\mathbf{A}}\mathbf{U}_G\right)^{-1}\mathbf{U}'_G$ $\widehat{\mathbf{A}}$ was described in Chap. 6, and is also valid for estimating RGESIM parameters.

The estimated CRLGSI vector of coefficients is $\widehat{\boldsymbol{\beta}}_{CR} = \widehat{\mathbf{K}}_{RG}\widehat{\boldsymbol{\beta}}$, where $\widehat{\boldsymbol{\beta}} = \widehat{\boldsymbol{\Phi}}^{-1}\widehat{\mathbf{A}}$ **w** is the estimated CLGSI vector of coefficients (Chap. 6). Let $\mathbf{w}' = [5 \quad -0.1 \quad -0.1 \quad 0 \quad 0 \quad 0]$ be the vector of economic weights and suppose that we restrict trait GY and its associated GEBV$_{GY}$; in this case, $\mathbf{U}'_{C1} = \begin{bmatrix} 1 & 0 & 0 & 0 & 0 & 0 \\ 0 & 0 & 0 & 1 & 0 & 0 \end{bmatrix}$, and according to matrices $\widehat{\mathbf{P}}$, $\widehat{\mathbf{C}}$, and $\widehat{\boldsymbol{\Gamma}}$ described earlier, $\widehat{\boldsymbol{\beta}}'_{CR} = [0.076 \quad -0.004 \quad -0.018 \quad 2.353 \quad -0.096 \quad -0.082]$ was the estimated CRLGSI vector of coefficients and the estimated CRLGSI was

$$\widehat{I}_{CR} = 0.076\text{GY} - 0.004\text{EHT} - 0.018\text{PHT} + 2.353\text{GEBV}_{\text{GY}} - 0.096\text{GEBV}_{\text{EHT}} - 0.082\text{GEBV}_{\text{PHT}}$$

where GEBV$_{GY}$, GEBV$_{EHT}$, and GEBV$_{PHT}$ are the GEBVs associated with the traits GY, EHT, and PHT respectively. The same procedure is valid for two or more restrictions.

The estimated CRLGSI selection response and expected genetic gain per trait were $\widehat{R}_{CR} = k_I\sqrt{\widehat{\boldsymbol{\beta}}'_{CR}\widehat{\boldsymbol{\Phi}}\widehat{\boldsymbol{\beta}}_{CR}} = 0.96$ and $\widehat{\mathbf{E}}'_{CR} = k_I = \frac{\widehat{\boldsymbol{\beta}}'_{CR}\widehat{\mathbf{A}}}{\sqrt{\widehat{\boldsymbol{\beta}}'_{CR}\widehat{\boldsymbol{\Phi}}\widehat{\boldsymbol{\beta}}_{CR}}}$ $[0 \quad -3.53 \quad -6.03 \quad 0 \quad -2.93 \quad -4.87]$ respectively, whereas the estimated CRLGSI accuracy was $\widehat{\rho}_{HI_{CR}} = \frac{\widehat{\sigma}_{I_{CR}}}{\widehat{\sigma}_H} = 0.51$. Note that in $\widehat{\mathbf{E}}'_{CR}$, the trait GY and its associated GEBV$_{GY}$ have null values, as we would expect.

The estimated RGESIM vector of coefficients was $\widehat{\boldsymbol{\beta}}'_{CR} = [0.015 \quad -0.001 \quad -0.004 \quad 0.998 \quad -0.029 \quad -0.045]$, and the estimated RGESIM index was $\widehat{I}_{RG} = 0.015\text{GY} - 0.001\text{EHT} - 0.004\text{PHT} + 0.998\text{GEBV}_{\text{GY}} - 0.029\text{GEBV}_{\text{EHT}} - 0.045\text{GEBV}_{\text{PHT}}$ where GEBV$_{GY}$, GEBV$_{EHT}$, and GEBV$_{PHT}$

are the GEBVs associated with traits GY, EHT, and PHT respectively. The same procedure is valid for two or more restrictions.

The estimated RGESIM selection response and expected genetic gain per trait were $\widehat{R}_{RG} = k_I\sqrt{\widehat{\boldsymbol{\beta}}'_{RG}\widehat{\boldsymbol{\Phi}}\widehat{\boldsymbol{\beta}}_{RG}} = 0.37$ and $\widehat{\mathbf{E}}'_{RG} = k_I = \frac{\widehat{\boldsymbol{\beta}}'_{RG}\widehat{\mathbf{A}}}{\sqrt{\widehat{\boldsymbol{\beta}}'_{RG}\widehat{\boldsymbol{\Phi}}\widehat{\boldsymbol{\beta}}_{RG}}}$ $[0 \quad -3.28 \quad -6.03 \quad 0 \quad -2.93 \quad -5.40]$ respectively, whereas the estimated RGESIM accuracy was $\widehat{\rho}_{\widehat{H}_{RG}\widehat{I}_{RG}} = \frac{\widehat{\sigma}_{\widehat{I}_{RG}}}{\widehat{\sigma}_{\widehat{H}_{RG}}} = 0.86$.

Fig. 8.3 presents the frequency distribution of the 244 estimated RGESIM index values for two null restrictions on traits GY and EHT and their associated $GEBV_{GY}$ and $GEBV_{EHT}$, for one selection cycle in an environment for a real maize (*Zea mays*) F_2 population with 233 molecular markers. Note that the frequency distribution of the estimated RGESIM index values approaches the normal distribution.

Now we compare the estimated CRLGSI and RGESIM selection responses and expected genetic gains per trait using the simulated data set described in Sect. 2.8.1 of Chap. 2. We used that data set for four phenotypic selection cycles (C2, C3, C4, and C5), each with four traits (T_1, T_2, T_3, and T_4), 500 genotypes, and four replicates per genotype. The economic weights for T_1, T_2, T_3, and T_4 were 1, −1, 1, and

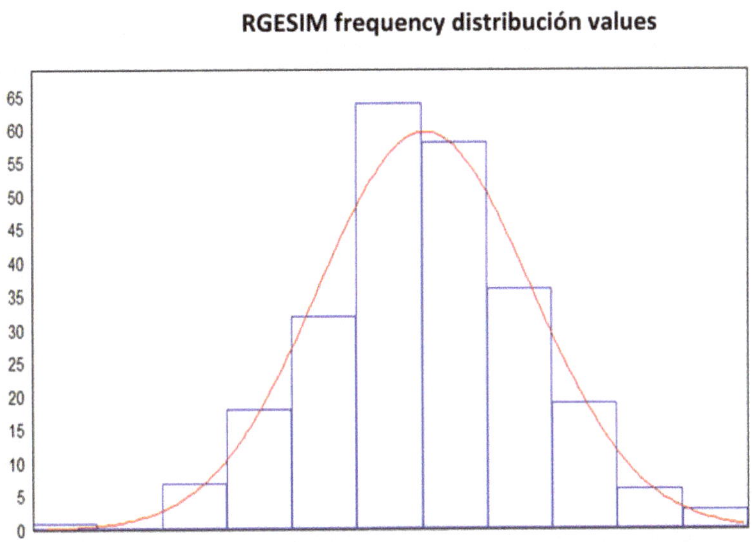

Fig. 8.3 Frequency distribution of the 244 estimated restricted genomic eigen selection index method (RGESIM) values for two null restrictions on traits grain yield (GY) and EHT and their associated genomic estimated breeding values (GEBVs), $GEBV_{GY}$ and $GEBV_{EHT}$ respectively, for one selection cycle in an environment for a real maize (*Zea mays*) F_2 population with 233 molecular markers. Note that the frequency distribution of the estimated RGESIM index values approaches normal distribution

1 respectively. For this data set, matrix **F** was an identity matrix of size 8×8 for all four selection cycles.

Columns 2, 3, and 4 (from left to right) of Table 8.4 present the estimated CRLGSI selection responses when their vectors of coefficients are normalized and the estimated RGESIM and selection responses for one, two, and three restrictions for four simulated selection cycles. The averages of the estimated CRLGSI selection responses of the traits and their associated GEBVs for each of the three null restrictions were 3.24 for one restriction, 4.08 for two restrictions, and 5.06 for three restrictions, whereas the averages of the estimated RGESIM selection responses were 3.08 for one restriction, 2.79 for two restrictions, and 3.23 for three restrictions. Note that although for one restriction the selection response was similar for both indices, for two and three restrictions the CRLGSI selection responses were greater than the RGESIM selection responses.

Table 8.5 presents the estimated CRLGSI and RGESIM expected genetic gains per trait for four traits (T1, T2, T3, and T4) and their associated GEBVs (in this case denoted by G1, G2, G3, and G4 to simplify the notation) in four simulated selection cycles and for one, two, and three null restrictions in four simulated selection cycles. Note that the null values of the traits and their restricted GEBVs are not shown in Table 8.5 with the aim of simplifying the table. The averages of the estimated CRLGSI expected genetic gains for the three traits and their associated GEBVs were −2.60, 2.16, 2.84, −1.21, 0.67, and 1.02 for one restriction; 2.74, 3.23, 0.78,

Table 8.4 Estimated combined null restricted linear genomic selection index (CRLGSI) and estimated combined predetermined proportional gain linear genomic selection index (CPPG-LGSI) selection responses for one, two, and three restrictions when their vectors of coefficients are normalized for four simulated selection cycles

	CRLGSI response for one, two and three null restrictions			CPPG-LGSI response for one, two and three predetermined restrictions		
Cycle	1	2	3	1	2	3
1	3.25	4.09	4.89	5.36	2.80	1.81
2	3.28	4.19	5.21	5.07	3.64	1.99
3	2.91	3.89	4.97	5.37	3.86	1.42
4	3.53	4.17	5.15	4.52	3.38	1.20
Average	3.24	4.08	5.06	5.08	3.42	1.60
	RGESIM response for one, two, and three null restrictions			PPG-GESIM response for one, two, and three predetermined restrictions		
Cycle	1	2	3	1	2	3
1	3.21	2.78	3.47	1.95	4.07	4.26
2	3.11	2.86	3.06	1.85	4.12	5.49
3	2.93	2.76	3.20	2.04	4.18	6.30
4	3.07	2.76	3.21	2.02	4.17	5.82
Average	3.08	2.79	3.23	1.96	4.14	5.47

Estimated null restricted genomic eigen selection index method (RGESIM) and predetermined proportional gain genomic eigen selection index method (PPG-GESIM) selection responses for one, two, and three restrictions for four simulated selection cycles. The selection intensity was 10% ($k_I = 1.755$)

Table 8.5 Estimated CRLGSI and estimated null RGESIM expected genetic gains per trait for four traits (T1, T2, T3, and T4) and their associated genomic estimated breeding values (G1, G2, G3, and G4) for four simulated selection cycles and for one, two, and three null restrictions for four simulated selection cycles. The selection intensity was 10% ($k_I = 1.755$)

	CRLGSI expected genetic gains for one, two and three null restrictions											
	One restriction[a]						Two restrictions[b]				Three restrictions[c]	
Cycle	T2	T3	T4	G2	G3	G4	T3	T4	G3	G4	T4	G4
1	−2.32	2.17	2.87	−1.48	0.73	1.24	2.60	3.38	0.86	1.15	4.08	1.50
2	−2.76	2.14	2.89	−1.19	0.76	0.96	2.81	3.30	0.87	0.98	3.95	1.25
3	−2.22	2.27	2.98	−1.15	0.62	0.97	2.77	3.14	0.69	0.90	3.93	1.33
4	−3.09	2.08	2.64	−1.05	0.58	0.92	2.80	3.08	0.70	0.93	4.13	1.24
Mean	−2.60	2.16	2.84	−1.21	0.67	1.02	2.74	3.23	0.78	0.99	4.02	1.33
	RGESIM expected genetic gains for one, two and three null restrictions											
	One restriction[a]						Two restrictions[b]				Three restrictions[c]	
Cycle	T2	T3	T4	G2	G3	G4	T3	T4	G3	G4	T4	G4
1	3.27	−1.52	−1.24	2.48	−0.88	−1.00	3.18	0.93	1.88	0.43	3.66	2.21
2	3.30	−1.79	−1.41	2.10	−1.09	−0.82	3.26	1.34	1.82	0.66	3.41	2.00
3	2.98	−1.62	−1.44	2.13	−0.83	−0.75	3.31	0.86	1.70	0.21	3.45	2.05
4	3.56	−1.73	−1.23	1.92	−0.89	−0.78	3.40	0.96	1.62	0.53	3.58	2.02
Mean	3.27	−1.67	−1.33	2.16	−0.92	−0.84	3.29	1.02	1.76	0.46	3.53	2.07

[a]All T1 and G1 expected genetic gains were null
[b]All T1, T2, G1, and G2 expected genetic gains were null
[c]All T1, T2, T3, G1, G2, and G3 expected genetic gains were null

and 0.99 for two restrictions; and 4.02 and 1.33 for three restrictions. On the other hand, the averages of the estimated RGESIM expected genetic gains for the three traits and their associated GEBVs were 3.27, −1.67, −1.33, 2.16, −0.92, and −0.84 for one restriction; 3.29, 1.02, 1.76, and 0.46 for two restrictions; and 3.53 and 2.07 for three restrictions. These results indicate that in terms of absolute values, the estimated expected genetic gains for the traits and their associated GEBVs were similar for both indices.

8.5 The Predetermined Proportional Gain Linear Genomic Eigen Selection Index Method

The predetermined proportional gain linear genomic eigen selection index method (PPG-GESIM) theory is based on the predetermined proportional gain linear phenotypic ESIM (PPG-ESIM) described in Chap. 7. In the PPG-ESIM, the vector of PPG (predetermined proportional gain) imposed by the breeder was $\mathbf{d}' = \begin{bmatrix} d_1 & d_2 & \cdots & d_r \end{bmatrix}$. However, because the PPG-GESIM uses phenotypic and GEBV information jointly to predict the net genetic merit, the vector of PPG imposed by the breeder ($\mathbf{d}_{PG}$) should be twice the standard vector $\mathbf{d}'$, that is, $\mathbf{d}'_{PG} = \begin{bmatrix} d_1 & d_2 & \cdots & d_r & d_{r+1} & d_{r+2} & \cdots & d_{2r} \end{bmatrix}$, where we would expect that if d_1 is the PPG imposed on trait 1, then d_{r+1} should be the PPG imposed on the GEBV associated with trait 1, etc. Thus, in the PPG-GESIM we have three possible options for determining (for each trait and GEBV) the PPG: e.g., for trait 1, $d_1 = d_{r+1}$, $d_1 > d_{r+1}$ or $d_1 < d_{r+1}$. This is the main difference between the standard PPG-ESIM described in Chap. 7 and the PPG-GESIM.

8.5.1 The PPG-GESIM Parameters

Using the same procedure described for RGESIM and PPG-ESIM, the PPG-GESIM vector of coefficients ($\boldsymbol{\beta}_{PG}$), which maximizes the PPG-GESIM selection response and the expected genetic gain per trait, is the first eigenvector of the following equation

$$(\mathbf{T}_{PG} - \lambda^2_{PG}\mathbf{I}_{2t})\boldsymbol{\beta}_{PG} = \mathbf{0}, \tag{8.45}$$

where $\mathbf{T}_{PG} = \mathbf{K}_{RG}\boldsymbol{\Phi}^{-1}\mathbf{A} + \mathbf{B}$, $\mathbf{K}_{PG} = [\mathbf{I}_{2t} - \mathbf{Q}_{RG}]$, $\mathbf{I}_{2t}$ is an identity matrix of size $2t \times 2t$, $\mathbf{Q}_{RG} = \boldsymbol{\Phi}^{-1}\mathbf{A}\mathbf{U}_G(\mathbf{U}'_G\mathbf{A}\boldsymbol{\Phi}^{-1}\mathbf{A}\mathbf{U}_G)^{-1}\mathbf{U}'_G\mathbf{A}$, $\mathbf{B} = \boldsymbol{\delta}\varphi'$, $\boldsymbol{\delta} = \boldsymbol{\Phi}^{-1}\mathbf{A}\mathbf{U}_G(\mathbf{U}'_G\mathbf{A}\boldsymbol{\Phi}^{-1}\mathbf{A}\mathbf{U}_G)^{-1}\mathbf{d}_{PG}$, and $\varphi' = \dfrac{\mathbf{d}'_{PG}(\mathbf{U}'_G\mathbf{A}\boldsymbol{\Phi}^{-1}\mathbf{A}\mathbf{U}_G)^{-1}\mathbf{U}'_G\mathbf{A}\boldsymbol{\Phi}^{-1}\mathbf{A}}{\mathbf{d}'_{PG}(\mathbf{U}'_G\mathbf{A}\boldsymbol{\Phi}^{-1}\mathbf{A}\mathbf{U}_G)^{-1}\mathbf{d}_{PG}}$.

When $\mathbf{B}$ is a null matrix, $\mathbf{T}_{PG} = \mathbf{K}_{RG}\boldsymbol{\Phi}^{-1}\mathbf{A}$ (matrix of the RGESIM), and when $\mathbf{U}'_G$ is a null matrix, $\mathbf{T}_{PG} = \boldsymbol{\Phi}^{-1}\mathbf{A}$ (matrix of the GESIM); this means that the PPG-GESIM includes the RGESIM and GESIM as particular cases. The optimized PPG-GESIM index can be written as $I_{PG} = \boldsymbol{\beta}'_{PG}\mathbf{f}$.

The vector of coefficients of $H = \mathbf{w}'_{PG}\boldsymbol{\alpha}$ in the PPG-GESIM can be written as

$$\mathbf{w}_{PG} = \mathbf{A}^{-1}[\boldsymbol{\Phi} + \mathbf{Q}'_{PG}\mathbf{A}]\boldsymbol{\beta}_{PG}, \tag{8.46}$$

where $\mathbf{Q}'_{PG} = \mathbf{A}\mathbf{U}_G\mathbf{D}_G(\mathbf{D}'_G\mathbf{U}'_G\mathbf{A}\boldsymbol{\Phi}^{-1}\mathbf{A}\mathbf{U}_G\mathbf{D}_G)^{-1}\mathbf{D}'_G\mathbf{U}'_G\mathbf{A}\boldsymbol{\Phi}^{-1}$, and $\mathbf{D}'_G = \begin{bmatrix} d_{2r} & 0 & \cdots & 0 & -d_1 \\ 0 & d_{2r} & \cdots & 0 & -d_2 \\ \vdots & \vdots & \ddots & \vdots & \vdots \\ 0 & 0 & \cdots & d_{2r} & -d_{2r-1} \end{bmatrix}$. Similar to RGESIM, it can be shown that the covariance between $H_{RG} = \mathbf{w}'_{PG}\boldsymbol{\alpha}$ and $I_{PG} = \boldsymbol{\beta}'_{PG}\mathbf{f}$ ($\sigma_{H_{PG}I_{PG}}$) is equal to the variance of $I_{PG} = \boldsymbol{\beta}'_{PG}\mathbf{f}$ ($\sigma^2_{I_{PG}} = \boldsymbol{\beta}'_{PG}\boldsymbol{\Phi}\boldsymbol{\beta}_{PG}$), that is, $\sigma_{H_{PG}I_{PG}} = \mathbf{w}'_{PG}\mathbf{A}\boldsymbol{\beta}_{PG} = \boldsymbol{\beta}'_{PG}\boldsymbol{\Phi}\boldsymbol{\beta}_{PG} = \sigma^2_{I_{PG}}$.

The maximized correlation between I_{PG} and H_{PG}, or PPG-GESIM accuracy, is

$$\rho_{H_{PG}I_{PG}} = \frac{\sqrt{\boldsymbol{\beta}'_{PG}\boldsymbol{\Phi}\boldsymbol{\beta}_{PG}}}{\sqrt{\mathbf{w}'_{PG}\mathbf{A}\mathbf{w}_{PG}}} \tag{8.47}$$

where $\mathbf{w}'_{PG}\mathbf{A}\mathbf{w}_{PG}$ is the variance of H_{PG}. Hereafter, to simplify the notation, we write Eq. (8.47) as λ_{PG}.

The maximized selection response and the expected genetic gain per trait of the PPG-GESIM are

$$R_{PG} = k_I\sqrt{\boldsymbol{\beta}'_{PG}\boldsymbol{\Phi}\boldsymbol{\beta}_{PG}} \tag{8.48}$$

and

$$\mathbf{E}_{PG} = k_I\frac{\mathbf{A}\boldsymbol{\beta}_{PG}}{\sqrt{\boldsymbol{\beta}'_{PG}\boldsymbol{\Phi}\boldsymbol{\beta}_{PG}}}, \tag{8.49}$$

respectively, where $\boldsymbol{\beta}_{PG}$ is the first eigenvector of Eq. (8.45).

8.5.2 Numerical Examples

The process for estimating PPG-ESIM parameters is similar to the method described for estimating RGESIM parameters. With a selection intensity of 10% ($k_I = 1.755$), we compare the combined predetermined proportional gain linear genomic selection index (CPPG-LGSI) and PPG-GESIM results using the real maize (*Zea mays*) F_2 population with 244 genotypes, 233 molecular markers, and three traits—GY (ton ha^{-1}), EHT (cm), and PHT (cm)—where $\widehat{\mathbf{P}} = \begin{bmatrix} 0.45 & 1.33 & 2.33 \\ 1.33 & 65.07 & 83.71 \\ 2.33 & 83.71 & 165.99 \end{bmatrix}$, $\widehat{\mathbf{G}} = \begin{bmatrix} 0.07 & 0.61 & 1.06 \\ 0.61 & 17.93 & 22.75 \\ 1.06 & 22.75 & 44.53 \end{bmatrix}$ and $\widehat{\boldsymbol{\Gamma}} = \begin{bmatrix} 0.07 & 0.65 & 1.05 \\ 0.65 & 10.62 & 14.25 \\ 1.05 & 14.25 & 26.37 \end{bmatrix}$ are the estimated matrices of $\mathbf{P}$, $\mathbf{G}$, and $\boldsymbol{\Gamma}$ respectively, whereas $\mathbf{w}' = [5 \quad -0.1 \quad -0.1 \quad 0 \quad 0 \quad 0]$ was the vector of economic weights.

The estimated CPPG-LGSI vector of coefficients was $\widehat{\boldsymbol{\beta}}_{CP} = \widehat{\boldsymbol{\beta}}_{CG} + \widehat{\theta}_{CP}\widehat{\boldsymbol{\delta}}$ (see Chap. 6 for additional details). Let $\widehat{\mathbf{A}} = \begin{bmatrix} \widehat{\mathbf{G}} & \widehat{\boldsymbol{\Gamma}} \\ \widehat{\boldsymbol{\Gamma}} & \widehat{\boldsymbol{\Gamma}} \end{bmatrix}$ and $\widehat{\boldsymbol{\Phi}} = \begin{bmatrix} \widehat{\mathbf{P}} & \widehat{\boldsymbol{\Gamma}} \\ \widehat{\boldsymbol{\Gamma}} & \widehat{\boldsymbol{\Gamma}} \end{bmatrix}$ be the estimated block matrices and $\mathbf{d}'_{PG} = [7 \quad -3 \quad 3.5 \quad -1.5]$ the vector of PPG imposed by the breeder on the traits GY and EHT, and their associated genomic estimated breeding values (GEBV_{GY} and GEBV_{EHT}), and let

$\mathbf{U}'_C = \begin{bmatrix} 1 & 0 & 0 & 0 & 0 & 0 \\ 0 & 1 & 0 & 0 & 0 & 0 \\ 0 & 0 & 0 & 1 & 0 & 0 \\ 0 & 0 & 0 & 0 & 1 & 0 \end{bmatrix}$ be the matrix of null restrictions on the CPPG-LGSI and $\mathbf{w}' = [5 \ -0.1 \ -0.1 \ 0 \ 0 \ 0]$ the vector of economic weights. It can be shown that $\widehat{\theta}_{\text{CP}} = -0.00009$ is the estimated value of the proportionality constant, $\widehat{\boldsymbol{\delta}}' = [-112.92 \quad -72.16 \quad 61.35 \quad 231.79 \quad 64.75 \quad -61.35]$, $\widehat{\boldsymbol{\beta}}'_{CP} = [-0.01 \quad 0.01 \quad -0.01 \quad 0.59 \quad 0.09 \quad -0.09]$ is the estimated CPPG-LGSI vector of coefficients, and the estimated CPPG-LGSI can be written as

$$\widehat{I}_{CP} = -0.01\text{GY} + 0.01\text{EHT} - 0.01\text{PHT} + 0.59\text{GEBV}_{\text{GY}} + 0.09\text{GEBV}_{\text{EHT}} - 0.09\text{GEBV}_{\text{PHT}}$$

where GEBV_{GY}, GEBV_{EHT}, and GEBV_{PHT} are the GEBVs associated with traits GY, EHT, and PHT respectively. The same procedure is valid for more than two predetermined restrictions. The estimated CPPG-LGSI selection response and expected genetic gain per trait were $\widehat{R}_{CP} = k_I\sqrt{\widehat{\boldsymbol{\beta}}'_{CP}\widehat{\boldsymbol{\Phi}}\widehat{\boldsymbol{\beta}}_{CP}} = 0.443$ and $\widehat{\mathbf{E}}'_{CP} = k_I \dfrac{\widehat{\boldsymbol{\beta}}'_{CP}\widehat{\mathbf{A}}}{\sqrt{\widehat{\boldsymbol{\beta}}'_{CP}\widehat{\boldsymbol{\Phi}}\widehat{\boldsymbol{\beta}}_{CP}}} = [-0.004 \quad 0.002 \quad -4.639 \quad -0.002 \quad 0.001 \quad -4.326]$ respectively, whereas the estimated CPPG-LGSI accuracy is $\widehat{\rho}_{HI_{CP}} = \dfrac{\widehat{\sigma}_{I_{CP}}}{\widehat{\sigma}_H} = 0.234$.

Because the estimated value of the proportionality constant was negative ($\widehat{\theta}_{\text{CP}} = -0.00009$), the expected genetic gains of the traits GY and EHT, and their associated genomic estimated breeding values (GEBV_{GY} and GEBV_{EHT}), which appeared in the $\widehat{\mathbf{E}}'_{CP}$ values, were not in accordance with the values of the vector of PPG imposed by the breeder, $\mathbf{d}'_{PG} = [7 \quad -3 \quad 3.5 \quad -1.5]$, as we would expect, and CPPG-LGSI accuracy (0.234) was low. These results indicate that in the CPPG-LGSI, it is very important for the estimated values of $\widehat{\theta}_{\text{CP}}$ to be positive (see Chaps. 3 and 6 for details).

In the PPG-GESIM, we need to find the solutions to equation $(\widehat{\mathbf{T}}_{PG} - \widehat{\lambda}^2_{PG_j}\mathbf{I}_{2t})\widehat{\boldsymbol{\beta}}_{PG_j} = \mathbf{0}$, for $\widehat{\lambda}^2_{PG_j}$ and $\widehat{\boldsymbol{\beta}}_{PG_j}$ (see Eq. 8.45). The estimated PPG-GESIM vector of coefficients was $\widehat{\boldsymbol{\beta}}'_{PG} = [0.001 \quad -0.050 \quad 0.029 \quad 0.975 \quad 0.154 \quad -0.157]$, which was transformed using matrix $\mathbf{F} = \begin{bmatrix} -0.1 & 0 & 0 & 0 & 0 & 0 \\ 0 & 3 & 0 & 0 & 0 & 0 \\ 0 & 0 & 2 & 0 & 0 & 0 \\ 0 & 0 & 0 & -1 & 0 & 0 \\ 0 & 0 & 0 & 0 & -1 & 0 \\ 0 & 0 & 0 & 0 & 0 & -1 \end{bmatrix}$, that is, we changed the direction of the original vector. With the $\widehat{\boldsymbol{\beta}}'_{PG}$ values, we can estimate the PPG-GESIM index as

$$\widehat{I}_{PG} = 0.001\text{GY} - 0.05\text{EHT} + 0.029\text{PHT} + 0.975\text{GEBV}_{\text{GY}} + 0.154\text{GEBV}_{\text{EHT}} - 0.157\text{GEBV}_{\text{PHT}}$$

where GEBV_{GY}, GEBV_{EHT}, and GEBV_{PHT} are the GEBVs associated with the traits GY, EHT, and PHT respectively. The estimated PPG-GESIM selection response, accuracy, and expected genetic gain per trait were $\widehat{R}_{PG} = k_I\sqrt{\widehat{\boldsymbol{\beta}}'_{PG}\widehat{\boldsymbol{\Phi}}\widehat{\boldsymbol{\beta}}_{PG}} = 0.696$, $\widehat{\rho}_{\widehat{H}_{PG}\widehat{I}_{PG}} = \frac{\widehat{\sigma}_{\widehat{I}_{PG}}}{\widehat{\sigma}_{\widehat{H}_{PG}}} = 0.843$, and $\widehat{\mathbf{E}}'_{PG} = k_I\frac{\widehat{\boldsymbol{\beta}}'_{PG}\widehat{\mathbf{A}}}{\sqrt{\widehat{\boldsymbol{\beta}}'_{PG}\widehat{\boldsymbol{\Phi}}\widehat{\boldsymbol{\beta}}_{PG}}} = [0.01 \;\; -1.00 \;\; -3.56 \;\; 0 \;\; -0.46 \;\; -3.98]$ respectively.

Fig. 8.4 presents the frequency distribution of the 244 estimated PPG-GESIM index values for two predetermined restrictions on the traits GY and EHT and their associated GEBVs (GEBV_{GY} and GEBV_{EHT}), for one selection cycle in an environment for a real maize (*Zea mays*) F_2 population with 233 molecular markers. Note that the frequency distribution of the estimated PPG-GESIM index values approaches normal distribution.

Now, with a selection intensity of 10% ($k_I = 1.755$) and a vector of predetermined restrictions $\mathbf{d}'_{PG} = [7 \;\; -3 \;\; 5 \;\; 3.5 \;\; -1.5 \;\; 2.5]$, we compare the estimated CPPG-LGSI and PPG-GESIM selection responses and expected genetic gains per

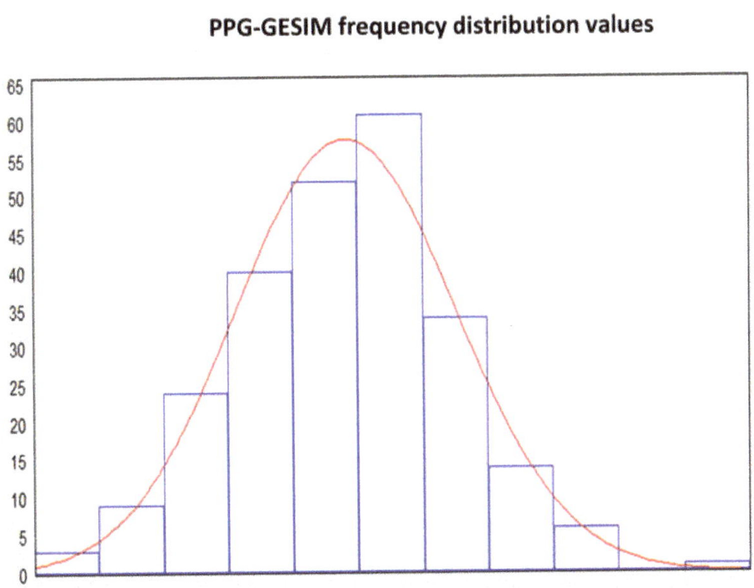

Fig. 8.4 Frequency distribution of the 244 estimated predetermined proportional gain genomic eigen selection index method (PPG-GESIM) values for two predetermined restrictions on the traits GY and EHT and their associated GEBVs, GEBV_{GY} and GEBV_{EHT}, for one selection cycle in an environment for a real maize (*Zea mays*) F_2 population with 233 molecular markers

trait using the simulated data set described in Sect. 2.8.1 of Chap. 2. Traits T1, T2, and T3 and their associated GEBVs (GEBV1, GEBV2, and GEBV3 respectively) were restricted, but trait T4 and its associated GEBV4 were not restricted. For this data set, matrix **F** was an identity matrix of size 8 × 8 for all four selection cycles.

Table 8.6 presents the estimated CPPG-LGSI selection responses when their vectors of coefficients are normalized, and the estimated PPG-GESIM selection responses for one, two, and three predetermined restrictions for four simulated selection cycles. The averages of the estimated CPPG-LGSI selection responses were 5.08 for one restriction, 3.42 for two restrictions, and 1.60 for three restrictions, whereas the averages of the estimated PPG-GESIM selection responses were 1.96 for one restriction, 4.14 for two restrictions, and 5.46 for three restrictions. For this data set, when the number of restrictions increases, the estimated CPPG-LGSI

Table 8.6 Estimated CPPG-LGSI expected genetic gains for one, two, and three restricted predetermined traits (T1, T2, and T3) and for one, two, and three restricted predetermined GEBVs (GEBV1, GEBV2, and GEBV3) for four simulated selection cycles

	CPPG-LGSI expected genetic gain for one predetermined restriction							
	Traits				Genomic estimated breeding values			
Cycle	T1	T2	T3	T4	GEBV1	GEBV2	GEBV3	GEBV4
1	8.24	−3.62	3.32	2.26	4.12	−2.33	1.75	1.09
2	7.98	−4.06	3.03	2.68	3.99	−2.24	1.79	1.04
3	8.61	−4.48	3.24	1.96	4.30	−2.32	1.70	0.98
4	8.30	−4.34	3.32	2.04	4.15	−2.16	1.62	0.92
Average	8.28	−4.12	3.23	2.23	4.14	−2.26	1.71	1.01
	CPPG-LGSI expected genetic gain for two predetermined restrictions							
	Traits				Genomic estimated breeding values			
Cycle	T1	T2	T3	T4	GEBV1	GEBV2	GEBV3	GEBV4
1	8.06	−3.46	3.30	2.06	4.03	−1.73	1.72	0.98
2	8.17	−3.50	3.08	2.65	4.09	−1.75	1.79	0.98
3	8.88	−3.81	3.31	1.83	4.44	−1.90	1.72	0.90
4	8.61	−3.69	3.43	1.99	4.30	−1.84	1.65	0.87
Average	8.43	−3.61	3.28	2.13	4.22	−1.81	1.72	0.93
	CPPG-LGSI expected genetic gain for three predetermined restrictions							
	Traits				Genomic estimated breeding values			
Cycle	T1	T2	T3	T4	GEBV1	GEBV2	GEBV3	GEBV4
1	5.77	−2.47	4.12	2.28	2.88	−1.24	2.06	0.98
2	5.68	−2.43	4.06	2.76	2.84	−1.22	2.03	0.97
3	5.87	−2.52	4.20	1.98	2.94	−1.26	2.10	0.79
4	5.91	−2.53	4.22	2.00	2.95	−1.27	2.11	0.83
Average	5.81	−2.49	4.15	2.26	2.90	−1.24	2.07	0.89

The selection intensity was 10% ($k_I = 1.755$) and the vector of predetermined restrictions was $\mathbf{d}'_{PG} = [7 \quad -3 \quad 5 \quad 3.5 \quad -1.5 \quad 2.5]$. Trait T4 and its associated GEBV4 were not restricted

selection response tends to decrease, whereas the estimated PPG-GESIM selection response increases.

Tables 8.7 presents the estimated CPPG-LGSI and PPG-GESIM expected genetic gains for one, two, and three predetermined restrictions respectively, for four simulated selection cycles. The averages of the estimated CPPG-LGSI expected genetic gains for the four traits and their four associated GEBVs were 8.28, −4.12, 3.23, 2.23, 4.14, −2.26, 1.71, and 1.01 for one restriction; 8.43, −3.61, 3.28, 2.13, 4.22, −1.81, 1.72, and 0.93 for two restrictions; and 5.81, −2.49, 4.15, 2.26, 2.90, −1.24, 2.07, and 0.89 for three restrictions. On the other hand, the averages of the estimated PPG-GESIM expected genetic gains for the four traits and their four associated GEBVs were 6.97, −1.31, 1.78, 0.52, 5.64, −1.74, 1.75, and 0.58 for one restriction; 6.93, −2.73, 1.29, 0.85, 5.75, −2.55, 1.49, and 0.79 for two restrictions, and 8.12, −3.27, 2.99, 1.13, 2.19, −1.15, 1.30, and 0.45 for three

Table 8.7 Estimated PPG-GESIM expected genetic gains for one, two, and three restricted traits (T1, T2, and T3) and for one, two, and three restricted GEBVs (GEBV1, GEBV2, and GEBV3) for four simulated selection cycles

	PPG-GESIM expected genetic gain for one predetermined restriction							
	Traits				Genomic estimated breeding values			
Cycle	T1	T2	T3	T4	GEBV1	GEBV2	GEBV3	GEBV4
1	6.89	−1.44	1.94	0.63	6.36	−1.89	2.04	0.62
2	6.71	−1.33	1.90	0.65	6.06	−2.00	1.97	0.75
3	7.09	−1.69	1.67	0.40	5.40	−1.72	1.63	0.55
4	7.18	−0.78	1.58	0.39	4.73	−1.34	1.35	0.39
Average	6.97	−1.31	1.78	0.52	5.64	−1.74	1.75	0.58
	PPG-GESIM expected genetic gain for two predetermined restrictions							
	Traits				Genomic estimated breeding values			
Cycle	T1	T2	T3	T4	GEBV1	GEBV2	GEBV3	GEBV4
1	6.61	−2.55	1.40	0.94	6.49	−2.80	1.75	0.87
2	5.67	−2.48	1.24	0.87	6.16	−2.84	1.70	0.91
3	7.35	−3.08	1.21	0.85	5.54	−2.49	1.37	0.82
4	8.10	−2.80	1.29	0.76	4.80	−2.08	1.16	0.56
Average	6.93	−2.73	1.29	0.85	5.75	−2.55	1.49	0.79
	PPG-GESIM expected genetic gain for three predetermined restrictions							
	Traits				Genomic estimated breeding values			
Cycle	T1	T2	T3	T4	GEBV1	GEBV2	GEBV3	GEBV4
1	7.21	−2.94	2.64	1.02	1.69	−1.10	1.07	0.45
2	7.71	−2.97	2.41	1.46	2.22	−1.15	1.21	0.45
3	8.72	−3.43	3.17	0.93	2.21	−1.06	1.34	0.42
4	8.85	−3.73	3.72	1.09	2.63	−1.29	1.60	0.48
Average	8.12	−3.27	2.99	1.13	2.19	−1.15	1.30	0.45

The selection intensity was 10% ($k_I = 1.755$) and the vector of predetermined restrictions was $\mathbf{d}'_{PG} = [7 \quad -3 \quad 5 \quad 3.5 \quad -1.5 \quad 2.5]$. Trait T4 and its associated GEBV4 were not restricted

restrictions. These results indicate that the estimated CPPG-LGSI expected genetic gains for the four traits and their four associated GEBVs were generally higher than the estimated PPG-GESIM expected genetic gains for the four traits and their four associated GEBVs.

References

Crossa J, Cerón-Rojas JJ (2011) Multi-trait multi-environment genome-wide molecular marker selection indices. J Indian Soc Agric Stat 62(2):125–142

Meyer CD (2000) Matrix analysis and applied linear algebra. Society for Industrial and Applied Mathematics (SIAM), Philadelphia, PA

9

Parameters and Numerical Examples of MRLPSI, MLGSI, MRLGSI and OMPPG-LGSI

Abstract Multistage linear selection indices select individual traits available at different times or stages and are applied mainly in animals and tree breeding, where the traits under consideration become evident at different ages. The main indices are: the unrestricted, the restricted, and the predetermined proportional gain selection index. The restricted and predetermined proportional gain indices allow null and predetermined restrictions to be imposed on the trait expected genetic gain (or multi-trait selection response) values, whereas the rest of the traits remain changed without any restriction. The three indices can use phenotypic, genomic, or both sets of information to predict the unobservable net genetic merit values of the candidates for selection and all of them maximize the selection response, the expected genetic gain for each trait, have maximum accuracy, are the best predictor of the net genetic merit, and provide the breeder with an objective rule for evaluating and selecting several traits simultaneously. The theory of the foregoing indices is based on the independent culling method and on the linear phenotypic selection index, and is described in this chapter in the phenotypic and genomic selection context. Their theoretical results are validated in a two-stage breeding selection scheme using real and simulated data.

9.1 Multistage Linear Phenotypic Selection Index

In a similar manner to the linear phenotypic selection index (LPSI, Chap. 2), the objectives of the multistage linear phenotypic selection index (MLPSI) are:

1. To predict the net genetic merit $H = \mathbf{w}'\mathbf{g}$, where $\mathbf{g}' = [g_1 \ g_2 \ \dots \ g_t]$ is the vector of true breeding values of an individual for t traits and $\mathbf{w}' = [w_1 \quad w_2 \quad \dots \quad w_t]$ is the vector of economic weights.
2. To select individuals with the highest H values at each stage as parents of the next generation.
3. To maximize the MLPSI selection response and its expected genetic gain per trait.

4. To provide the breeder with an objective rule for evaluating and selecting several traits simultaneously.

When selection is based on all the individual traits of interest jointly, the LPSI vector of coefficients that maximizes the selection response $R = k\sqrt{\mathbf{b}'\mathbf{P}\mathbf{b}}$ and the expected genetic gain per trait $\mathbf{E} = k\frac{\mathbf{C}\mathbf{b}}{\sqrt{\mathbf{b}'\mathbf{P}\mathbf{b}}}$ is $\mathbf{b} = \mathbf{P}^{-1}\mathbf{C}\mathbf{w}$, where $\mathbf{C}$ and $\mathbf{P}$ are the covariance matrices of the true breeding values ($\mathbf{g}$) and trait phenotypic values ($\mathbf{y}$) respectively, and k is the selection intensity. In MLPSI terminology, the LPSI is called a one-stage selection index. The MLPSI is an extension of the LPSI theory to the multistage selection context and, as we shall see, the MLPSI theoretical results are very similar to the LPSI theoretical results described in Chap. 2.

9.1.1 The MLPSI Parameters for Two Stages

Let $\mathbf{y}' = [y_1 \quad y_2 \quad \cdots \quad y_t]$ be a vector with t traits of interest and suppose that we can select only n_i of them ($n_i < t$) at stage i ($i = 1, 2, \cdots, N$), such that after N stages ($N < t$), $\sum_{i=1}^{N} n_i = t$. Thus, for each stage we should have a selection index with a different number of traits. For example, at stage i the index would be $I_i = \sum_{j=1}^{n_i} b_{ij}y_{ij}$, and at stage N the index would be $I_N = \sum_{j=1}^{n_1} b_{1j}y_{1j} + \sum_{j=1}^{n_2} b_{2j}y_{2j} + \cdots + \sum_{j=1}^{n_N} b_{Nj}y_{Nj} = \sum_{i=1}^{N} I_i$, where the double subscript of y_{ij} indicates that the jth trait is measured at stage i, so that at each sub-index I_i, all the n_i traits are measured at the same age.

Suppose that there are four traits of interest and that $\mathbf{y}' = [y_1 \quad y_2 \quad y_3 \quad y_4]$ is the vector of observable phenotypic values and $\mathbf{g}' = [g_1 \quad g_2 \quad g_3 \quad g_4]$ is the vector of unobservable breeding values. If at the first and second stages we select two traits, then $n_1 = n_2 = 2$ and $\mathbf{y}'$ can be partitioned as $\mathbf{y}' = [\mathbf{x}_1' \quad \mathbf{x}_2']$, where $\mathbf{x}_1' = [y_1 \quad y_2]$ and $\mathbf{x}_2' = [y_3 \quad y_4]$ are the vectors of traits that become evident at the first and second stages respectively. At the first stage, the phenotypic covariance matrix of $\mathbf{x}_1$ ($\mathbf{P}_1$) and the covariance matrix of $\mathbf{x}_1$ with the vector of true breeding values $\mathbf{g}$ ($\mathbf{G}_1$) can be written as $Var(\mathbf{x}_1) = \begin{bmatrix} Var(y_1) & Cov(y_1, y_2) \\ Cov(y_2, y_1) & Var(y_2) \end{bmatrix} = \mathbf{P}_1$ and

$$Cov(\mathbf{x}_1, \mathbf{g}) = \begin{bmatrix} Cov(y_1, g_1) & Cov(y_1, g_2) & Cov(y_1, g_3) & Cov(y_1, g_4) \\ Cov(y_2, g_1) & Cov(y_2, g_2) & Cov(y_2, g_3) & Cov(y_2, g_4) \end{bmatrix} = \mathbf{G}_1$$

respectively. For the second stage, in addition to matrix $\mathbf{P}_1$, we need the phenotypic covariance matrix between $\mathbf{x}_1$ and $\mathbf{x}_2$ ($\mathbf{P}_{12}$) and the phenotypic covariance matrix of $\mathbf{x}_2$ ($\mathbf{P}_2$); thus, the covariance matrix of phenotypic values at stage 2 is $\mathbf{P} = \begin{bmatrix} \mathbf{P}_1 & \mathbf{P}_{12} \\ \mathbf{P}_{21} & \mathbf{P}_2 \end{bmatrix}$. In a similar manner, in addition to matrix $\mathbf{G}_1$, at stage 2 we need the covariance between $\mathbf{x}_2$ and $\mathbf{g}$ ($\mathbf{G}_2$); that is, at stage 2 the covariance matrix between phenotypic and breeding values can be written as $\mathbf{G} = \begin{bmatrix} \mathbf{G}_1 \\ \mathbf{G}_2 \end{bmatrix}$. Matrices $\mathbf{G}$ and $\mathbf{C}$ are not exactly the same, because although $\mathbf{C} = Var(\mathbf{g})$, $\mathbf{G} = \begin{bmatrix} Cov(\mathbf{x}_1, \mathbf{g}) \\ Cov(\mathbf{x}_2, \mathbf{g}) \end{bmatrix} = \begin{bmatrix} \mathbf{G}_1 \\ \mathbf{G}_2 \end{bmatrix}$ and this latter matrix changes at each stage.

Let $\mathbf{w}' = [w_1 \quad w_2 \quad w_3 \quad w_4]$ be the vector of economic weights; then, at the first and second stages the MLPSI vectors of coefficients are $\mathbf{b}_1' = \mathbf{w}'\mathbf{G}'_1\mathbf{P}_1^{-1} = [b_{11} \quad b_{12}]$ and $\mathbf{b}_2' = \mathbf{w}'\mathbf{G}'\mathbf{P}^{-1} = [b_{21} \quad b_{22} \quad b_{23} \quad b_{24}]$ respectively. The selection indices at stages 1 and 2 can be written as $I_1 = b_{11}y_1 + b_{12}y_2 = \mathbf{b}_1'\mathbf{x}_1$ and $I_2 = b_{21}y_1 + b_{22}y_2 + b_{23}y_3 + b_{24}y_4 = \mathbf{b}_2'\mathbf{y}$, which could be correlated and then numerical integration would be required to find optimal truncation points and selection intensities (Xu and Muir 1992; Hicks et al. 1998) before obtaining the maximized MLPSI selection response and expected genetic gain per trait.

The accuracy of the MLPSI at stages 1 and 2 can be written as

$$\rho_{HI_1} = \sqrt{\frac{\mathbf{b}_1'\mathbf{P}_1\mathbf{b}_1}{\mathbf{w}'\mathbf{C}\mathbf{w}}} \quad \text{and} \quad \rho_{HI_2} = \sqrt{\frac{\mathbf{b}_2'\mathbf{P}^*\mathbf{b}_2}{\mathbf{w}'\mathbf{C}^*\mathbf{w}}}, \tag{9.1}$$

respectively. Let k_1 and k_2 be the selection intensities for stages 1 and 2; then, the maximized MLPSI expected genetic gains per trait can be written as

$$\mathbf{E}_1 = k_1\frac{\mathbf{G}_1'\mathbf{b}_1}{\sqrt{\mathbf{b}_1'\mathbf{P}_1\mathbf{b}_1}} \quad \text{and} \quad \mathbf{E}_2 = k_2\frac{\mathbf{b}_2'\mathbf{C}^*}{\sqrt{\mathbf{b}_2'\mathbf{P}^*\mathbf{b}_2}}, \tag{9.2}$$

and the total expected genetic gain per trait for the two stages is equal to $\mathbf{E}_1 + \mathbf{E}_2$. In a similar manner, the maximized selection responses for both stages are

$$R_1 = k_1\sqrt{\mathbf{b}_1'\mathbf{P}_1\mathbf{b}_1} \quad \text{and} \quad R_2 = k_2\sqrt{\mathbf{b}_2'\mathbf{P}^*\mathbf{b}_2}, \tag{9.3}$$

and the total selection response for the two stages is $R_1 + R_2$. In Eqs. (9.1) to (9.3), matrices $\mathbf{P}^*$ and $\mathbf{C}^*$ are matrices $\mathbf{P}$ and $\mathbf{C}$ respectively, adjusted for previous selection on $I_1 = \mathbf{b}_1'\mathbf{x}_1$. That is, the MLPSI accuracy, expected genetic gain per trait, and selection response at stage 2 are affected by previous selection on I_1 (Saxton 1983) and it is necessary to adjust $\mathbf{P}$ and $\mathbf{C}$.

One method for adjusting matrices $\mathbf{P}$ and $\mathbf{C}$ has been provided by Cochran (1951) and Cunningham (1975). Suppose that X, Y, and W are three jointly normally distributed random variables and that the covariance among them is known, then the covariance between X and Y adjusted for the effects of selection on W can be obtained as

$$Cov(X, Y)^* = Cov(X, Y) - u\frac{Cov(X, W)Cov(Y, W)}{Var(W)}, \tag{9.4}$$

where $u = k_1(k_1 - \tau)$, k_1 is the selection intensity at stage 1 and τ is the truncation point when $I_1 = \mathbf{b}_1'\mathbf{x}_1$ is applied. For example, if the selection intensity at the first stage is 5%, $k_1 = 2.063$, $\tau = 1.645$, and $u = 0.862$ (Falconer and Mackay 1996, Table A).

According to Dekkers (2014), with the result of Eq. (9.4), it is possible to obtain matrices $\mathbf{P}^*$ and $\mathbf{C}^*$ using the following two equations:

$$\mathbf{P}^* = Var(\mathbf{y})^* = \mathbf{P} - u\frac{Cov(\mathbf{y},\mathbf{x}_1)\mathbf{b}_1\mathbf{b}_1'Cov(\mathbf{x}_1,\mathbf{y})}{\mathbf{b}_1'Var(\mathbf{x}_1)\mathbf{b}_1}$$

$$= \mathbf{P} - u\frac{\begin{bmatrix}\mathbf{P}_1 \\ \mathbf{P}_{21}\end{bmatrix}\mathbf{b}_1\mathbf{b}_1'[\mathbf{P}_1 \quad \mathbf{P}_{21}]}{\mathbf{b}_1'\mathbf{P}_1\mathbf{b}_1} \tag{9.5}$$

and

$$\mathbf{C}^* = Var(\mathbf{g})^* = \mathbf{C} - u\frac{Cov(\mathbf{g},\mathbf{x}_1)\mathbf{b}_1\mathbf{b}_1'Cov(\mathbf{x}_1,\mathbf{g})}{\mathbf{b}_1'Var(\mathbf{x}_1)\mathbf{b}_1} = \mathbf{C} - u\frac{\mathbf{G}_1'\mathbf{b}_1\mathbf{b}_1'\mathbf{G}_1}{\mathbf{b}_1'\mathbf{P}_1\mathbf{b}_1}. \tag{9.6}$$

With the Eq. (9.5) result, the correlation between $I_1 = \mathbf{b}_1'\mathbf{x}_1$ and $I_2 = \mathbf{b}_2'\mathbf{y}$ is

$$Corr(I_1, I_2) = \frac{\mathbf{b}_1'[\mathbf{P}_1 \quad \mathbf{P}_{21}]\mathbf{b}_2}{\sqrt{\mathbf{b}_1'\mathbf{P}_1\mathbf{b}_1}\sqrt{\mathbf{b}_2'\mathbf{P}\mathbf{b}_2}} = \rho_{12}, \tag{9.7}$$

where $\sqrt{\mathbf{b}_1'\mathbf{P}_1\mathbf{b}_1}$ and $\sqrt{\mathbf{b}_2'\mathbf{P}\mathbf{b}_2}$ are the standard deviations of the variances of $I_1 = \mathbf{b}_1'\mathbf{x}_1$ and $I_2 = \mathbf{b}_2'\mathbf{y}$ respectively.

9.1.2 The Selection Intensities

Selection intensity k is related to the height of the ordinate of the normal curve (z) and the proportion selected (p) in the LPSI as $k = z/p$. In the multistage selection context, it is usual to fix the total proportion to be selected (p) before selection is carried out and then to determine the unknown proportion q_i (i=1, 2,···, N) for each stage under the restriction

$$p = \prod_{i=1}^{N} q_i, \tag{9.8}$$

where N is the number of stages. In the two-stage selection scheme, we would have $p = q_1 q_2$. Based on the fixed proportion p and the ρ_{12} value (Eq. 9.7), Young (1964) used the bivariate truncated normal distribution theory to obtain the selection intensity for two stages. A truncated distribution is a conditional distribution resulting when the domain of the parent distribution is restricted to a smaller region (Hattaway 2010). In the multistage selection context, a truncation occurs when a sample of individuals from the parent distribution are selected as parents for the next selection cycle, thus creating a new population of individuals that follow a truncated normal distribution.

Suppose that $I_1 = \mathbf{b}_1'\mathbf{x}_1$ and $I_2 = \mathbf{b}_2'\mathbf{y}$ have joint normal distribution and let I_1 and I_2 be transformed as $v_1 = \frac{I_1 - \mu_{I_1}}{\sigma_{I_1}}$ and $v_2 = \frac{I_2 - \mu_{I_2}}{\sigma_{I_2}}$ with a mean of zero and a variance of 1, where μ_{I_2} and μ_{I_2} are the means, whereas σ_{I_1} and σ_{I_2} are the standard deviations of the variances of I_1 and I_2 respectively. In this case, the method of selection is to retain animals or plants with $v_1 \geq c_1$ at stage 1 and $v_1 + v_2 \geq c_2$ at stage 2, where c_1 and c_2 are truncation points for I_1 and I_2 respectively.

The selected population has bivariate left truncated normal distribution with a probability density function given by $h(v_1, v_2) = \frac{f(v_1, v_2)}{p}$, where $f(v_1, v_2) = \frac{1}{2\pi\sqrt{1-\rho_{12}^2}} \exp\left\{ -\frac{1}{2(1-\rho_{12}^2)} \left[v_1^2 + v_2^2 - 2\rho_{12} v_1 v_2 \right] \right\}$ and ρ_{12} is the correlation between v_1 and v_2. The fixed total proportion (p) before selection can be written as $p = \int_{c_1}^{\infty} \int_{c_2 - v_1}^{\infty} f(v_1, v_2) dv_2 dv_1$, where c_1 and c_2 are truncation points for I_1 and I_2, respectively. Then, as p is fixed, Young (1964) integrated by parts (Thomas 2014)

$$\int_{c_1}^{\infty} \int_{c_2 - v_1}^{\infty} f(v_1, v_2) dv_1 dv_2 \tag{9.9}$$

and found the expectations of v_1 and v_2 in the selected population, writing the selection intensity values for stages 1 (k_1) and 2 (k_2) as

$$k_1 = \frac{z(c_1)Q(a)}{p} + \frac{z(c_3)Q(b)\sqrt{(1+\rho_{12})/2}}{p} \tag{9.10}$$

and

$$k_2 = \frac{\rho_{12} z(c_1)Q(a)}{p} + \frac{z(c_3)Q(b)\sqrt{(1+\rho_{12})/2}}{p} \tag{9.11}$$

respectively, where $z(c_1) = \frac{\exp\{-0.5c_1^2\}}{\sqrt{2\pi}}$ and $z(c_3) = \frac{\exp\{-0.5c_3^2\}}{\sqrt{2\pi}}$ are the heights of the ordinates of the standard normal distribution at the lowest value of c_1 and $c_3 = \frac{c_2}{\sqrt{2(1+\rho_{12})}}$ and p is the total proportion of the population of animal or plant lines selected; $a = \frac{c_2 - c_1(1+\rho_{12})}{\sqrt{1-\rho_{12}^2}}$ and $b = \frac{2c_1 - c_2}{\sqrt{2(1-\rho_{12})}}$, whereas $Q(a) = 1 - \Phi(a)$ and $Q(b) = 1 - \Phi(b)$ are the complement of the standard normal distribution; $\Phi(a) = \int_{-\infty}^{a} \frac{1}{\sqrt{2\pi}} \exp\{-0.5w^2\} dw$ and $\Phi(b) = \int_{-\infty}^{b} \frac{1}{\sqrt{2\pi}} \exp\{-0.5t^2\} dt$ are probabilities of the standard normal distribution, i.e., $\Phi(a) = P_r(W \leq a)$ and $\Phi(b) = P_r(T \leq b)$.

Young (1964) provided figures to obtain values of c_1 and c_2 when the ρ_{12} values are between -0.8 and 0.8, and the p values are between 0.05 and 0.8. For example, suppose that $\rho_{12} = 0.8$ and $p = 0.2$ (or 20%), then, according to Young (1964, Fig. 9), $c_1 = 0.80$ and $c_2 = 1.6$, and to find the selection intensities for the first (k_1) and second stages (k_2) we need to solve Eqs. (9.10) and (9.11). That is, as $c_1 = 0.80$, $c_2 = 1.6$, $\rho_{12} = 0.8$, and $p = 0.2$, then $z(c_1) = \frac{\exp\{-0.5(0.8)^2\}}{\sqrt{2\pi}} = 0.290$, $z(c_3) = \frac{\exp\{-0.5[(1.6)^2/2(1.8)]\}}{\sqrt{2\pi}} = 0.28$, $a = \frac{1.6-0.8(1.8)}{\sqrt{1-(0.8)^2}} = 0.27$, $b = \frac{2(0.8)-1.6}{\sqrt{2(0.2)}} = 0$, $\Phi(a) = 0.6064$, $\Phi(b) = 0.5$, $Q(a) = 1 - \Phi(a) = 0.3936$, and $Q(b) = 1 - \Phi(b) = 0.5$. Based on these results, the selection intensities for stages 1 and 2 are

$$k_1 = \frac{(0.29)(0.3936)}{0.2} + \frac{(0.28)(0.5)(0.9)}{0.2} = 0.744 \quad \text{and}$$

$$k_2 = \frac{(0.8)(0.29)(0.3936)}{0.2} + \frac{(0.28)(0.5)(0.9)}{0.2} = 0.721$$

respectively. Note that the values of $\Phi(a) = 0.6064$ and $\Phi(b) = 0.5$ can be obtained from any table with values showing the area under the curve of the standard normal distribution (e.g., Rausand and Hϕyland 2004, Table F.1).

One problem with Eqs. (9.10) and (9.11) is that they tend to overestimate the selection intensities values and also overestimate the selection response when the total proportion retained p is lower than 10%. Cochran (1951) have given two equations to obtain selection intensities in the two stages context but his equations also overestimate the selection intensities values when p is lower than 10%. Up to now, there is not an accurate method to estimate selection intensities for two or more stages in the MLPSI context. Mi et al. (2014) have developed an R package called *selectiongain* that enables calculation of the OMLPSI selection response for up to 20 selection stages. *Selectiongain* uses raw integration to obtain the first moment of a lower truncated multivariate standard normal distribution and then it estimates the OMLPSI selection response at each stage; however, this integral requires complex numerical algorithms with no convergence criteria (Arismendi 2013) and could also overestimate the selection intensity at each stage.

9.1.3 Numerical Example

To illustrate the two-stage selection theory, we use the poultry data of Xu and Muir (1992). This data set contains four traits: age at sexual maturity, defined as the age (in days) at which the first trap-nested egg was laid (y_1); rate of lay, defined as 100 times (total eggs in the laying period)/(total days in the laying period) (y_2); body weight (in pounds) measured at 32 weeks of age (y_3); and average egg weight (in ounces per dozen) of all the eggs laid up to 32 weeks of age (y_4). The estimated phenotypic and genetic covariance matrices were $\widehat{\mathbf{P}} = \begin{bmatrix} 137.178 & -90.957 & 0.136 & 0.564 \\ -90.957 & 201.558 & 1.103 & -1.231 \\ 0.136 & 1.103 & 0.202 & 0.104 \\ 0.564 & -1.231 & 0.104 & 2.874 \end{bmatrix}$ and $\widehat{\mathbf{C}} = \begin{bmatrix} 14.634 & -18.356 & -0.109 & 1.233 \\ -18.356 & 32.029 & 0.103 & -2.574 \\ -0.109 & 0.103 & 0.089 & 0.023 \\ 1.233 & -2.574 & 0.023 & 1.225 \end{bmatrix}$ respectively, whereas the vector of economic weights for the four traits was $\mathbf{w}' = [-3.555 \quad 19.536 \quad -113.746 \quad 48.307]$.

Suppose that at the first and second stages we select two traits ($n_1 = n_2 = 2$); then, $\mathbf{y}' = [\mathbf{x}_1' \quad \mathbf{x}_2']$, where $\mathbf{x}_1' = [y_1 \quad y_2]$ and $\mathbf{x}_2' = [y_3 \quad y_4]$. The estimated phenotypic ($\widehat{\mathbf{P}}_1$) and genetic ($\widehat{\mathbf{G}}_1$) covariance matrices for the first stage were $\widehat{\mathbf{P}}_1 = \begin{bmatrix} 137.178 & -90.957 \\ -90.957 & 1.103 \end{bmatrix}$ and $\widehat{\mathbf{G}}_1 = \begin{bmatrix} 14.634 & -18.356 & -0.109 & 1.233 \\ -18.356 & 32.029 & 0.103 & -2.574 \end{bmatrix}$ respectively. For the first and second stages, the estimated MLPSI vector of coefficients were $\widehat{\mathbf{b}}_1' = \mathbf{w}'\widehat{\mathbf{G}}'_1\widehat{\mathbf{P}}_1 = [-0.918 \quad 2.339]$ and $\widehat{\mathbf{b}}_2' = \widehat{\mathbf{w}}'\widehat{\mathbf{C}}\widehat{\mathbf{P}}^{-1} = [-0.59 \quad 2.78 \quad -49.45 \quad 3.75]$ respectively.

The estimated correlation value between the estimated indices $\widehat{I}_1 = \widehat{\mathbf{b}}_1'\mathbf{x}_1$ and $\widehat{I}_2 = \widehat{\mathbf{b}}_2'\mathbf{y}$ was $\widehat{\rho}_{12} = \dfrac{\widehat{\mathbf{b}}_1'[\widehat{\mathbf{P}}_1 \quad \widehat{\mathbf{P}}_{21}]\widehat{\mathbf{b}}_2}{\sqrt{\widehat{\mathbf{b}}_1'\widehat{\mathbf{P}}_1\widehat{\mathbf{b}}_1}\sqrt{\widehat{\mathbf{b}}_2'\widehat{\mathbf{P}}\widehat{\mathbf{b}}_2}} = 0.88$, where $\sqrt{\widehat{\mathbf{b}}_1'\widehat{\mathbf{P}}_1\widehat{\mathbf{b}}_1}$ and $\sqrt{\widehat{\mathbf{b}}_2'\widehat{\mathbf{P}}\widehat{\mathbf{b}}_2}$ were the estimated standard deviations of the variance of $\widehat{I}_1$ and $\widehat{I}_2$ respectively. Assuming that $p = 0.2$ (or 20%), an approximate selection intensity for the first stage was $k_1 = 0.744$, whence the estimated MLPSI selection response, expected genetic gain per trait, and accuracy were $\widehat{R}_1 = k_1\sqrt{\widehat{\mathbf{b}}_1'\widehat{\mathbf{P}}_1\widehat{\mathbf{b}}_1} = 29.85$, $\widehat{\mathbf{E}}'_1 = k_1\dfrac{\widehat{\mathbf{G}}_1'\widehat{\mathbf{b}}_1}{\sqrt{\widehat{\mathbf{b}}_1'\widehat{\mathbf{P}}_1\widehat{\mathbf{b}}_1}} = [-1.046 \quad 1.702 \quad 0.006 \quad -0.133]$, and $\widehat{\rho}_{HI_1} = \sqrt{\dfrac{\widehat{\mathbf{b}}_1'\widehat{\mathbf{P}}_1\widehat{\mathbf{b}}_1}{\mathbf{w}'\widehat{\mathbf{C}}\mathbf{w}}} = 0.353$ respectively.

According to the $k_1 = 0.744$ value, the approached value of u was $u = 0.554$, and by Eqs. (9.5) and (9.6), the estimated and adjusted phenotypic ($\widehat{\mathbf{P}}^*$) and genetic ($\widehat{\mathbf{C}}^*$) covariance matrices for the second stage were $\widehat{\mathbf{P}}^* = \begin{bmatrix} 97.682 & -26.241 & 0.422 & 0.168 \\ -26.241 & 95.518 & 0.634 & -0.582 \\ 0.422 & 0.634 & 0.200 & 0.107 \\ 0.168 & -0.582 & 0.107 & 2.870 \end{bmatrix}$ and

$$\widehat{\mathbf{C}}^* = \begin{bmatrix} 13.540 & -16.575 & -0.102 & 1.094 \\ -16.575 & 29.129 & 0.092 & -2.348 \\ -0.102 & 0.092 & 0.089 & 0.024 \\ 1.094 & -2.384 & 0.024 & 1.207 \end{bmatrix}, \text{ respectively.}$$

For the second stage, the approximated selection intensity was $k_2 = 0.721$, whereas the estimated MLPSI selection response, expected genetic gain per trait and accuracy, were $\widehat{R}_2 = k_{I_2}\sqrt{\widehat{\mathbf{b}}_2'\widehat{\mathbf{P}}_2^*\widehat{\mathbf{b}}_2} = 24.84$, $\widehat{\mathbf{E}}_2' = k_{I_2}\frac{\widehat{\mathbf{C}}^{*\prime}\widehat{\mathbf{b}}_2}{\sqrt{\widehat{\mathbf{b}}_2'\widehat{\mathbf{P}}_2^*\widehat{\mathbf{b}}_2}} = [-0.443 \quad 0.804 \quad -0.087 \quad -0.087]$, and $\widehat{\rho}_{HI_2} = \sqrt{\frac{\widehat{\mathbf{b}}_2'\widehat{\mathbf{P}}_2^*\widehat{\mathbf{b}}_2}{\mathbf{w}'\widehat{\mathbf{C}}^*\mathbf{w}}} = 0.314$ respectively. Finally, the total estimated MLPSI selection response and expected genetic gain per trait were $\widehat{R}_1 + \widehat{R}_2 = 54.69$ and $\widehat{\mathbf{E}}_1' + \widehat{\mathbf{E}}_2' = [-1.488 \quad 2.506 \quad -0.081 \quad -0.219]$.

9.2 The Multistage Restricted Linear Phenotypic Selection Index

The multistage restricted linear phenotypic selection index (MRLPSI) is an extension of the null restricted linear phenotypic selection index (RLPSI) described in Chap. 3 to the multistage case; thus, the theoretical results of the MRLPSI are very similar to those of the RLPSI. The MRLPSI allows restrictions equal to zero to be imposed on the expected genetic gains of some traits, whereas other traits increase (or decrease) their expected genetic gains without any restrictions being imposed.

9.2.1 *The MRLPSI Parameters for Two Stages*

In Chap. 3, we indicated that vector $\mathbf{b}_R = \mathbf{Kb}$ is a linear transformation of the LPSI vector of coefficients ($\mathbf{b}$) made by the projector matrix $\mathbf{K}$, and that matrix $\mathbf{K}$ is idempotent ($\mathbf{K} = \mathbf{K}^2$) and projects $\mathbf{b}$ into a space smaller than the original space of $\mathbf{b}$. The reduction of the space into which matrix $\mathbf{K}$ projects $\mathbf{b}$ is equal to the number of zeros that appears on the expected genetic gain per trait. Hence, the MRLPSI vector of coefficients for stages 1 and 2 should be a linear transformation of the MLPSI vector of coefficients at stages 1 ($\mathbf{b}_1 = \mathbf{P}_1^{-1}\mathbf{G}_1\mathbf{w}$) and 2 ($\mathbf{b}_2 = \mathbf{P}^{-1}\mathbf{Cw}$) described in Sect. 9.1.1 of this chapter, and should be written as

$$\mathbf{b}_{R_1} = \mathbf{K}_1\mathbf{b}_1 \tag{9.12}$$

and

$$\mathbf{b}_{R_2} = \mathbf{K}_2\mathbf{b}_2, \tag{9.13}$$

respectively, where, at stage 1, $\mathbf{K}_1 = [\mathbf{I}_1 - \mathbf{Q}_1]$, $\mathbf{Q}_1 = \mathbf{P}_1^{-1}\boldsymbol{\Psi}_1\left(\boldsymbol{\Psi}_1'\mathbf{P}_1^{-1}\boldsymbol{\Psi}_1\right)^{-1}\boldsymbol{\Psi}_1'$, $\boldsymbol{\Psi}_1' = \mathbf{U}'\mathbf{G}'_1$, $\mathbf{I}_1$ is an identity matrix of the same size as $\mathbf{P}_1$, and $\mathbf{P}_1^{-1}$ is the inverse of matrix $\mathbf{P}_1$. At stage 2, $\mathbf{K}_2 = [\mathbf{I}_2 - \mathbf{Q}_2]$, $\mathbf{Q}_2 = \mathbf{P}^{-1}\boldsymbol{\Psi}_2\left(\boldsymbol{\Psi}_2'\mathbf{P}^{-1}\boldsymbol{\Psi}_2\right)^{-1}\boldsymbol{\Psi}_2'$, $\boldsymbol{\Psi}_2' = \mathbf{U}'\mathbf{C}$, $\mathbf{I}_2$ is an identity matrix of the same size as $\mathbf{P}$, and $\mathbf{P}^{-1}$ is the inverse of matrix $\mathbf{P}$. By Eqs. (9.12) and (9.13), the MRLPSI for stages 1 and 2 can be written as $I_1 = \mathbf{b}_{R_1}'\mathbf{x}_1$ and $I_2 = \mathbf{b}_{R_2}'\mathbf{y}$, where $\mathbf{y}' = \begin{bmatrix}\mathbf{x}_1' & \mathbf{x}_2'\end{bmatrix}$; $\mathbf{x}_1'$ and $\mathbf{x}_2'$ are the vectors of traits that become evident at the first and second stages respectively.

Let k_1 and k_2 be the selection intensities for stages 1 and 2 (Eqs. 9.10 and 9.11) respectively, and let $\mathbf{P}^*$ and $\mathbf{C}^*$ be the covariance matrices adjusted in the MRLPSI context according to Eqs. (9.5) and (9.5) respectively. The maximized MRLPSI selection response, expected genetic gain per trait, and accuracy at stages 1 and 2 can be written as

$$R_{R_1} = k_1\sqrt{\mathbf{b}_{R_1}'\mathbf{P}_1\mathbf{b}_{R_1}} \quad \text{and} \quad R_{R_1} = k_2\sqrt{\mathbf{b}_{R_2}'\mathbf{P}^*\mathbf{b}_{R_2}}, \tag{9.14}$$

$$\mathbf{E}_{R_1} = k_1\frac{\mathbf{G}_1'\mathbf{b}_{R_1}}{\sqrt{\mathbf{b}_{R_1}'\mathbf{P}_1\mathbf{b}_{R_1}}} \quad \text{and} \quad \mathbf{E}_{R_2} = k_2\frac{\mathbf{b}_{R_2}'\mathbf{C}^*}{\sqrt{\mathbf{b}_{R_2}'\mathbf{P}^*\mathbf{b}_{R_2}}} \tag{9.15}$$

and

$$\rho_{R_1} = \sqrt{\frac{\mathbf{b}_{R_1}'\mathbf{P}_1\mathbf{b}_{R_1}}{\mathbf{w}'\mathbf{C}\mathbf{w}}} \quad \text{and} \quad \rho_{R_2} = \sqrt{\frac{\mathbf{b}_{R_2}'\mathbf{P}^*\mathbf{b}_{R_2}}{\mathbf{w}'\mathbf{C}^*\mathbf{w}}}, \tag{9.16}$$

respectively, whereas the total MRLPSI selection response and expected genetic gain per trait for both stages are equal to $R_{R_1} + R_{R_2}$ and $\mathbf{E}_{R_1} + \mathbf{E}_{R_2}$.

9.2.2 Numerical Examples

To illustrate the MRLPSI theory for a two-stage selection breeding scheme, we use the real data set of the White Leghorn chickens of Hicks et al. (1998). This data set is conformed with six traits (y_1 to y_6) that correspond to records consisting of the number of eggs laid during different periods: from week 0 through 4 (y_1), 4 through 8 (y_2), 8 through 28 (y_3), 28 through 32 (y_4), 32 through 36 (y_5), and 36 through 52 (y_6) respectively. The estimated phenotypic and genotypic covariance matrices were

$$\widehat{\mathbf{P}} = \begin{bmatrix} 102 & 32 & 14 & 4 & 3 & -1 \\ 32 & 80 & 80 & 16 & 17 & 7 \\ 14 & 80 & 298 & 78 & 112 & 62 \\ 4 & 16 & 78 & 66 & 80 & 51 \\ 3 & 17 & 112 & 80 & 135 & 49 \\ -1 & 7 & 62 & 51 & 49 & 98 \end{bmatrix} \quad \text{and} \quad \widehat{\mathbf{C}} = \begin{bmatrix} 44 & 11 & -11 & -3 & -8 & -3 \\ 11 & 26 & 24 & 7 & 7 & 3 \\ -11 & 24 & 62 & 23 & 37 & 20 \\ -3 & 7 & 23 & 14 & 23 & 14 \\ -8 & 7 & 37 & 23 & 42 & 25 \\ -3 & 3 & 20 & 14 & 25 & 18 \end{bmatrix},$$

respectively, and $\mathbf{w}' = \begin{bmatrix} 0.08 & 0.08 & 0.38 & 0.08 & 0.08 & 0.31 \end{bmatrix}$ was the vector of economic weights.

Let $\mathbf{y}' = \begin{bmatrix} y_1 & y_2 & y_3 & y_4 & y_5 & y_6 \end{bmatrix}$ and $\mathbf{g}' = \begin{bmatrix} g_1 & g_2 & g_3 & g_4 & g_5 & g_6 \end{bmatrix}$ be the vectors of observed phenotypic and unobserved genotypic values respectively, and suppose that at stage 1 we select four traits and at stage 2 we select two traits, then $\mathbf{x}_1' = \begin{bmatrix} y_1 & y_2 & y_3 & y_4 \end{bmatrix}$ and $\mathbf{x}_2' = \begin{bmatrix} y_5 & y_6 \end{bmatrix}$ are the vector of observations at stages 1 and 2 respectively, whereas $\mathbf{y}' = \begin{bmatrix} \mathbf{x}_1' & \mathbf{x}_2' \end{bmatrix}$ is the vector of total observations at stage 2. We need to estimate vectors $\mathbf{b}_{R_1}' = \mathbf{b}_1'\mathbf{K}_1'$ and $\mathbf{b}_{R_2}' = \mathbf{b}_2'\mathbf{K}_2'$, where $\mathbf{b}_1' = \mathbf{w}'\mathbf{G}'_1 \mathbf{P}_1^{-1}$ and $\mathbf{b}_2' = \mathbf{w}'\mathbf{G}'\mathbf{P}^{-1}$. In Chap. 3, we described methods of estimating matrices $\mathbf{K}_1 = [\mathbf{I}_1 - \mathbf{Q}_1]$, $\mathbf{Q}_1 = \mathbf{P}_1^{-1}\boldsymbol{\Psi}_1\left(\boldsymbol{\Psi}_1'\mathbf{P}_1^{1}\boldsymbol{\Psi}_1\right)^{-1}\boldsymbol{\Psi}_1'$, $\boldsymbol{\Psi}_1' = \mathbf{U}'\mathbf{G}'_1$, $\mathbf{K}_2 = [\mathbf{I}_2 - \mathbf{Q}_2]$, $\mathbf{Q}_2 = \mathbf{P}^{-1}\boldsymbol{\Psi}_2\left(\boldsymbol{\Psi}_2'\mathbf{P}^{-1}\boldsymbol{\Psi}_2\right)^{-1}\boldsymbol{\Psi}_2'$, and $\boldsymbol{\Psi}_2' = \mathbf{U}'\mathbf{C}$, which are used in this subsection.

At stage 1, the estimated phenotypic and genotypic covariance matrices were

$$\widehat{\mathbf{P}}_1 = \begin{bmatrix} 102 & 32 & 14 & 4 \\ 32 & 80 & 80 & 16 \\ 14 & 80 & 298 & 78 \\ 4 & 16 & 78 & 66 \end{bmatrix} \quad \text{and} \quad \mathbf{G}_1 = \begin{bmatrix} 44 & 11 & -11 & -3 & -8 & -3 \\ 11 & 26 & 24 & 7 & 7 & 3 \\ -11 & 24 & 62 & 23 & 37 & 20 \\ -3 & 7 & 23 & 14 & 22 & 14 \end{bmatrix}$$

respectively. At both stages, traits y_1 and y_2 are restricted. Matrix $\mathbf{U}$ can be written as $\mathbf{U}' = \begin{bmatrix} 1 & 0 & 0 & 0 & 0 & 0 \\ 0 & 1 & 0 & 0 & 0 & 0 \end{bmatrix}$, whence the estimated matrix of restrictions was $\widehat{\boldsymbol{\Psi}}_1' = \mathbf{U}\widehat{\mathbf{G}}_1' = \begin{bmatrix} 44 & 11 & -11 & -3 \\ 11 & 26 & 24 & 7 \end{bmatrix}$; therefore, the estimated matrices of $\mathbf{Q}_1 = \mathbf{P}_1^{-1}\boldsymbol{\Psi}_1\left(\boldsymbol{\Psi}_1'\mathbf{P}_1^{-1}\boldsymbol{\Psi}_1\right)^{-1}\boldsymbol{\Psi}_1'$ and $\mathbf{K}_1 = [\mathbf{I}_4 - \mathbf{Q}_1]$ were

$$\widehat{\mathbf{Q}}_1 = \widehat{\mathbf{P}}_1^{-1}\widehat{\boldsymbol{\Psi}}_1\left(\widehat{\boldsymbol{\Psi}}_1'\widehat{\mathbf{P}}_1^{-1}\widehat{\boldsymbol{\Psi}}_1\right)^{-1}\widehat{\boldsymbol{\Psi}}_1' = \begin{bmatrix} 0.923 & -0.013 & -0.511 & -0.144 \\ 0.164 & 1.026 & 1.093 & 0.317 \\ -0.145 & -0.069 & -0.001 & -0.001 \\ 0.010 & 0.159 & 0.178 & 0.052 \end{bmatrix} \text{ and}$$

$$\widehat{\mathbf{K}}_1 = \left[\mathbf{I}_4 - \widehat{\mathbf{Q}}_1\right] = \begin{bmatrix} 0.077 & 0.013 & 0.511 & 0.144 \\ 0.164 & -0.026 & -1.093 & -0.317 \\ 0.145 & 0.069 & 1.001 & 0.001 \\ -0.010 & -0.159 & -0.178 & 0.948 \end{bmatrix} \text{ respectively, where}$$

$\mathbf{I}_4$ is an identity matrix of size 4×4.

The estimated vector $\mathbf{b}_{R_1}' = \mathbf{b}_1'\mathbf{K}_1'$ was $\widehat{\mathbf{b}}'_{R_1} = \widehat{\mathbf{b}}_1'\widehat{\mathbf{K}}_1' = [0.044 \quad -0.095 \quad 0.0450.131]$, where $\widehat{\mathbf{b}}_1' = \mathbf{w}'\widehat{\mathbf{G}}_1'\widehat{\mathbf{P}}_1^{-1} = \begin{bmatrix} -0.067 & 0.125 & 0.045 & 0.167 \end{bmatrix}$, and $\widehat{I}_{R_1} = \widehat{\mathbf{b}}'_{R_1}\mathbf{x}_1$ was the estimated MRLPSI at stage 1. The estimated MRLPSI vector of coefficients at stage 2 was $\widehat{\mathbf{b}}'_{R_2} = \widehat{\mathbf{b}}_2'\widehat{\mathbf{K}}_2' = [0.045 \quad -0.068 \quad 0.028 \; -0.057 \; 0.099 \; 0.106]$ and $\widehat{I}_{R_2} = \widehat{\mathbf{b}}'_{R_2}\mathbf{y}$ was the estimated MRLPSI at stage 2.

The estimated correlation value ($\widehat{\rho}_{R_{12}}$) between $\widehat{I}_{R_1} = \widehat{\mathbf{b}}'_{R_1}\mathbf{x}_1$ and $\widehat{I}_{R_2} = \widehat{\mathbf{b}}'_{R_2}\mathbf{y}$ was $\widehat{\rho}_{R_{12}} = \dfrac{\widehat{\mathbf{b}}'_{R_1}\left[\widehat{\mathbf{P}}_1 \quad \widehat{\mathbf{P}}_{21}\right]\widehat{\mathbf{b}}_{R_2}}{\sqrt{\widehat{\mathbf{b}}'_{R_1}\widehat{\mathbf{P}}_1\widehat{\mathbf{b}}_{R_1}}\sqrt{\widehat{\mathbf{b}}'_{R_2}\widehat{\mathbf{P}}\widehat{\mathbf{b}}_{R_2}}} = 0.564$, where $\sqrt{\widehat{\mathbf{b}}'_{R_1}\widehat{\mathbf{P}}_1\widehat{\mathbf{b}}_{R_1}}$ and $\sqrt{\widehat{\mathbf{b}}'_{R_2}\widehat{\mathbf{P}}\widehat{\mathbf{b}}_{R_2}}$ are the estimated standard deviations of the variance of $\widehat{I}_{R_1} = \widehat{\mathbf{b}}'_{R_1}\mathbf{x}_1$ and $\widehat{I}_{R_2} = \widehat{\mathbf{b}}'_{R_2}\mathbf{y}$ respectively. According to Young (1964, Fig. 8), and Eqs. (9.10) and (9.11), the selection intensities for stages 1 and 2 were $k_1 = 0.641$ and $k_2 = 0.593$

respectively. The estimated selection responses and expected genetic gains per traits for both stages were $\widehat{R}_{R_1} = k_1\sqrt{\widehat{\mathbf{b}}'_{R_1}\widehat{\mathbf{P}}_1\widehat{\mathbf{b}}_{R_1}} = 0.973$ and $\widehat{R}_{R_2} = k_2\sqrt{\widehat{\mathbf{b}}'_{R_2}\widehat{\mathbf{P}}^*\widehat{\mathbf{b}}_{R_2}} = 0.930$, $\widehat{\mathbf{E}}'_{R_1} = k_1\frac{\widehat{\mathbf{G}}'_1\widehat{\mathbf{b}}_{R_1}}{\sqrt{\widehat{\mathbf{b}}'_{R_1}\widehat{\mathbf{P}}_1\widehat{\mathbf{b}}_{R_1}}} = [0 \quad 0 \quad 1.271 \quad 0.870 \quad 1.482 \quad 0.974]$ and $\widehat{\mathbf{E}}'_{R_2} = k_2\frac{\widehat{\mathbf{C}}^{*\prime}\widehat{\mathbf{b}}_{R_2}}{\sqrt{\widehat{\mathbf{b}}'_{R_2}\widehat{\mathbf{P}}^*\widehat{\mathbf{b}}_{R_2}}} = [0 \quad 0 \quad 1.419 \quad 1.014 \quad 2.037 \quad 1.349]$, whereas $\widehat{R}_{R_1} + \widehat{R}_{R_2} = 1.903$ and $\widehat{\mathbf{E}}'_{R_1} + \widehat{\mathbf{E}}'_{R_2} = [0 \quad 0 \quad 2.691 \quad 1.884 \quad 3.519 \quad 2.322]$ were the total estimated MRLPSI selection response and expected genetic gain per trait respectively.

Finally, the estimated MRLPSI accuracy at stage 1 was $\widehat{\rho}_{R_1} = \sqrt{\frac{\widehat{\mathbf{b}}'_{R_1}\widehat{\mathbf{P}}_1\widehat{\mathbf{b}}_{R_1}}{\mathbf{w}'\widehat{\mathbf{C}}\mathbf{w}}} = 0.320$ and at stage 2 it was $\widehat{\rho}_{R_2} = \sqrt{\frac{\widehat{\mathbf{b}}'_{R_2}\widehat{\mathbf{P}}^*\widehat{\mathbf{b}}_{R_2}}{\mathbf{w}'\widehat{\mathbf{C}}^*\mathbf{w}}} = 0.334$. In this case, $\widehat{\rho}_{R_2} > \widehat{\rho}_{R_1}$. We can explain these results considering that although $\widehat{\rho}_{R_2}$ was obtained with six traits, $\widehat{\rho}_{R_1}$ was obtained only with four traits, two of them restricted.

9.3 The Multistage Predetermined Proportional Gain Linear Phenotypic Selection Index

The main objectives of the multistage predetermined proportional gain linear phenotypic selection index (MPPG-LPSI) are the same as those of the predetermined proportional gain linear phenotypic selection index (PPG-LPSI) described in Chap. 3, i.e., to optimize, under some predetermined restrictions, the expected genetic gains per trait, to predict the net genetic merit, and to select the individual with the highest net genetic merit values as parents of the next generation under some predetermined restrictions. The MPPG-LPSI allows restrictions different from zero to be imposed on the expected genetic gains of some traits, whereas other traits increase (or decrease) their expected genetic gains without any restrictions being imposed.

9.3.1 The MPPG-LPSI Parameters

In a similar manner to the MRLPSI, the MPPG-LPSI vector of coefficients for stages 1 and 2 should be a linear transformation of the MLPSI vector of coefficients at stages 1 ($\mathbf{b}_1 = \mathbf{P}_1^{-1}\mathbf{G}_1\mathbf{w}$) and 2 ($\mathbf{b}_2 = \mathbf{P}^{-1}\mathbf{C}\mathbf{w}$), and should be written as

$$\mathbf{b}_{M_1} = \mathbf{K}_{M_1}\mathbf{b}_1 \tag{9.17}$$

and

$$\mathbf{b}_{M_2} = \mathbf{K}_{M_2}\mathbf{b}_2, \tag{9.18}$$

respectively, where, at stage 1, $\mathbf{K}_{M_1} = [\mathbf{I}_1 - \mathbf{Q}_{M_1}]$, $\mathbf{Q}_{M_1} = \mathbf{P}_1^{-1}\mathbf{M}_1(\mathbf{M}_1'\mathbf{P}_1^{-1}\mathbf{M}_1)^{-1}\mathbf{M}_1'$, $\mathbf{M}_1' = \mathbf{D}'\mathbf{\Psi}'_1$, $\mathbf{\Psi}_1' = \mathbf{U}'\mathbf{G}'_1$, $\mathbf{I}_1$ is an identity matrix of the same size as $\mathbf{P}_1$, and $\mathbf{P}_1^{-1}$ is the inverse of matrix $\mathbf{P}_1$. At stage 2, $\mathbf{K}_M = [\mathbf{I} - \mathbf{Q}_M]$, $\mathbf{Q}_M = \mathbf{P}^{-1}\mathbf{M}(\mathbf{M}'\mathbf{P}^{-1}\mathbf{M})^{-1}\mathbf{M}'$, $\mathbf{M}' = \mathbf{D}'\mathbf{\Psi}'$, $\mathbf{\Psi}' = \mathbf{U}'\mathbf{C}$, $\mathbf{I}$ is an identity matrix of the same size as $\mathbf{P}$, $\mathbf{P}^{-1}$ is the inverse of matrix $\mathbf{P}$, and $\mathbf{D}' = \begin{bmatrix} d_r & 0 & \cdots & 0 & -d_1 \\ 0 & d_r & \cdots & 0 & -d_2 \\ \vdots & \vdots & \ddots & \vdots & \vdots \\ 0 & 0 & \cdots & d_r & -d_{r-1} \end{bmatrix}$, where d_q ($q = 1, 2\ldots, r$) is the qth element of $\mathbf{d}' = [d_1 \;\; d_2 \;\; \cdots \;\; d_r]$, the vector PPG (predetermined proportional gains) imposed by the breeder (see Chap. 3 for details).

By Eqs. (9.17) and (9.18), the MPPG-LPSI for stages 1 and 2 can be written as $I_{M_1} = \mathbf{b}_{M_1}\mathbf{x}_1$ and $I_{M_2} = \mathbf{b}_{M_2}\mathbf{y}$ respectively, where, assuming that at stage 1 we select four traits and at stage 2 we select two traits, $\mathbf{x}_1' = [y_1 \;\; y_2 \;\; y_3 \;\; y_4]$ and $\mathbf{x}_2' = [y_5 \;\; y_6]$ are the vectors of phenotypic observations at stages 1 and 2 respectively, and $\mathbf{y}' = [\mathbf{x}_1' \;\; \mathbf{x}_2']$ is the vector of total phenotypic observations at stage 2.

Let k_1 and k_2 be the selection intensities for stages 1 and 2 (Eqs. 9.10 and 9.11) respectively and let $\mathbf{P}^*$ and $\mathbf{C}^*$ be the adjusted matrices according to Eqs. (9.5) and (9.6) in the MPPG-LPSI context. Then, the MPPG-LPSI selection response and expected genetic gain per trait for both stages can be written as

$$R_{M_1} = k_1\sqrt{\mathbf{b}_{M_1}'\mathbf{P}_1\mathbf{b}_{M_1}} \quad \text{and} \quad R_{M_2} = k_2\sqrt{\mathbf{b}_{M_2}'\mathbf{P}^*\mathbf{b}_{M_2}} \tag{9.19}$$

and

$$\mathbf{E}_{M_1} = k_1\frac{\mathbf{G}_1'\mathbf{b}_{M_1}}{\sqrt{\mathbf{b}_{M_1}'\mathbf{P}_1\mathbf{b}_{M_1}}} \quad \text{and} \quad \mathbf{E}_{M_2} = k_2\frac{\mathbf{b}_{M_2}'\mathbf{C}^*}{\sqrt{\mathbf{b}_{M_2}'\mathbf{P}^*\mathbf{b}_{M_2}}}, \tag{9.20}$$

respectively, whereas the total MPPG-LPSI selection response and expected genetic gain per trait for both stages are equal to $R_{M_1} + R_{M_2}$ and $\mathbf{E}_{M_1} + \mathbf{E}_{M_2}$. In addition, the MPPG-LPSI accuracy for both stages can be written as

$$\rho_{M_1} = \sqrt{\frac{\mathbf{b}_{M_1}'\mathbf{P}_1\mathbf{b}_{M_1}}{\mathbf{w}'\mathbf{C}\mathbf{w}}} \quad \text{and} \quad \rho_{M_2} = \sqrt{\frac{\mathbf{b}_{M_2}'\mathbf{P}^*\mathbf{b}_{M_2}}{\mathbf{w}'\mathbf{C}^*\mathbf{w}}}. \tag{9.21}$$

9.3.2 Numerical Examples

We use the real data set described in Sect. 9.2.2 to illustrate the theoretical results of the MPPG-LPSI in the same form as we did with those of the MRLPSI. We need to estimate vectors $\mathbf{b}'_{M_1} = \mathbf{b}'_1\mathbf{K}'_{M_1}$ and $\mathbf{b}'_{M_2} = \mathbf{b}'_2\mathbf{K}'_{M_2}$, where $\mathbf{b}'_1 = \mathbf{w}'\mathbf{G}'_1\mathbf{P}_1^{-1}$ and $\mathbf{b}'_2 = \mathbf{w}'\mathbf{G}'\mathbf{P}^{-1}$. In Chap. 3 we have given methods to estimates $\mathbf{K}_M = [\mathbf{I} - \mathbf{Q}_M]$, $\mathbf{Q}_M = \mathbf{P}^{-1}\mathbf{M}(\mathbf{M}'\mathbf{P}^{-1}\mathbf{M})^{-1}\mathbf{M}'$, $\mathbf{M}' = \mathbf{D}'\mathbf{\Psi}'$, and $\mathbf{\Psi}' = \mathbf{U}'\mathbf{C}$, which will be used in this subsection.

The estimated phenotypic and genotypic covariance matrices at stage 1 were $\widehat{\mathbf{P}}_1 = \begin{bmatrix} 102 & 32 & 14 & 4 \\ 32 & 80 & 80 & 16 \\ 14 & 80 & 298 & 78 \\ 4 & 16 & 78 & 66 \end{bmatrix}$ and $\mathbf{G}_1 = \begin{bmatrix} 44 & 11 & -11 & -3 & -8 & -3 \\ 11 & 26 & 24 & 7 & 7 & 3 \\ -11 & 24 & 62 & 23 & 37 & 20 \\ -3 & 7 & 23 & 14 & 22 & 14 \end{bmatrix}$ respectively, whereas $\mathbf{w}' = [0.08 \quad 0.08 \quad 0.38 \quad 0.08 \quad 0.08 \quad 0.31]$ was the vector of economic weights. The traits restricted at both stages are y_1, y_2, and y_3. The vector of PPG was $\mathbf{d}' = [2 \quad 3 \quad 5]$, whence $\mathbf{D}' = \begin{bmatrix} 5 & 0 & -2 \\ 0 & 5 & -3 \end{bmatrix}$ and $\mathbf{U}' = \begin{bmatrix} 1 & 0 & 0 & 0 & 0 & 0 \\ 0 & 1 & 0 & 0 & 0 & 0 \\ 0 & 0 & 1 & 0 & 0 & 0 \end{bmatrix}$ were matrices $\mathbf{D}'$ and $\mathbf{U}$. The estimated matrices of $\mathbf{M}'_1$ and $\mathbf{K}_{M_1} = [\mathbf{I} - \mathbf{Q}_{M_1}]$ were $\widehat{\mathbf{M}}'_1 = \mathbf{D}'\mathbf{\Psi}'_1 = \begin{bmatrix} 242 & 7 & -178 & -61 \\ 88 & 58 & -66 & -34 \end{bmatrix}$ and

$$\widehat{\mathbf{K}}_{M_1} = \begin{bmatrix} 0.176 & 0.205 & 0.606 & 0.159 \\ 0.031 & 0.032 & -0.007 & 0.199 \\ 0.195 & 0.235 & 0.852 & -0.098 \\ 0.130 & 0.130 & -0.098 & 0.940 \end{bmatrix}$$

respectively, where $\widehat{\mathbf{\Psi}}'_1 = \mathbf{U}'\widehat{\mathbf{G}}'_1$.

At stages 1 and 2, the estimated MPPG-LPSI vector of coefficients were $\widehat{\mathbf{b}}'_{M_1} = \widehat{\mathbf{b}}'_1\widehat{\mathbf{K}}'_{M_1} = [0.068 \quad 0.035 \quad 0.039 \quad 0.160]$ and $\widehat{\mathbf{b}}'_1 = \mathbf{w}'\widehat{\mathbf{G}}'_1\widehat{\mathbf{P}}_1^{-1} = [-0.067 \quad 0.125 \quad 0.045 \quad 0.167]$, whence the estimated MPPG-LGSI were $\widehat{I}_{M_1} = \widehat{\mathbf{b}}'_{M_1}\mathbf{x}_1$ and $\widehat{I}_{M_2} = \widehat{\mathbf{b}}'_{M_2}\mathbf{y}$. The estimated correlation value ($\widehat{\rho}_{M_{12}}$) between $\widehat{I}_{M_1} = \widehat{\mathbf{b}}'_{M_1}\mathbf{x}_1$ and $\widehat{I}_{M_2} = \widehat{\mathbf{b}}'_{M_2}\mathbf{y}$ was $\widehat{\rho}_{M_{12}} = \frac{\widehat{\mathbf{b}}'_{M_1}[\widehat{\mathbf{P}}_1 \quad \widehat{\mathbf{P}}_{21}]\widehat{\mathbf{b}}_{M_2}}{\sqrt{\widehat{\mathbf{b}}'_{M_1}\widehat{\mathbf{P}}_1\widehat{\mathbf{b}}_{M_1}}\sqrt{\widehat{\mathbf{b}}'_{M_2}\widehat{\mathbf{P}}\widehat{\mathbf{b}}_{M_2}}} = 0.870$, where $\sqrt{\widehat{\mathbf{b}}'_{M_1}\widehat{\mathbf{P}}_1\widehat{\mathbf{b}}_{M_1}}$ and $\sqrt{\widehat{\mathbf{b}}'_{M_2}\widehat{\mathbf{P}}\widehat{\mathbf{b}}_{M_2}}$ were the estimated standard deviations of variance of $\widehat{I}_{M_1} = \widehat{\mathbf{b}}'_{M_1}\mathbf{x}_1$ and $\widehat{I}_{M_2} = \widehat{\mathbf{b}}'_{M_2}\mathbf{y}$ respectively. According to Young (1964, Fig. 8), the selection intensities for stages 1 and 2 were $k_1 = 0.744$ and $k_2 = 0.721$ (Eqs. 9.10 and 9.11) respectively.

The estimated selection responses and expected genetic gains per traits for both stages were $\widehat{R}_{M_1} = k_1\sqrt{\widehat{\mathbf{b}}'_{M_1}\widehat{\mathbf{P}}_1\widehat{\mathbf{b}}_{M_1}} = 1.553$ and $\widehat{R}_{M_2} = k_2\sqrt{\widehat{\mathbf{b}}'_{M_2}\widehat{\mathbf{P}}^*\widehat{\mathbf{b}}_{M_2}} = 1.401$, $\widehat{\mathbf{E}}'_{M_1} = k_1\frac{\widehat{\mathbf{G}}'_1\widehat{\mathbf{b}}_{M_1}}{\sqrt{\widehat{\mathbf{b}}'_{M_1}\widehat{\mathbf{P}}_1\widehat{\mathbf{b}}_{M_1}}} = [0.877 \quad 1.316 \quad 2.193 \quad 1.128 \quad 1.655 \quad 1.037]$, and

$\widehat{\mathbf{E}}'_{M_2} = k_2 \frac{\widehat{\mathbf{C}}^{*\prime}\widehat{\mathbf{b}}_{M_2}}{\sqrt{\widehat{\mathbf{b}}'_{M_2}\widehat{\mathbf{P}}^*\widehat{\mathbf{b}}_{M_2}}} = [0.878 \quad 1.346 \quad 2.604 \quad 1.433 \quad 2.506 \quad 1.602]$, whereas $\widehat{R}_{M_1} + \widehat{R}_{M_2} = 2.954$ and $\widehat{\mathbf{E}}'_{M_1} + \widehat{\mathbf{E}}'_{M_2} = [1.755 \quad 2.662 \quad 4.797 \; 2.561 \; 4.161 \; 2.639]$ were the total estimated MPPGLPSI selection response and expected genetic gain per trait respectively. Note that the vector of predetermined restriction was $\mathbf{d}' = [2 \quad 3 \quad 5]$. This means that the MPPG-LPSI efficiency at predicting the total expected genetic gain per trait was high because the difference between each predetermined value (2, 3, and 5) and the total of each predicted value (1.755, 2.662, and 4.797) were 0.245, 0.338, and 0.203 respectively.

Finally, the estimated MPPG-LPSI accuracy at stage 1 was $\widehat{\rho}_{M_1} = \sqrt{\frac{\widehat{\mathbf{b}}'_{M_1}\widehat{\mathbf{P}}_1\widehat{\mathbf{b}}_{M_1}}{\mathbf{w}'\widehat{\mathbf{C}}\mathbf{w}}}$ $= 0.435$, and at stage 2 it was $\widehat{\rho}_{M_2} = \sqrt{\frac{\widehat{\mathbf{b}}'_{M_2}\widehat{\mathbf{P}}^*\widehat{\mathbf{b}}_{M_2}}{\mathbf{w}'\widehat{\mathbf{C}}^*\mathbf{w}}} = 0.428$; that is, both were very similar.

9.4 The Multistage Linear Genomic Selection Index

We describe the multistage linear genomic selection indices (MLGSI) as an extension of the linear genomic selection index (LGSI, Chap. 5) theory to the multistage genomic selection context; thus, the theoretical results of the MLGSI are very similar to those of the LGSI. The MLGSI is a linear combination of genomic estimated breeding values (GEBVs) and is useful for predicting individual net genetic merit and for selecting individuals from a nonphenotyped testing population as parents of the next selection cycle.

9.4.1 *The MLGSI Parameters*

The objective of the MLGSI is to predict the net genetic merit $H = \mathbf{w}'\mathbf{g}$, where $\mathbf{g}$ is a vector of true breeding values and $\mathbf{w}'$ is the vector of economic weights, using only GEBVs. In Chap. 5, we indicated that the covariance between $\boldsymbol{\gamma}_i$ and $\mathbf{g}_i$ is equal to the variance of $\boldsymbol{\gamma}_i$, i.e., $Cov(\mathbf{g}_i, \boldsymbol{\gamma}_i) = s_i^2$, and that the GEBV associated with the ith trait is a predictor of the ith vector of genomic breeding values ($\boldsymbol{\gamma}_i$). In the testing population, the only observable information is $\mathbf{w}'$ and the GEBV associated with the traits of interest. For this reason, in practice, we construct a linear combination of GEBVs, which should be a good predictor of $H = \mathbf{w}'\mathbf{g}$.

Suppose that the breeder is interested in four traits, and that $\boldsymbol{\gamma}' = [\gamma_1 \quad \gamma_2 \quad \gamma_3 \quad \gamma_4]$, $\mathbf{g}' = [g_1 \quad g_2 \quad g_3 \quad g_4]$, and $\mathbf{w}' = [w_1 \quad w_2 \quad w_3 \quad w_4]$ are the vectors of genomic breeding values ($\boldsymbol{\gamma}$), true breeding values ($\mathbf{g}$), and economic weights ($\mathbf{w}$) respectively. Let $\boldsymbol{\Gamma} = Var(\boldsymbol{\gamma}) = \begin{bmatrix} s_1^2 & s_{12} & s_{13} & s_{14} \\ s_{21} & s_2^2 & s_{23} & s_{24} \\ s_{31} & s_{32} & s_3^2 & s_{34} \\ s_{41} & s_{42} & s_{43} & s_4^2 \end{bmatrix}$ and $\mathbf{C} = (\mathbf{g}) = \begin{bmatrix} \sigma_1^2 & \sigma_{12} & \sigma_{13} & \sigma_{14} \\ \sigma_{21} & \sigma_2^2 & \sigma_{23} & \sigma_{24} \\ \sigma_{31} & \sigma_{32} & \sigma_3^2 & \sigma_{34} \\ \sigma_{41} & \sigma_{42} & \sigma_{43} & \sigma_4^2 \end{bmatrix}$ be the covariance matrix of $\mathbf{g}$ and $\boldsymbol{\gamma}$. At a two-stage selection breeding scheme, $\boldsymbol{\gamma}' = [\gamma_1 \quad \gamma_2 \quad \gamma_3 \quad \gamma_4]$ can be partitioned into $\boldsymbol{\gamma}_1' = [\gamma_1 \quad \gamma_2]$ and $\boldsymbol{\gamma}_2' = [\gamma_3 \quad \gamma_4]$; therefore, at stage 1, $\boldsymbol{\Gamma}_1 = Var(\boldsymbol{\gamma}_1) = \begin{bmatrix} s_1^2 & s_{12} \\ s_{21} & s_2^2 \end{bmatrix}$ is the genomic covariance matrix of $\boldsymbol{\gamma}_1' = [\gamma_1 \quad \gamma_2]$ and $Cov(\boldsymbol{\gamma}_1, \mathbf{g}) = \begin{bmatrix} s_1^2 & s_{12} & s_{13} & s_{14} \\ s_{12} & s_2^2 & s_{23} & s_{24} \end{bmatrix} = \mathbf{A}_1$ is the covariance matrix of $\boldsymbol{\gamma}_1' = [\gamma_1 \quad \gamma_2]$ with $\mathbf{g}' = [g_1 \quad g_2 \quad g_3 \quad g_4]$. Matrix $\mathbf{A}_1$ indicates that we are assuming that the covariance between $\boldsymbol{\gamma}_i$ and $\mathbf{g}_j$ ($i, j = 1, 2, \cdots, g$; g= number of genotypes) is equal to the covariance between $\boldsymbol{\gamma}_i$ and $\boldsymbol{\gamma}_j$. This is because, in practice, in the testing population, we can only estimate matrix $\boldsymbol{\Gamma}$.

At stage 2, $\boldsymbol{\Gamma} = Var(\boldsymbol{\gamma})$ is the covariance matrix of $\boldsymbol{\gamma}$ and $\mathbf{A} = \boldsymbol{\Gamma}$ is the covariance matrix of the vector of genomic breeding values $\boldsymbol{\gamma}$ with the vector of breeding values $\mathbf{g}$. The MLGSI vector of coefficients at stages 1 and 2 are $\boldsymbol{\beta}_1' = \mathbf{w}'\mathbf{A}'_1\boldsymbol{\Gamma}_1^{-1} = [\beta_{11} \quad \beta_{12}]$ and $\boldsymbol{\beta}_2' = \mathbf{w}'\mathbf{A}\boldsymbol{\Gamma}^{-1} = \mathbf{w}' = [w_1 \quad w_2 \quad w_3 \quad w_4]$ respectively, and the MLGSI for both stages can be written as $I_1 = \beta_{11}\gamma_1 + \beta_{12}\gamma_2 = \boldsymbol{\beta}_1'\boldsymbol{\gamma}_1$ and $I_2 = w_1\gamma_1 + w_2\gamma_2 + w_3\gamma_3 + w_4\gamma_4 = \mathbf{w}'\boldsymbol{\gamma}$.

Let k_1 and k_2 be the MLGSI selection intensities for stages 1 and 2. For both stages, the MLGSI accuracies (ρ_{HI_1} and ρ_{HI_2}), expected genetic gains per trait ($\mathbf{E}_1$ and $\mathbf{E}_2$) and selection responses (R_1 and R_2) can be written as

$$\rho_{HI_1} = \sqrt{\frac{\boldsymbol{\beta}_1'\boldsymbol{\Gamma}_1\boldsymbol{\beta}_1}{\mathbf{w}'\mathbf{C}\mathbf{w}}} \quad \text{and} \quad \rho_{HI_2} = \sqrt{\frac{\mathbf{w}'\boldsymbol{\Gamma}^*\mathbf{w}}{\mathbf{w}'\mathbf{C}^*\mathbf{w}}}, \tag{9.22}$$

$$\mathbf{E}_1 = k_1\frac{\mathbf{A}_1'\boldsymbol{\beta}_1}{\sqrt{\boldsymbol{\beta}_1'\boldsymbol{\Gamma}_1\boldsymbol{\beta}_1}} \quad \text{and} \quad \mathbf{E}_2 = k_2\frac{\boldsymbol{\Gamma}^*\mathbf{w}}{\sqrt{\mathbf{w}'\boldsymbol{\Gamma}^*\mathbf{w}}} \tag{9.23}$$

and

$$R_1 = k_1\sqrt{\boldsymbol{\beta}_1'\boldsymbol{\Gamma}_1\boldsymbol{\beta}_1} \quad \text{and} \quad R_2 = k_2\sqrt{\mathbf{w}'\boldsymbol{\Gamma}^*\mathbf{w}}. \tag{9.24}$$

The total MLGSI expected genetic gain per trait and selection response at both stages are equal to $\mathbf{E}_1 + \mathbf{E}_2$ and $R_1 + R_2$. To simplify notation, in Eqs. (9.23) and (9.24), we have omitted the intervals between stages or selection cycles (L_G). Matrices $\mathbf{C}^*$ and $\boldsymbol{\Gamma}^*$ in Eqs. (9.22) to (9.23) are matrices $\boldsymbol{\Gamma}$ and $\mathbf{C}$ adjusted for previous selection on I_1.

We adjust matrices $\boldsymbol{\Gamma}$ and $\mathbf{C}$ for previous selection on I_1 as

$$\boldsymbol{\Gamma}^* = \boldsymbol{\Gamma} - u\frac{\mathbf{A}_1'\boldsymbol{\beta}_1\boldsymbol{\beta}_1'\mathbf{A}_1}{\boldsymbol{\beta}_1'\boldsymbol{\Gamma}_1\boldsymbol{\beta}_1} \tag{9.25}$$

and

$$\mathbf{C}^* = \mathbf{C} - u\frac{\mathbf{G}_1'\mathbf{b}_1\mathbf{b}_1'\mathbf{G}_1}{\mathbf{b}_1'\mathbf{P}_1\mathbf{b}_1}, \tag{9.26}$$

respectively, where $u = k_1(k_1 - \tau)$, k_1 is the standardized selection differential, and τ is the truncation point when $I_1 = \boldsymbol{\beta}_1'\boldsymbol{\gamma}_1$ is applied. All the terms in Eq. (9.26) were defined in Eq. (9.6).

The correlation between $I_1 = \boldsymbol{\beta}_1'\boldsymbol{\gamma}_1$ and $I_2 = \mathbf{w}'\boldsymbol{\gamma}$ can be written as

$$Corr(I_1, I_2) = \frac{\boldsymbol{\beta}_1'\mathbf{A}_1\mathbf{w}}{\sqrt{\boldsymbol{\beta}_1'\boldsymbol{\Gamma}_1\boldsymbol{\beta}_1}\sqrt{\mathbf{w}'\boldsymbol{\Gamma}\mathbf{w}}} = \rho_{I_1 I_2}, \tag{9.27}$$

where $\sqrt{\boldsymbol{\beta}_1'\boldsymbol{\Gamma}_1\boldsymbol{\beta}_1}$ and $\sqrt{\mathbf{w}'\boldsymbol{\Gamma}\mathbf{w}}$ are the standard deviations of the variances of $I_1 = \boldsymbol{\beta}_1'\boldsymbol{\gamma}_1$ and $I_2 = \mathbf{w}'\boldsymbol{\gamma}$ respectively. In Eq. (9.27), matrix $\boldsymbol{\Gamma}$ was not adjusted according to Eq. (9.25).

9.4.2 Estimating the Genomic Covariance Matrix

All the MLGSI parameters are associated with matrix $\boldsymbol{\Gamma}$; thus, the estimation of this matrix in the testing population is very important. We estimate matrix $\boldsymbol{\Gamma}$ according to the estimation method described in Chap. 5 (Eq. 5.25), that is, as

$$\widehat{\boldsymbol{\Gamma}}_l = \left\{\widehat{\sigma}_{\gamma_{qq'}}\right\}, \tag{9.28}$$

where $\widehat{\sigma}_{\gamma_{qq'}} = \frac{1}{g}\left(\widehat{\boldsymbol{\gamma}}_{ql} - \mathbf{1}\widehat{\mu}_{\gamma_{ql}}\right)'\mathbf{G}_l^{-1}\left(\widehat{\boldsymbol{\gamma}}_{q'l} - \mathbf{1}\widehat{\mu}_{\gamma_{q'l}}\right)$ is the estimated covariance between $\widehat{\boldsymbol{\gamma}}_{ql} = \mathbf{X}_l\widehat{\mathbf{u}}_q$ and $\widehat{\boldsymbol{\gamma}}_{q'l} = \mathbf{X}_l\widehat{\mathbf{u}}_{q'}$ at stage l or selection cycle of the testing population; g is the number of genotypes; $\widehat{\mu}_{\gamma_{ql}}$ and $\widehat{\mu}_{\gamma_{q'l}}$ are the estimated arithmetic means of the values of $\widehat{\boldsymbol{\gamma}}_{ql}$ and $\widehat{\boldsymbol{\gamma}}_{q'l}$; $\mathbf{1}$ is an $g \times 1$ vector of 1s and $\mathbf{G}_l = c^{-1}\mathbf{X}_l\mathbf{X}_l'$ is the additive genomic relationship matrix at stage l or selection cycle in the testing population (see Chap. 5 for details).

9.4.3 Numerical Examples

We illustrate the MLGSI theoretical results using the data described in Chap. 2, Sect. 2.8.1 simulated for eight phenotypic and seven genomic selection cycles, each with four traits (T_1, T_2, T_3 and T_4), 500 genotypes, four replicates for each genotype, 2500 molecular markers, and 315 quantitative trait loci in one environment. The economic weights of T_1, T_2, T_3, and T_4 were 1, −1, 1, and 1 respectively. In this subsection, and only for illustrative purposes, we use the data set from cycle 1.

The genotypic and genomic estimated covariance matrices in cycle 1 were $\widehat{\mathbf{C}} = \begin{bmatrix} 36.21 & -12.93 & 8.35 & 2.74 \\ -12.93 & 13.04 & -3.4 & -2.24 \\ 8.35 & -3.4 & 9.96 & 0.16 \\ 2.74 & -2.24 & 0.16 & 6.64 \end{bmatrix}$ and $\widehat{\mathbf{\Gamma}} = \begin{bmatrix} 16.26 & -6.51 & 5.60 & 2.29 \\ -6.51 & 5.79 & -2.23 & -1.62 \\ 5.60 & -2.23 & 3.75 & 0.94 \\ 2.29 & -1.62 & 0.94 & 2.62 \end{bmatrix}$ respectively, whereas $\mathbf{w}' = [1 \;\; -1 \;\; 1 \;\; 1]$ was the vector of economic weights. Matrices $\widehat{\mathbf{P}}$ and $\widehat{\mathbf{C}}$ were obtained according to Eqs. (2.22) to (2.24), whereas matrix $\widehat{\mathbf{\Gamma}}$ was obtained according to Eq. (9.28).

Suppose that we select two traits at stages 1 and 2. Then, at stage 1, $\widehat{\mathbf{\Gamma}}_1 = \begin{bmatrix} 16.26 & -6.51 \\ -6.51 & 5.79 \end{bmatrix}$ and $\widehat{\mathbf{A}}_1 = \begin{bmatrix} 16.26 & -6.51 & 5.60 & 2.29 \\ -6.51 & 5.79 & -2.33 & -1.62 \end{bmatrix}$ are the estimated covariance matrices of $\mathbf{\Gamma}_1$ and $\mathbf{A}_1$ respectively, and the estimated MLGSI vector of coefficients was $\widehat{\boldsymbol{\beta}}'_1 = \mathbf{w}'\widehat{\mathbf{A}'}_1\widehat{\mathbf{\Gamma}}_1^{-1} = [1.39 \quad -1.25]$. Because at stage 2 $\boldsymbol{\beta}'_2 = \mathbf{w}'\mathbf{A}\mathbf{\Gamma}^{-1} = \mathbf{w}' = [w_1 \quad w_2 \quad w_3 \quad w_4]$, the estimated MLGSI vector of coefficients is the vector of economic weights. Thus, $\widehat{\rho}_{I_1 I_2} = \dfrac{\widehat{\boldsymbol{\beta}}'_1\widehat{\mathbf{A}}_1\mathbf{w}}{\sqrt{\widehat{\boldsymbol{\beta}}'_1\widehat{\mathbf{\Gamma}}_1\widehat{\boldsymbol{\beta}}_1}\sqrt{\mathbf{w}'\widehat{\mathbf{\Gamma}}\mathbf{w}}} =$ 0.97 was the estimated correlation between $\widehat{I}_1 = \widehat{\boldsymbol{\beta}}'_1\widehat{\boldsymbol{\gamma}}_1$ and $\widehat{I}_2 = \mathbf{w}'\widehat{\boldsymbol{\gamma}}$, and assuming that the fixed proportion was 0.2 (20%), $k_1 = 0.744$ and $k_2 = 0.721$ were the approximated selection intensities for stages 1 and 2 respectively. The adjusted matrices $\mathbf{\Gamma}^*$ and $\mathbf{C}^*$ for previous selection on $\widehat{I}_1 = \widehat{\boldsymbol{\beta}}'_1\widehat{\boldsymbol{\gamma}}_1$ were $\widehat{\mathbf{\Gamma}}^* = \begin{bmatrix} 7.96 & -2.11 & 2.71 & 0.88 \\ -2.11 & 3.46 & -0.80 & -0.87 \\ 2.71 & -0.80 & 2.75 & 0.45 \\ 0.88 & -0.87 & 0.45 & 2.38 \end{bmatrix}$ and $\widehat{\mathbf{C}}^* = \begin{bmatrix} 24.40 & -5.65 & 5.47 & 1.39 \\ -5.65 & 8.55 & -1.63 & -1.41 \\ 5.47 & -1.63 & 9.26 & -0.17 \\ 1.39 & -1.41 & -0.17 & 6.49 \end{bmatrix}$.

The estimated MLGSI accuracy, selection response, and expected genetic gain for stage 1 in the testing population were $\widehat{\rho}_{HI_1} = \sqrt{\dfrac{\widehat{\boldsymbol{\beta}}'_1\widehat{\mathbf{\Gamma}}_1\widehat{\boldsymbol{\beta}}_1}{\mathbf{w}'\widehat{\mathbf{C}}\mathbf{w}}} = 0.71$, $\widehat{R}_1 = k_1\sqrt{\widehat{\boldsymbol{\beta}}'_1\widehat{\mathbf{\Gamma}}_1\widehat{\boldsymbol{\beta}}_1} = 5.90$, and $\widehat{\mathbf{E}}'_1 = k_1\dfrac{\widehat{\mathbf{A}}'_1\widehat{\boldsymbol{\beta}}_1}{\sqrt{\widehat{\boldsymbol{\beta}}'_1\widehat{\mathbf{\Gamma}}_1\widehat{\boldsymbol{\beta}}_1}} = [2.88 \;\; -1.53 \;\; 1.00 \;\; 0.49]$ respectively, whereas at stage 2, the estimated MLGSI accuracy, selection response, and expected genetic gain were $\widehat{\rho}_{HI_2} = \sqrt{\dfrac{\mathbf{w}'\widehat{\mathbf{\Gamma}}^*\mathbf{w}}{\mathbf{w}'\widehat{\mathbf{C}}^*\mathbf{w}}} = 0.64$, $\widehat{R}_2 = k_2\sqrt{\mathbf{w}'\widehat{\mathbf{\Gamma}}^*\mathbf{w}} = 4.10$, and $\widehat{\mathbf{E}}'_2 = k_2\dfrac{\widehat{\mathbf{\Gamma}}^*\mathbf{w}}{\sqrt{\mathbf{w}'\widehat{\mathbf{\Gamma}}^*\mathbf{w}}} = [1.74 \;\; -0.92 \;\; 0.85 \;\; 0.58]$ respectively. The estimated MLGSI accuracy, selection response, and expected genetic gain at stage 2 were lower than at stage 1. This means that the adjusted matrices $\widehat{\boldsymbol{\Gamma}}^*$ and $\widehat{\mathbf{C}}^*$ negatively affected the estimated MLPSI parameters at stage 2. The total estimated MLGSI selection response and expected genetic gain for stages 1 and 2 were $\widehat{R}_1 + \widehat{R}_2 = 9.99$ and $\widehat{\mathbf{E}}_1' + \widehat{\mathbf{E}}_2' = [4.62 \quad -2.45 \quad 1.85 \quad 1.07]$.

9.5 The Multistage Restricted Linear Genomic Selection Index (MRLGSI)

The restricted linear genomic selection index (RLGSI) described in Chap. 3 is extended to the multistage restricted linear genomic selection index (MRLGSI) context in a two-stage breeding selection scheme.

9.5.1 The MRLGSI Parameters

In Sect. 9.4.1, we indicated that the MLGSI vector of coefficients at stage 1 can be written as $\boldsymbol{\beta}_1' = \mathbf{w}'\mathbf{A'}_1\boldsymbol{\Gamma}_1^{-1} = [\beta_{11} \quad \beta_{12}]$ and at stage 2 as $\boldsymbol{\beta}_2' = \mathbf{w}'\mathbf{A}\boldsymbol{\Gamma}^{-1} = \mathbf{w}' = [w_1 \quad w_2 \quad w_3 \quad w_4]$. It can be shown that the MRLGSI vector of coefficients is a linear transformation of vectors $\boldsymbol{\beta}_1$ and $\boldsymbol{\beta}_2$ made by matrix $\mathbf{K}_G$, which is a projector (see Chaps. 3 and 6 for details) that projects $\boldsymbol{\beta}_1$ and $\boldsymbol{\beta}_2$ into a space smaller than the original space of $\boldsymbol{\beta}_1$ and $\boldsymbol{\beta}_2$. Thus, at stages 1 and 2, the MRLGSI vector of coefficients is

$$\boldsymbol{\beta}_{R_1} = \mathbf{K}_{G_1}\boldsymbol{\beta}_1 \tag{9.29}$$

and

$$\boldsymbol{\beta}_{R_2} = \mathbf{K}_{G_2}\boldsymbol{\beta}_2 = \mathbf{K}_{G_2}\mathbf{w}, \tag{9.30}$$

respectively, where $\mathbf{K}_{G_1} = [\mathbf{I} - \mathbf{Q}_{G_1}]$, $\mathbf{Q}_{G_1} = \mathbf{U}_1(\mathbf{U}_1'\boldsymbol{\Gamma}_1\mathbf{U}_1)^{-1}\mathbf{U}_1'\boldsymbol{\Gamma}_1$, $\mathbf{K}_{G_2} = [\mathbf{I} - \mathbf{Q}_{G_2}]$, and $\mathbf{Q}_{G_2} = \mathbf{U}_2(\mathbf{U}_2'\boldsymbol{\Gamma}\mathbf{U}_2)^{-1}\mathbf{U}_2'\boldsymbol{\Gamma}$ are matrix projectors. By Eqs. (9.29) and (9.30), the MRLGSI at stages 1 and 2 can be written as $I_{R_1} = \boldsymbol{\beta}_{R_1}'\boldsymbol{\gamma}_1$ and $I_{R_2} = \boldsymbol{\beta}_{R_2}'\boldsymbol{\gamma}$ respectively, where $\boldsymbol{\gamma}_1' = [\gamma_1 \quad \gamma_2]$ and $\boldsymbol{\gamma}' = [\gamma_1 \quad \gamma_2 \quad \gamma_3 \quad \gamma_4]$ are vectors of genomic breeding values, which can be estimated using GEBVs, as described in Chap. 5. In Chap. 6 we described methods for constructing matrix $\mathbf{U}'$ and estimating matrix $\mathbf{K}_G$; those methods are also valid in the MRLGSI context.

In a similar manner to the MLGSI context, MRLGSI accuracies, expected genetic gains per trait, and selection responses for stages 1 and 2 in the testing population can be written as

$$\rho_{HI_1} = \sqrt{\frac{\boldsymbol{\beta}'_{R_1}\boldsymbol{\Gamma}_1\boldsymbol{\beta}_{R_1}}{\mathbf{w}'\mathbf{C}\mathbf{w}}} \quad \text{and} \quad \rho_{HI_2} = \sqrt{\frac{\boldsymbol{\beta}'_{R_2}\boldsymbol{\Gamma}^*\boldsymbol{\beta}_{R_2}}{\mathbf{w}'\mathbf{C}^*\mathbf{w}}}, \tag{9.31}$$

$$\mathbf{E}_{R_1} = k_1\frac{\mathbf{A}'_1\boldsymbol{\beta}_{R_1}}{\sqrt{\boldsymbol{\beta}'_{R_1}\boldsymbol{\Gamma}_1\boldsymbol{\beta}_{R_1}}} \quad \text{and} \quad \mathbf{E}_{R_2} = k_2\frac{\boldsymbol{\Gamma}^*\boldsymbol{\beta}_{R_2}}{\sqrt{\boldsymbol{\beta}'_{R_2}\boldsymbol{\Gamma}^*\boldsymbol{\beta}_{R_2}}} \tag{9.32}$$

and

$$R_{R_1} = k_1\sqrt{\boldsymbol{\beta}'_{R_1}\boldsymbol{\Gamma}_1\boldsymbol{\beta}_{R_1}} \quad \text{and} \quad R_{R_2} = k_2\sqrt{\boldsymbol{\beta}'_{R_2}\boldsymbol{\Gamma}^*\boldsymbol{\beta}_{R_2}}, \tag{9.33}$$

respectively. The total MRLGSI expected genetic gain per trait and selection response for both stages are equal to $\mathbf{E}_{R_1} + \mathbf{E}_{R_2}$ and $R_{R_1} + R_{R_2}$. To simplify the notation, in Eqs. (9.32) and (9.33), we have omitted the intervals between stages or selection cycles (L_G). Matrices $\boldsymbol{\Gamma}^*$ and $\mathbf{C}^*$ in Eqs. (9.31) to (9.33) are matrices $\boldsymbol{\Gamma}$ and $\mathbf{C}$ adjusted for previous selection.

In the MRLGSI context, matrices $\boldsymbol{\Gamma}^*$ and $\mathbf{C}^*$ can be obtained as

$$\boldsymbol{\Gamma}^* = \boldsymbol{\Gamma} - u\frac{\mathbf{A}'_1\boldsymbol{\beta}_{R_1}\boldsymbol{\beta}'_{R_1}\mathbf{A}_1}{\boldsymbol{\beta}'_{R_1}\boldsymbol{\Gamma}_1\boldsymbol{\beta}_{R_1}} \tag{9.34}$$

and

$$\mathbf{C}^* = \mathbf{C} - u\frac{\mathbf{G}'_1\mathbf{b}_{R_1}\mathbf{b}'_{R_1}\mathbf{G}_1}{\mathbf{b}'_{R_1}\mathbf{P}_1\mathbf{b}_{R_1}}, \tag{9.35}$$

where $\boldsymbol{\beta}_{R_1}$ was defined in Eq. (9.29) and vector $\mathbf{b}_{R_1}$ can be obtained according to the RLPSI as described in Chap. 3. The term $u = k(k - \tau)$ was defined earlier.

The correlation between $I_{R_1} = \boldsymbol{\beta}'_{R_1}\boldsymbol{\gamma}_1$ and $I_{R_2} = \boldsymbol{\beta}'_{R_2}\boldsymbol{\gamma}$ can be written as

$$\rho_{I_{R_1}I_{R_2}} = \frac{\boldsymbol{\beta}'_{R_1}\mathbf{A}_1\boldsymbol{\beta}_{R_2}}{\sqrt{\boldsymbol{\beta}'_{R_1}\boldsymbol{\Gamma}_1\boldsymbol{\beta}_{R_1}}\sqrt{\boldsymbol{\beta}'_{R_2}\boldsymbol{\Gamma}\boldsymbol{\beta}_{R_2}}}, \tag{9.36}$$

where $\sqrt{\boldsymbol{\beta}'_{R_1}\boldsymbol{\Gamma}_1\boldsymbol{\beta}_{R_1}}$ and $\sqrt{\boldsymbol{\beta}'_{R_2}\boldsymbol{\Gamma}\boldsymbol{\beta}_{R_2}}$ are the standard deviations of the variances of $I_{R_1} = \boldsymbol{\beta}'_{R_1}\boldsymbol{\gamma}_1$ and $I_{R_2} = \boldsymbol{\beta}'_{R_2}\boldsymbol{\gamma}$ respectively. In Eq. (9.36), matrix $\boldsymbol{\Gamma}$ was not adjusted for previous selection on $I_{R_1} = \boldsymbol{\beta}'_{R_1}\boldsymbol{\gamma}_1$.

9.5.2 Numerical Examples

To illustrate the MRLGSI theory in a two-stage breeding selection scheme, we use the simulated data described in Sect. 9.4.3. In that subsection we indicated that the estimated covariance matrices of $\mathbf{\Gamma}_1$ and $\mathbf{A}_1$ were $\widehat{\mathbf{\Gamma}}_1 = \begin{bmatrix} 16.26 & -6.51 \\ -6.51 & 5.79 \end{bmatrix}$ and $\widehat{\mathbf{A}}_1 = \begin{bmatrix} 16.26 & -6.51 & 5.60 & 2.29 \\ -6.51 & 5.79 & -2.33 & -1.62 \end{bmatrix}$, and that $\widehat{\boldsymbol{\beta}}'_1 = \mathbf{w}'\widehat{\mathbf{A}}'_1\widehat{\mathbf{\Gamma}}_1^{-1} = [1.39 \;\; -1.25]$ was the estimated MLGSI vector of coefficients at stage 1. At stage 2, the estimated MLGSI vector of coefficients was $\mathbf{w}' = [1 \;\; -1 \;\; 1 \;\; 1]$, the vector of economic weights.

Suppose that we restrict only trait 2; then at stages 1 and 2, matrix $\mathbf{U}'_1 = [0 \;\; 1]$ and matrix $\mathbf{U}'_2 = [0 \;\; 1 \;\; 0 \;\; 0]$ respectively. In addition, $\widehat{\mathbf{Q}}_{G_1} = \mathbf{U}_1\left(\mathbf{U}'_1\widehat{\mathbf{\Gamma}}_1\mathbf{U}_1\right)^{-1}\mathbf{U}'_1\widehat{\mathbf{\Gamma}}_1$, $\widehat{\mathbf{Q}}_{G_2} = \mathbf{U}_2\left(\mathbf{U}'_2\widehat{\mathbf{\Gamma}}\mathbf{U}_2\right)^{-1}\mathbf{U}'_2\widehat{\mathbf{\Gamma}}$, $\widehat{\mathbf{K}}_{G_1} = \left[\mathbf{I} - \widehat{\mathbf{Q}}_{G_1}\right]$, and $\widehat{\mathbf{K}}_{G_2} = \left[\mathbf{I} - \widehat{\mathbf{Q}}_{G_2}\right]$ are the estimated matrices described in Eqs. (9.29) and (9.30) for stages 1 and 2. It can be shown that, at stages 1 and 2, $\widehat{\boldsymbol{\beta}}'_{R_1} = \widehat{\boldsymbol{\beta}}'_1\widehat{\mathbf{K}}'_{G_1} = [1.39 \;\; 1.558]$ and $\widehat{\boldsymbol{\beta}}'_{R_2} = \mathbf{w}'\widehat{\mathbf{K}}'_{G_2} = [1.0 \;\; 1.81 \;\; 1.01.0]$ are the MRLGSI vectors of coefficients respectively.

Suppose that the total proportion retained for the two stages was 20%, then at stage 1, $k_1 = 0.744$ is an associated approximated selection intensity and the estimated MRLGSI selection response, expected genetic gain per trait, and accuracy were $\widehat{R}_{R_1} = k_1\sqrt{\widehat{\boldsymbol{\beta}}'_{R_1}\widehat{\mathbf{\Gamma}}_1\widehat{\boldsymbol{\beta}}_{R_1}} = 3.083$, $\widehat{\mathbf{E}}_{R_1} = [2.225 \;\; 0 \;\; 0.742 \;\; 0.117]$, and $\widehat{\rho}_{HI_1} = \sqrt{\dfrac{\widehat{\boldsymbol{\beta}}'_{R_1}\widehat{\mathbf{\Gamma}}_1\widehat{\boldsymbol{\beta}}_{R_1}}{\mathbf{w}'\widehat{\mathbf{C}}\mathbf{w}}} = 0.370$ respectively. The estimated MRLGSI expected genetic gain, accuracy, and selection response at stage 2 were $\widehat{\mathbf{E}}_{R_2} = k_2\dfrac{\widehat{\boldsymbol{\beta}}'_{R_2}\widehat{\mathbf{\Gamma}}^*}{\sqrt{\widehat{\boldsymbol{\beta}}'_{R_2}\widehat{\mathbf{\Gamma}}^*\widehat{\boldsymbol{\beta}}_{R_2}}} = [1.156 \;\; 0 \;\; 0.793 \;\; 0.536]$, $\widehat{\rho}_{HI_2} = \sqrt{\dfrac{\widehat{\boldsymbol{\beta}}'_{R_2}\widehat{\mathbf{\Gamma}}^*\widehat{\boldsymbol{\beta}}_{R_2}}{\mathbf{w}'\widehat{\mathbf{C}}^*\mathbf{w}}} = 0.32$, and $\widehat{R}_{R_2} = k_2\sqrt{\widehat{\boldsymbol{\beta}}'_{R_2}\widehat{\mathbf{\Gamma}}^*\widehat{\boldsymbol{\beta}}_{R_2}} = 2.485$ respectively, where $k_2 = 0.721$ was the approximated selection intensity value for stage 2.

The estimated total MRLGSI selection response and expected genetic gain at stages 1 and 2 were $\widehat{R}_{R_1} + \widehat{R}_{R_2} = 5.568$ and $\mathbf{E}'_{R_1} + \mathbf{E}'_{R_2} = [3.380 \;\; 0 \;\; 1.535 \; 0.653]$ respectively. Note that, in effect, the expected genetic gain for trait 2 was 0, as expected.

9.6 The Multistage Predetermined Proportional Gain Linear Genomic Selection Index

The MPPG-LGSI is an adaptation of the predetermined proportional gain linear genomic selection index (PPG-LGSI) described in Chap. 6; thus, the theoretical results, properties, and objectives of both indices are similar. The MPPG-LGSI objective is to change μ_q to $\mu_q + d_q$, where d_q is a predetermined change in μ_q. We solve this problem by minimizing the mean squared difference between $\mathrm{I} = \boldsymbol{\beta}'\boldsymbol{\gamma}$ and $H = \mathbf{w}'\mathbf{g}$ ($E[(H - \mathrm{I})^2]$) under the restriction $\mathbf{U}'\boldsymbol{\Gamma}\boldsymbol{\beta} = \theta_G\mathbf{d}$, where θ_G is a proportionality constant, $\mathbf{d}' = [d_1\ d_2 \ldots d_r]$ is the vector of predetermined restrictions, $\mathbf{U}'$ is a matrix $(t-1) \times t$ of 1s and 0s, and $\boldsymbol{\Gamma}$ is a covariance matrix of additive genomic breeding values, $\boldsymbol{\gamma}' = [\gamma_1\ \gamma_2 \ldots \gamma_t]$, where r is the number of predetermined restrictions and t the number of traits.

9.6.1 The OMPPG-LGSI Parameters

According to the results in Chap. 6, at stages 1 and 2, the MPPG-LGSI vector of coefficients can be written as

$$\boldsymbol{\beta}_{P_1} = \boldsymbol{\beta}_{R_1} + \theta_1\mathbf{U}_1\left(\mathbf{U}_1'\boldsymbol{\Gamma}_1\mathbf{U}_1\right)^{-1}\mathbf{d} \tag{9.37}$$

and

$$\boldsymbol{\beta}_{P_2} = \boldsymbol{\beta}_{R_2} + \theta_2\mathbf{U}_2\left(\mathbf{U}_2'\boldsymbol{\Gamma}\mathbf{U}_2\right)^{-1}\mathbf{d}, \tag{9.38}$$

respectively, where $\boldsymbol{\beta}_{R_1} = \mathbf{K}_{G_1}\boldsymbol{\beta}_1$, $\boldsymbol{\beta}_{R_2} = \mathbf{K}_{G_2}\boldsymbol{\beta}_2 = \mathbf{K}_{G_2}\mathbf{w}$, $\mathbf{K}_{G_1} = \left[\mathbf{I} - \mathbf{Q}_{G_1}\right]$, $\mathbf{Q}_{G_1} = \mathbf{U}_1\left(\mathbf{U}_1'\boldsymbol{\Gamma}_1\mathbf{U}_1\right)^{-1}\mathbf{U}_1'\boldsymbol{\Gamma}_1$, $\mathbf{K}_{G_2} = \left[\mathbf{I} - \mathbf{Q}_{G_2}\right]$, and $\mathbf{Q}_{G_2} = \mathbf{U}_2\left(\mathbf{U}_2'\boldsymbol{\Gamma}\mathbf{U}_2\right)^{-1}\mathbf{U}_2'\boldsymbol{\Gamma}$ were described in Eqs. (9.29) and (9.30). Also, it can be shown that the proportionality constants for stages 1 (θ_1) and 2 (θ_2) are

$$\theta_1 = \frac{\mathbf{d}'\left(\mathbf{U}_1'\boldsymbol{\Gamma}_1\mathbf{U}_1\right)^{-1}\mathbf{U}_1'\mathbf{A}_1\mathbf{w}}{\mathbf{d}'\left(\mathbf{U}_1'\boldsymbol{\Gamma}_1\mathbf{U}_1\right)^{-1}\mathbf{d}} \quad \text{and} \quad \theta_2 = \frac{\mathbf{d}'\left(\mathbf{U}_2'\boldsymbol{\Gamma}\mathbf{U}_2\right)^{-1}\mathbf{U}_2'\boldsymbol{\Gamma}\mathbf{w}}{\mathbf{d}'\left(\mathbf{U}_2'\boldsymbol{\Gamma}\mathbf{U}_2\right)^{-1}\mathbf{d}}, \tag{9.39}$$

respectively. By Eqs. (9.37) to (9.39), the MPPG-LGSI for stages 1 and 2 can be written as $I_{P_1} = \boldsymbol{\beta}_{P_1}'\boldsymbol{\gamma}_1$ and $I_{P_2} = \boldsymbol{\beta}_{P_2}'\boldsymbol{\gamma}$ respectively, where $\boldsymbol{\gamma}_1$ and $\boldsymbol{\gamma}$ are vectors of genomic breeding values, which can be estimated using GEBVs (see Chap. 5 for details).

For stages 1 and 2, the MPPG-LGSI accuracies (ρ_{HI_1} and ρ_{HI_2}), expected genetic gains per trait ($\mathbf{E}_{P_1}$ and $\mathbf{E}_{P_2}$), and selection responses (R_{P_1} and R_{P_2}) can be written as

$$\rho_{HI_1} = \sqrt{\frac{\boldsymbol{\beta}_{P_1}'\boldsymbol{\Gamma}_1\boldsymbol{\beta}_{P_1}}{\mathbf{w}'\mathbf{C}\mathbf{w}}} \quad \text{and} \quad \rho_{HI_2} = \sqrt{\frac{\boldsymbol{\beta}_{P_2}'\boldsymbol{\Gamma}^*\boldsymbol{\beta}_{P_2}}{\mathbf{w}'\mathbf{C}^*\mathbf{w}}}, \tag{9.40}$$

$$\mathbf{E}_{P_1} = k_1\frac{\mathbf{A}_1'\boldsymbol{\beta}_{P_1}}{\sqrt{\boldsymbol{\beta}_{P_1}'\boldsymbol{\Gamma}_1\boldsymbol{\beta}_{P_1}}} \quad \text{and} \quad \mathbf{E}_{P_2} = k_2\frac{\boldsymbol{\Gamma}^*\boldsymbol{\beta}_{P_2}}{\sqrt{\boldsymbol{\beta}_{P_2}'\boldsymbol{\Gamma}^*\boldsymbol{\beta}_{P_2}}} \tag{9.41}$$

and

$$R_{P_1} = k_1\sqrt{\boldsymbol{\beta}'_{P_1}\boldsymbol{\Gamma}_1\boldsymbol{\beta}_{P_1}} \quad \text{and} \quad R_{P_2} = k_2\sqrt{\boldsymbol{\beta}'_{P_2}\boldsymbol{\Gamma}^*\boldsymbol{\beta}_{P_2}}, \tag{9.42}$$

respectively. The total MPPG-LGSI expected genetic gain per trait and selection response at both stages are equal to $\mathbf{E}_{P_1} + \mathbf{E}_{P_2}$ and $R_{P_1} + R_{P_2}$. To simplify the notation, in Eqs. (9.41) and (9.42), we omitted the intervals between stages or selection cycles (L_G). Matrices $\boldsymbol{\Gamma}^*$ and $\mathbf{C}^*$ are matrices $\boldsymbol{\Gamma}$ and $\mathbf{C}$ adjusted for previous selection on I_{P_1} according to Eqs. (9.34) and (9.35) respectively in the MPPG-LGSI context.

The correlation between $I_{P_1} = \boldsymbol{\beta}'_{P_1}\boldsymbol{\gamma}_1$ and $I_{P_2} = \boldsymbol{\beta}'_{P_2}\boldsymbol{\gamma}$ can be written as

$$\rho_{12} = \frac{\boldsymbol{\beta}'_{P_1}\mathbf{A}_1\boldsymbol{\beta}_{P_2}}{\sqrt{\boldsymbol{\beta}'_{P_1}\boldsymbol{\Gamma}_1\boldsymbol{\beta}_{P_1}}\sqrt{\boldsymbol{\beta}'_{P_2}\boldsymbol{\Gamma}\boldsymbol{\beta}_{P_2}}}. \tag{9.43}$$

In Eq. (9.43), matrix $\boldsymbol{\Gamma}$ was not adjusted for previous selection on $I_{P_1} = \boldsymbol{\beta}'_{P_1}\boldsymbol{\gamma}_1$.

9.6.2 *Numerical Examples*

To illustrate the MPPG-LGSI theory, we use the simulated data described in Sect. 9.4.3. Suppose that we select two traits at stages 1 and 2; then, at stage 1, $\widehat{\boldsymbol{\Gamma}}_1 = \begin{bmatrix} 16.26 & -6.51 \\ -6.51 & 5.79 \end{bmatrix}$ and $\widehat{\mathbf{A}}_1 = \begin{bmatrix} 16.26 & -6.51 & 5.60 & 2.29 \\ -6.51 & 5.79 & -2.33 & -1.62 \end{bmatrix}$ are the estimated covariance matrices of $\boldsymbol{\Gamma}_1$ and $\mathbf{A}_1$ respectively. We restricted trait 2 with $\mathbf{d} = -2$; then, at the stage 1 matrix $\mathbf{U}'_1 = [0 \quad 1]$ and at the stage 2 matrix $\mathbf{U}'_2 = [0\ 1\ 0\ 0]$. In addition, $\widehat{\mathbf{Q}}_{G_1} = \mathbf{U}_1\left(\mathbf{U}'_1\widehat{\boldsymbol{\Gamma}}_1\mathbf{U}_1\right)^{-1}\mathbf{U}'_1\widehat{\boldsymbol{\Gamma}}_1$, $\widehat{\mathbf{Q}}_{G_2} = \mathbf{U}_2\left(\mathbf{U}'_2\widehat{\boldsymbol{\Gamma}}\mathbf{U}_2\right)^{-1}\mathbf{U}'_2\widehat{\boldsymbol{\Gamma}}$, $\widehat{\mathbf{K}}_{G_1} = \left[\mathbf{I} - \widehat{\mathbf{Q}}_{G_1}\right]$, and $\widehat{\mathbf{K}}_{G_2} = \left[\mathbf{I} - \widehat{\mathbf{Q}}_{G_2}\right]$ are the estimates of matrix projectors associated with stages 1 and 2 (Eqs. 9.37 and 9.38 for details).

In Sect. 9.4.3, we showed that the estimated MRLGSI vector of coefficients for stage 1 was $\widehat{\boldsymbol{\beta}}'_{R_1} = \boldsymbol{\beta}'_1\widehat{\mathbf{K}}'_{G_1} = [1.386 \quad 1.550]$. Thus, by Eq. (9.37), to obtain $\widehat{\boldsymbol{\beta}}_{P_1} = \widehat{\boldsymbol{\beta}}_{R_1} + \widehat{\theta}_1\mathbf{U}_1\left(\mathbf{U}'_1\widehat{\boldsymbol{\Gamma}}_1\mathbf{U}_1\right)^{-1}\mathbf{d}$, we only need to obtain $\widehat{\theta}_1$ and $\mathbf{U}_1\left(\mathbf{U}'_1\widehat{\boldsymbol{\Gamma}}_1\mathbf{U}_1\right)^{-1}\mathbf{d}$, where $\mathbf{d} = -2$ and $\widehat{\theta}_1 = \dfrac{\mathbf{d}'\left(\mathbf{U}'_1\widehat{\boldsymbol{\Gamma}}_1\mathbf{U}_1\right)^{-1}\mathbf{U}'_1\widehat{\mathbf{A}}_1\mathbf{w}}{\mathbf{d}'\left(\mathbf{U}'_1\widehat{\boldsymbol{\Gamma}}_1\mathbf{U}_1\right)^{-1}\mathbf{d}}$. It can be shown that $\mathbf{U}_1\left(\mathbf{U}'_1\widehat{\boldsymbol{\Gamma}}_1\mathbf{U}_1\right)^{-1}$ $\mathbf{d} = \begin{bmatrix} 0 \\ -0.345 \end{bmatrix}$ and $\widehat{\theta}_1 = 8.125$; therefore, $\widehat{\boldsymbol{\beta}}'_{P_1} = [1.39 \quad -1.25]$ is the MPPG-LGSI vector of coefficients at stage 1.

Suppose that the total proportion retained for the two stages was 20%; then, $k_1 = 0.744$ is an approximate selection intensity associated with MPPG-LGSI and the estimated MPPG-LGSI accuracy, selection response, and expected genetic gain at stage 1 were $\widehat{\rho}_{HI_1} = \sqrt{\frac{\widehat{\boldsymbol{\beta}}'_{P_1}\widehat{\boldsymbol{\Gamma}}_1\widehat{\boldsymbol{\beta}}_{P_1}}{\mathbf{w}'\widehat{\mathbf{C}}\mathbf{w}}} = 0.71$, $\widehat{R}_{P_1} = k_1\sqrt{\widehat{\boldsymbol{\beta}}'_{P_1}\widehat{\boldsymbol{\Gamma}}_1\widehat{\boldsymbol{\beta}}_{P_1}} = 5.90$ and $\widehat{\mathbf{E}}'_{P_1} = k_1\frac{\widehat{\mathbf{A}}'_1\widehat{\boldsymbol{\beta}}_{P_1}}{\sqrt{\widehat{\boldsymbol{\beta}}'_{P_1}\widehat{\boldsymbol{\Gamma}}_1\widehat{\boldsymbol{\beta}}_{P_1}}} = [2.88 \quad -1.53 \quad 1.00 \quad 0.49]$ respectively.

It can be shown that at stage 2, $\mathbf{d}'\left(\mathbf{U}'_1\widehat{\boldsymbol{\Gamma}}_1\mathbf{U}_1\right)^{-1}\mathbf{U}'_1 = [0 \quad -0.345 \quad 0 \quad 0]$, $\widehat{\theta}_2 = 8.125$ and $\widehat{\boldsymbol{\beta}}'_{P_2} = \mathbf{w}' = [1 \quad -1 \quad 1 \quad 1]$. Thus, the estimated MPPG-LGSI accuracy, selection response, and expected genetic gain at this stage were $\widehat{\rho}_{HI_2} = \sqrt{\frac{\mathbf{w}'\widehat{\boldsymbol{\Gamma}}^*\mathbf{w}}{\mathbf{w}'\widehat{\mathbf{C}}^*\mathbf{w}}} = 0.64$, $\widehat{R}_{P_2} = k_2\sqrt{\mathbf{w}'\widehat{\boldsymbol{\Gamma}}^*\mathbf{w}} = 4.10$, and $\widehat{\mathbf{E}}'_{P_2} = k_2\frac{\widehat{\boldsymbol{\Gamma}}^*\mathbf{w}}{\sqrt{\mathbf{w}'\widehat{\boldsymbol{\Gamma}}^*\mathbf{w}}} = [1.74 \quad -0.92 \quad 0.85 \quad 0.58]$ respectively, where $k_2 = 0.721$. The estimated total MPPG-LGSI selection response and expected genetic gain for both stages were $\widehat{R}_{P_1} + \widehat{R}_{P_2} = 9.99$ and $\widehat{\mathbf{E}}'_{P_1} + \widehat{\mathbf{E}}'_{P_2} = [4.62 \quad -2.45 \quad 1.85 \quad 1.07]$ respectively. Note that the total expected genetic gain for trait 2 was -2.45, which is similar to $\mathbf{d} = -2$, the PPG imposed by the breeder. Finally, to simplify the notation, we omitted the intervals between stages or selection cycles (L_G) in the estimated MPPG-LPSI selection response and expected genetic gain for both stages.

References

Arismendi JC (2013) Multivariate truncated moments. J Multivar Anal 117:41–75

Cochran WG (1951) Improvement by means of selection. In: Neyman J (ed) Proc. 2nd Berkeley Symp. on Math., Stat. and Probability, pp 449–470

Cunningham EP (1975) Multi-stage index selection. Theor Appl Genet 46:55–61

Dekkers JCM (2014) Multiple stage selection. Armidale Animal Breeding Summer Course 2014, University of New England, Armidale, NSW, Australia

Falconer DS, Mackay TFC (1996) Introduction to quantitative genetics. Longman, New York

Hattaway JT (2010) Parameter estimation and hypothesis testing for the truncated normal distribution with applications to introductory statistics grades. All Theses and Dissertations. Paper 2053

Hicks C, Muir WM, Stick DA (1998) Selection index updating for maximizing rate of annual genetic gain in laying hens. Poult Sci 77:1–7

Mi X, Utz HF, Technow F, Melchinger AE (2014) Optimizing resource allocation for multistage selection in plant breeding with R package *Selectiongain*. Crop Sci 54:1413–1418

Rausand M, Hφyland A (2004) System reliability theory: models, statistical methods, and applications, 2nd edn. Wiley, Hoboken, NJ

Saxton AM (1983) A comparison of exact and sequential methods in multi-stage index selection. Theor Appl Genet 66:23–28

Thomas GB (2014) Thomas' calculus: early transcendentals, 3r edn. Pearson Education, Inc., Boston, MA

Xu S, Muir WM (1992) Selection index updating. Theor Appl Genet 83:451–458

Young SSY (1964) Multi-stage selection for genetic gain. Heredity 19:131–143

10

Stochastic Simulation using LPSI, ESIM, Restrictive LPSI and Restrictive ESIM

Fernando H. Toledo, José Crossa and Juan Burgueño

Abstract Stochastic simulation can contribute to a better understanding of the problem, and has already been successfully applied to evaluate other breeding scenarios. Despite all the theories developed in this book concerning different types of indices, including phenotypic data and/or data on molecular markers, no examples have been presented showing the long-term behavior of different indices. The objective of this chapter is to present some results and insights into the in silico (computer simulation) performance comparison of over 50 selection cycles of a recurrent and generic population breeding program with different selection indices, restricted and unrestricted. The selection indices included in this stochastic simulation were the linear phenotypic selection index (LPSI), the eigen selection index method (ESIM), the restrictive LPSI, and the restrictive ESIM.

10.1 Stochastic Simulation

Simulations were used to evaluate the accuracy, effectiveness, response to selection, and the decrease in the overall genetic variance in a recurrent selection scheme under the use of the Smith (1936) and Hazel (1943) index (or linear phenotypic selection index, LPSI, see Chap. 2 for details); the eigen selection index method (ESIM, see Chap. 7 for details); the Kempthorne and Nordskog (1959) restricted index (K&N or restricted phenotypic selection index, RLPSI, see Chap. 3 for details); and the restricted eigen selection index method (RESIM, see Chap. 3 for details). The different scenarios are described below and encompass variations in the nature of the genetic correlation between traits in addition to their expected heritabilities.

F. H. Toledo (✉) · J. Crossa · J. Burgueño
Biometrics and Statistics Unit, International Maize and Wheat Improvement Center (CIMMYT), Mexico, Mexico
e-mail: f.toledo@cgiar.org

10.1.1 Breeding Design

A total of 50 forward recurrent selection cycles of modern breeding were simulated, in which the breeder has the ability to select based on breeding value estimates of genetically correlated traits, and to apply the various above-mentioned selection indices. All simulated scenarios (described below) followed a common general breeding design. In each cycle, 350 full sib progenies (S_1) were generated taking 700 parents at random from the base population. From each progeny, 100 double-haploid lines were randomly derived (which shortened the cycle interval by five inbreeding generations). The simulated phenotypic values of the 35,000 resulting lines were then evaluated in simulated trials. The selection was made by means of the progeny average performance. The selected progenies (top quarter) according to each index were then recombined by random mating a sample of the lines within the progeny to recover the population for the next cycle.

10.1.2 Simulating Quantitative Traits

Genetically correlated quantitative traits were simulated assuming a full pleiotropic model. This was carried out by randomly sampling genetic effects for all segregating sites from a multivariate normal distribution with zero mean and a previously stated variance–covariance. The genetic effects were in turn used to compute true breeding values (TBVs). An individual's phenotype was obtained by taking its TBV and adding a zero mean normally random term with variance consistent with the expected heritability (h^2) for the trait at which phenotyping occurred. The genetic variance in each cycle was calculated as the variance of the TBV of the individuals in that generation. However, it was expressed as relative values of the genetic variance in the initial cycle. The realized response to selection was also standardized in units of the genetic standard deviation in cycle 0. Cycle 0 was used as the base generation because it represents the available genetic variability, and also to observe, from the start, the genetic changes in future breeding generations.

An empirical genome was considered comprising a set of 10 linkage groups (chromosomes), each 200 cM in length, and 1000 uniformly distributed segregating sites. To represent the historical evolution and recent breeding efforts up to the present day in addition to incorporating a steady state of known linkage disequilibrium (LD) structure existing in crops, the starting populations (cycle 0) were taken after 200 generations of random mating within an effective population size of 1000 segregating for all loci in which the allele frequency was 0.5.

The in silico meiosis reflected the Mendelian laws of segregation for diploid species, by a count-location process that mimics the Haldane map function (Haldane 1919). Thus, homologous chromosomes are paired into bivalents and recombined through randomly positioned chiasmata. The number of chiasmata follows a Poisson distribution, where the λ parameter represents the chromosome length in Morgans and their positions are uniformly distributed, i.e., without interference between crossovers or any mutagenesis process.

10.1.3 Simulated Scenarios

Three traits were considered, one with low heritability (the first, $h^2 = 0.2$) and two with high heritability (the second and the third, $h^2 = 0.5$). The correlations between the first and second trait vary from positive ($\rho_G = 0.5$) to negative ($\rho_G = -0.5$). The third trait was always considered with segregation independent from the two others.

The selection process involved two unrestricted indices: the LPSI (see Chap. 2), which ranks the progenies based on the average merit of their lines considering equal economic weights for all traits, and the ESIM (see Chap. 7), where the progenies were ranked in terms of ESIM values. Regarding the restricted selection indices, the RLPSI (or K&N) was employed (see Chap. 3) with equal economic weights for the traits in addition to the RESIM (see Chap. 7). Because of the restrictions, two different situations were evaluated in the latter cases, i.e., where the restrictions were applied for each of the first and second traits separately.

Thus, all simulated scenarios encompass a three-way factorial: four selection procedures (the LPSI, the ESIM, the RLPSI or K&N, and the RESIM); two correlation scenarios, positive ($\rho_G = 0.5$) and negative correlations ($\rho_G = -0.5$) between the first and second trait; and two constraint situations, where the restrictions were applied separately for the first and second traits.

To simulate genetically correlated traits a full pleiotropic model was assumed. Gene effects were sampled from a multivariate normal distribution with zero mean and a previously stated variance–covariance matrix. In that sense it is possible to represent a quantitative and infinitesimal model. Each genes has its own effect varying according to a probabilistic density i.e., genes with positive and negative effects varying its effects sizes; alleles with large effects at lower frequency (major genes) and alleles with modest effects at higher frequency (minor genes).

10.1.4 Inferences

Results are presented as summaries of 100 Monte Carlo replicates for each scenario and include the response to selection, decreases in the genetic variance, selection accuracy, and observed heritabilities. The meiosis routine was implemented in C++, and compiled, linked, and through the facilities provided by the Rccp R package (Eddelbuettel 2013). All simulations were performed, analyzed, and summarized in R version 3.3.3 (R Development Core Team 2017).

10.2 Results

Overriding the results of the simulations regarding the four selection indices under the different trait genetic correlations and restrictions, scenarios are presented in terms of the consistency of the observed heritabilities of the traits; the response to selection and changes in genetic variance for each trait; and the accuracy of the indices' selection.

First of all, the results show the stability of the Monte Carlo replicates in terms of possible deviations in the observed heritability from that expected, which in turn may affect further inferences (Table 10.1). The *type I* error (α) of the *t test* comparing expected and observed heritabilities for all simulated scenarios did not show important and significant departures. Slight departures that may be due to Monte Carlo error ($P < 0.05$) were found, namely: for both high and low heritability traits of the LPSI at cycle 5 when they were negatively correlated; for the independent trait also with the LPSI at cycle 50, but, when the other traits are positively correlated; for the high heritable trait at the first and last cycles, both under positive correlation in the ESIM and RESIM indices respectively; and for the low heritability trait in both restricted indices (RLPSI and RESIM) in cycles 0 and 5 for respective and negative and positive correlations.

A complementary estimate of the power (*type II* error or β) of the tests was performed considering departures from the expected heritabilities of 1%. It was verified that the average power if the observed estimates was around 70%, which reinforces the appropriateness of the simulation findings.

10.2.1 Realized Genetic Gains

Figure 10.1 shows the average genetic gains (expressed as standard deviations from the mean of cycle 0) for cycles 0–50 for the traits (low and high heritabilities and the independent trait); the four selection indices (unrestricted: LPSI and ESIM and restricted: RLPSI and RESIM) when the correlations are positive and negative.

It is important to note that even after 50 recurrent cycles none of the scenarios has shown any indication that the selection plateau has been reached (Fig. 10.1). It is considered that even with the variation of the gains in the scenarios, there were increases in the merit of the target traits. Thus, the employment of selection indices is an effective way of achieving progress in long-term multi-trait selection.

As expected, the unrestricted selection indices have shown genetic gains higher than their restricted counterparts (Fig. 10.1). It must be highlighted that the restrictions proved their properties because when any trait was restricted, no gains were obtained for that trait (data not shown). The higher gains obtained with unrestricted indices is well known and justified in comparison with their restricted homologous because the net genetic merit is beneficiated by the gains in all traits, while, with gains constrained to zero in some traits, there are no indirect gains that may be highlighted especially because of positive correlations.

The independent trait has presented the higher gains in comparison with the other traits for all correlation and selection process scenarios. The higher gains, however, were for the RESIM followed by the RLPSI in both positive and negative correlations (Fig. 10.1e and f). These findings may be understood both under the nature of the trait (independent inheritance) and over the properties of the restricted indices.

Table 10.1 Mean (μ), standard deviation (σ), *type I* error (α), and *type II*/power error (β) of the observed heritability in 100 Monte Carlo replicates for traits with low and high heritability (h^2) and independent at cycles 0, 5, and 50 of a simulated selection given four indices, the linear phenotypic selection index (LPSI), the ESIM, the restricted linear phenotypic selection index (RLPSI), and the RESIM, with positive and negative correlations between the traits low h^2 and high h^2

Scenario/cycle		Low h^2				High h^2				Independent			
				Errors				Errors				Errors	
		μ	σ	α	β	μ	σ	α	β	μ	σ	α	β
LPSI													
Negative	0	0.201	0.016	0.565	0.993	0.502	0.035	0.536	0.513	0.502	0.035	0.524	0.531
	5	0.204	0.016	0.016	0.993	0.493	0.031	0.028	0.612	0.504	0.034	0.191	0.540
	50	0.200	0.015	0.852	0.996	0.501	0.035	0.672	0.524	0.505	0.034	0.124	0.547
Positive	0	0.200	0.014	0.782	0.999	0.500	0.037	0.902	0.485	0.503	0.033	0.347	0.556
	5	0.200	0.016	0.886	0.995	0.499	0.032	0.758	0.601	0.502	0.033	0.561	0.562
	50	0.200	0.015	0.950	0.997	0.504	0.035	0.214	0.513	0.507	0.035	0.046	0.511
ESIM													
Negative	0	0.200	0.013	0.711	1.000	0.502	0.035	0.552	0.529	0.503	0.038	0.371	0.464
	5	0.201	0.015	0.677	0.997	0.497	0.033	0.349	0.559	0.502	0.031	0.445	0.611
	50	0.200	0.015	0.749	0.997	0.503	0.034	0.309	0.548	0.499	0.035	0.815	0.514
Positive	0	0.199	0.015	0.559	0.996	0.493	0.032	0.029	0.593	0.502	0.030	0.530	0.642
	5	0.200	0.014	0.892	0.999	0.498	0.032	0.490	0.589	0.505	0.037	0.222	0.472
	50	0.202	0.018	0.405	0.975	0.504	0.029	0.132	0.679	0.500	0.033	0.888	0.560
RLPSI													
Negative	0	0.198	0.016	0.178	0.994	0.497	0.033	0.405	0.557	0.503	0.032	0.438	0.587
	5	0.203	0.014	0.042	0.999	0.505	0.033	0.110	0.567	0.501	0.034	0.471	0.541
	50	0.202	0.015	0.173	0.996	0.501	0.032	0.742	0.602	0.500	0.033	0.397	0.572
Positive	0	0.201	0.014	0.295	0.999	0.504	0.033	0.239	0.570	0.502	0.034	0.130	0.539
	5	0.201	0.016	0.709	0.990	0.507	0.035	0.059	0.510	0.502	0.034	0.689	0.557
	50	0.201	0.013	0.331	1.000	0.504	0.032	0.195	0.591	0.507	0.035	0.061	0.533

(continued)

Table 10.1 (continued)

Scenario/cycle		Low h^2				High h^2				Independent			
				Errors				Errors				Errors	
		μ	σ	α	β	μ	σ	α	β	μ	σ	α	β
RESIM													
Negative	0	0.201	0.017	0.504	0.987	0.506	0.034	0.075	0.545	0.502	0.033	0.646	0.567
	5	0.202	0.016	0.230	0.993	0.506	0.032	0.090	0.585	0.500	0.031	0.757	0.618
	50	0.199	0.015	0.490	0.996	0.507	0.034	0.051	0.549	0.501	0.032	0.516	0.604
Positive	0	0.203	0.014	0.018	0.999	0.506	0.035	0.075	0.510	0.505	0.033	0.172	0.578
	5	0.202	0.014	0.171	0.998	0.502	0.034	0.489	0.555	0.501	0.031	0.734	0.626
	50	0.200	0.017	0.911	0.988	0.508	0.036	0.034	0.495	0.501	0.032	0.694	0.601

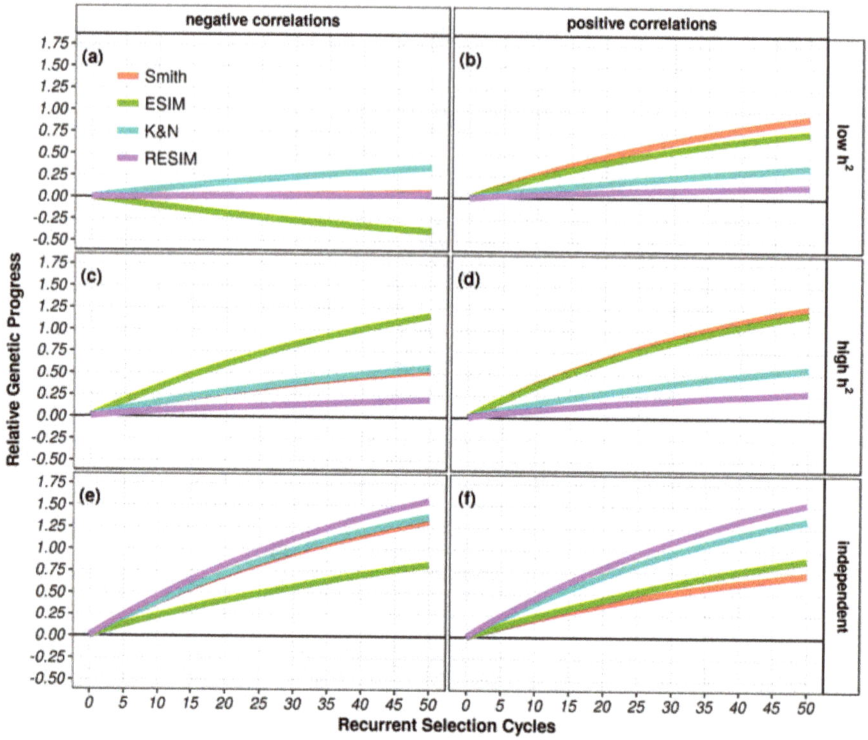

Fig. 10.1 Average genetic gains in 100 Monte Carlo replicates for traits with low and high heritability (h^2 0.2 and 0.5) and independent along cycles 0–50 of a simulated selection given four indices, the linear phenotypic selection index (LPSI), the ESIM, the restricted linear phenotypic selection index (RLPSI), and the RESIM with positive (0.5) and negative (−0.5) correlations between the traits low h^2 and high h^2. (**a**) Gains for the trait with low heritability when it is negatively correlated with the high heritability trait. (**b**) Gains for the trait with low heritability when it is positively correlated with the high heritability trait. (**c**) Gains for the trait with high heritability when it is negatively correlated with the low heritability trait. (**d**) Gains for the trait with high heritability when it is negatively correlated with the low heritability trait. (**e**) Gains for the independent trait when the other traits are negatively correlated. (**f**) Gains for the independent trait when the other traits are positively correlated

As the third trait becomes independent from the others, there are no indirect effects owing to the constraints in the gains of the other traits. With regard to the technical features of the RESIM, it must be emphasized that because of the eigen decomposition, the largest eigenvector obtains higher weight from the most variable trait and consequently ends in distinct gains, which in this case is the independent trait.

The Smith (or LPSI) and ESIM produce similar genetic gains for highly heritable traits when the genetic correlations are positive (Fig. 10.1d). The ESIM is simply another way of obtaining the LPSI based on the eigen decomposition theory, which avoids the assignment of economic weights. Thus, the results prove that the same results may be found with both indices. However, the ESIM is the preferred index because of its advantages over the LPSI: no subjective decision for selecting economic weights, and better statistical sampling properties.

When the traits are negatively correlated, the trait with greater heritability has shown important realized genetic gains based on the ESIM and similar gains for the LPSI and its restricted analogous, i.e., the RLPSI (Fig. 10.1a and c). In addition, when traits are negatively correlated, restricting the traits with low heritability is an alternative, to ensure similar progress to the use of unrestricted indices for highly heritable traits. On the contrary, it is also interesting to note that the ESIM has the worst performance when the traits are negatively correlated for trait with lower heritability (Fig. 10.1a).

On the other hand, as already pointed out, the ESIM performance surpasses all the others with regard to the highly heritable trait (Fig. 10.1c and d). The reason for this is similar to the above-mentioned regarding the properties of the eigen decomposition. When the first trait is negatively correlated with the second one, heavier weight is given to the trait with higher heritability than to the trait with low heritability. However, when the traits are positively correlated, synergic and indirect effects increase both traits, one positively affecting the other.

When the traits are positively correlated but with low heritability, the LPSI and the ESIM have similar realized genetic gains until cycle 25; after this selection cycle, the LPSI is superior to the ESIM (Fig. 10.1b). In this case, the two restrictive indices, the RLPSI and the RESIM, are given lower realized genetic gains than the LPSI and the ESIM (Fig. 10.1b). Finally, considering the third trait (the independent one), the RESIM provides the greater realized genetic gains (Fig. 10.1e and f).

10.2.2 Genetic Variances

In Fig. 10.2, the average relative decreases in the genetic variances along the 50 cycles of selection for the three traits (with low and high heritability traits in addition to the independent trait) under the selection system given by the four selection indices, restricted (the RLPSI and the RESIM) and unrestricted (the LPSI and the ESIM), both with negative and positive correlations between the first and second traits.

As a general result, it is clear that after selection there were decreases in the genetic variance along the recurrent cycles (Fig. 10.2). From the most conservative decrease (around 40% in Fig. 10.2a and b) to the sharp decrease (close to 10% in Fig. 10.2e and f) and in contrast to the trends in genetic gains, it is possible to conceive that the genetic variability was not yet exhausted by selection. This observation endorses what was said regarding the effectiveness of the selection indices as a criterion for long-term multi-trait selection.

As expected, the restricted indices are more conservative, maintaining greater genetic variance (Fig. 10.2). Their feature is to prevent the restricted trait from changing its genetic merit. Thus, they tend to keep its genetic variance unchanged,

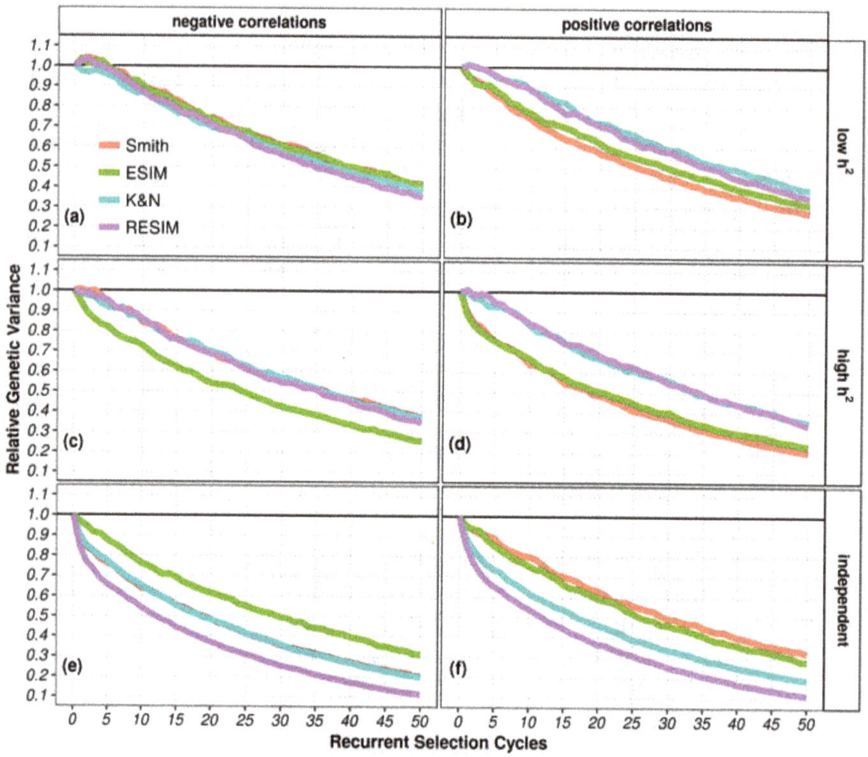

Fig. 10.2 Average genetic variances in 100 Monte Carlo replicates for traits with low and high heritability (h^2 0.2 and 0.5) and independent along cycles 0–50 of a simulated selection given four selection indices, the LPSI, the ESIM, the RLPSI, and the RESIM, with positive (0.5) and negative (−0.5) correlations between the traits low h^2 and high h^2. (**a**) Genetic variance of the low heritability trait when it is negatively correlated with the high heritability trait. (**b**) Genetic variance of the low heritability trait when it is positively correlated with the high heritability trait. (**c**) Genetic variance of the high heritability trait when it is negatively correlated with the low heritability trait. (**d**) Genetic variance of the high heritability trait when it is negatively correlated with the low heritability trait. (**e**) Genetic variance of the independent trait when the other traits are negatively correlated. (**f**) Genetic variance of the independent trait when the other traits are positively correlated

which is reflected in the lower decreases in the genetic variance, even under the indirect effects of the other traits.

It should be noted that there was a slight increase in variance in the short term (up to cycle 3) for the trait with lower heritability when negatively correlated with the highly heritable one (Fig. 10.2a and b). This is an outcome of the changes in allele frequencies of the first trait due to the indirect effects of the second trait and/or the release of genetic disequilibrium owing to the assortative mating of the individuals given higher weights regarding the second trait (highly heritable).

Reflecting the findings regarding the genetic gains (Fig. 10.1), the trait with strong decreases in genetic variance on average was the one in which the response to selection was more pronounced, i.e., the independent trait (Fig. 10.2e and f). This trait has shown stronger decreases over the selection through the ESIM index in both positive and negative correlation scenarios. As mentioned before, as the third trait is independent of the others, a greater response to selection was achieved in that trait and consequently strong changes in allele frequencies, which drove the decreases in genetic variance.

When the heritability is high, it is easy to differentiate the trends in the decrease in the genetic variance between restricted and unrestricted indices (Fig. 10.2c). It is more evident, especially when the traits are positively correlated (Fig. 10.2d). Thus, the ESIM has the highest decreases followed by the LPSI. Nevertheless, for the traits with low heritability, the decreases in genetic variance are indistinguishable between the indices, showing that the effectiveness of the response to selection is a function of the heritability (Fig. 10.2a and b).

10.2.3 Selection Accuracy

The accuracy of the selection was measured as the square root of the correlation between the net genetic merit and the estimated linear function of each index. Figure 10.3 shows the absolute accuracies (left axis) and relative values in relation to the mean accuracy of the first cycle (right axis) for all indices in both negative (Fig. 10.3a) and positive (Fig. 10.3b) correlation scenarios.

In all cases, a reduction in the selection precision of all the indices was observed. The effect of selection is the improvement in the genetic merit of the traits by means of changes in allele frequencies that also affect/decrease the genetic variance. However, as a side effect, the selection becomes harder and has lower precision.

The LPSI has shown greater accuracy in comparison with the other indices in any situation (Fig. 10.3a and b). Its main feature is precisely maximizing the correlation between the net genetic merit and the linear combination of the trait. It may be

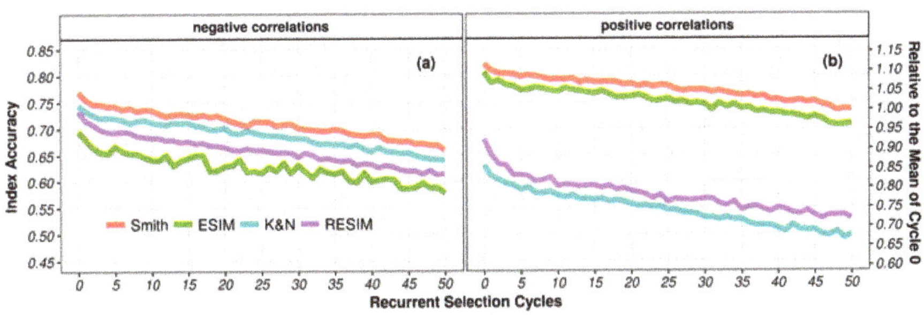

Fig. 10.3 Average absolute and relative accuracy of selection in 100 Monte Carlo replicates for traits with low and high heritability (h^2) and independent along cycles 0–50 of a simulated selection given four selection indices, the LPSI, the ESIM, the RLPSI, and the RESIM with positive and negative correlations between the traits low h^2 and high h^2

argued that the ESIM also does that; however, only when the phenotypic and genotypic variances and covariances are known are they the best linear predictors. Thus, according to what was found, it is possible to note that the ESIM was more affected by the sampling properties when estimating matrices of variance and covariance (Fig. 10.3a).

For the scenario with positive correlations, the differences between the two types of indices, the restricted ones and the unrestricted ones, were clear, as the unrestricted indices have shown greater selection accuracy (Fig. 10.3b). This reflects the fact that the restricted index constrains the gains by means of restrictions in the correlation between the net genetic merit and the linear combination of the traits.

References

Eddelbuettel D (2013) Seamless R and C++ Integration with Rcpp. Springer, New York
Haldane JBS (1919) The combination of linkage values and the calculation of distance between the loci of linked factors. J Genet 8:299–309
Hazel IN (1943) The genetics basis for constructing selection indexes. Genetics 28:476–490
Kempthorne O, Nordskog AW (1959) Restricted selection indices. Biometrics 15:10–19
R Core Team (2017) R: A language and environment for statistical computing
Smith HF (1936) A discriminant function for plant selection. Ann Eugenics 7:240–250

11

A Comprehensive Study of RIndSel

Gregorio Alvarado, Angela Pacheco, Sergio Pérez-Elizalde, Juan Burgueño, and Francisco M. Rodríguez

Abstract RIndSel is a graphical unit interface that uses selection index theory to select individual candidates as parents for the next selection cycle. The index can be a linear combination of phenotypic values, genomic estimated breeding values, or a linear combination of phenotypic values and marker scores. Based on the restriction imposed on the expected genetic gain per trait, the index can be unrestricted, null restricted, or predetermined proportional gain indices. RIndSel is compatible with any of the following versions of Windows: XP, 7, 8, and 10. Furthermore, it can be installed on 32-bit and 64-bit computers. In the context of fixed and mixed models, RIndSel estimates the phenotypic and genetic covariance using two main experimental designs: randomized complete block design and lattice or alpha lattice design. In the following, we explain how RIndSel can be used to determine individual candidates as parents for the next cycle of improvement.

11.1 Background

The linear selection index theory (see Chaps. 2 to 9 for details) can be difficult to apply without the use of specific codes developed in statistical analysis system (SAS) software. At the International Maize and Wheat Improvement Center (CIMMYT, for its Spanish acronym), codes were developed in SAS software version 9.4 (SAS institute 2017) that can help to determine individuals as parents for the next selection cycle. The SAS codes can be found at the following link: https://data.cimmyt.org/dataset.xhtml?persistentId=hdl:11529/10242.

G. Alvarado · A. Pacheco · J. Burgueño (✉) · F. M. Rodríguez
Biometrics and Statistics Unit, International Maize and Wheat Improvement Center (CIMMYT), Mexico, Mexico
e-mail: g.alvarado@cgiar.org; J.Burgueno@cgiar.org

S. Pérez-Elizalde
Departamento de Socioeconomía Estadística e Informática, Colegio de Postgraduados, Mexico, Mexico

Afterward, the SAS codes were translated to R language as scripts (Pacheco et al. 2017) and denoted by RIndSel (R software to analyze Selection Indices), with the objective of creating a user-friendly graphical unit interface (GUI) in JAVA. The link to download the software is: https://data.cimmyt.org/dataset.xhtml?persistentId=hdl:11529/10854.

11.2 Requirements, Installation, and Opening

RIndSel is compatible with a Windows platform, in any of the following versions: XP, 7, 8, and 10; furthermore, it can be installed on 32-bit and 64-bit computers. To install RIndSel on a computer, the user must double-click on the executable file downloaded over the link given above and then follow the instructions that appear in the installation box. Once RIndSel has been installed, it can be opened by:

1. Double-clicking on the shortcut located in the desktop.
2. Locating it in the Windows menu and clicking.
3. Locating the software via the pathway C:/RIndSel, and double-clicking on RIndSel.exe.

As we shall see, the software has been partitioned into two modules.

11.3 First Module: Data Reading and Helping

This module (Fig. 11.1) deploys two small boxes upper left denoted by "*Open File*" and "*Help*." With *Open File,* the user may access a set of files where he/she can open, for example, the file of phenotypic data, which should contain information

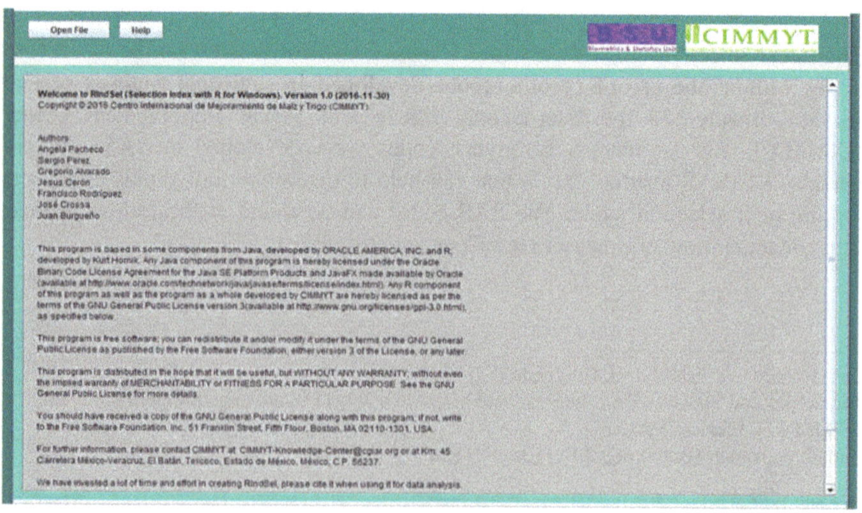

Fig. 11.1 Module for reading data

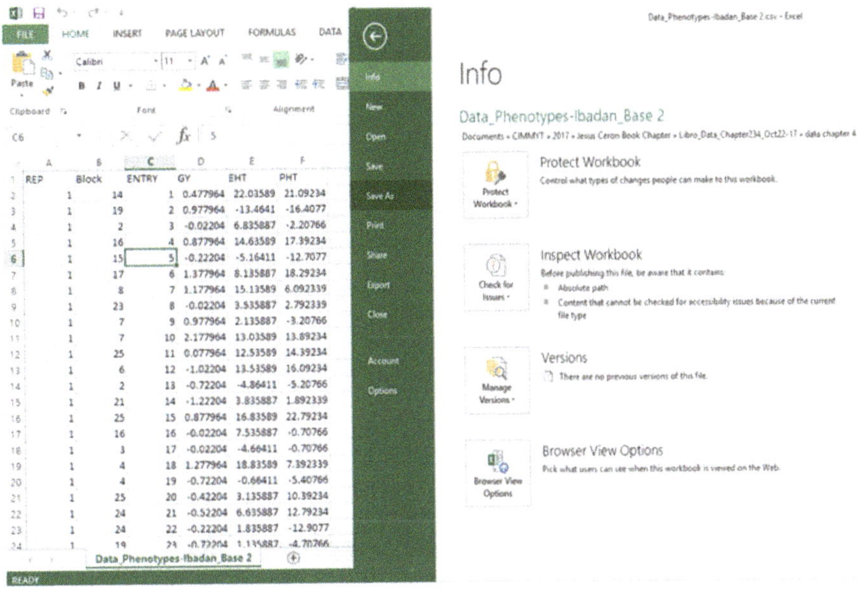

Fig. 11.2 Steps for saving a comma delimited file

associated with the experimental design. This file contains information about the field book where the experimental design variables can be identified in the first columns, whereas the remaining columns contain information about traits measured in the field; design variables and traits are connected by the plot number. Previously, the data set should have been captured in a spreadsheet using Excel or any other similar software and saved as a comma delimited file. To save the data as a comma delimited file in Excel, the following steps should be taken. In the Excel file that contains the data set (Fig. 11.2), select from the main menu: FILE → Save As → Browser View Options (look for the path were the data will be saved) → Save as type (look for CSV, comma separated values). The end of the file name should be ". csv," indicating that the file is ready to be used.

The small box "*Help*" (Fig. 11.1) shows basic features such as the installation manual and software licenses. The installation manual provides a brief description of the selection indices that can be calculated and the pathway to where the software is located (Fig. 11.3). Furthermore, it shows folders related to the software features such as how the software could be used. There is also a folder called "*Examples*," where the user can find data for test phenotypic selection indices, selection indices of coded score markers, and wide genome selection indices. The folders "*Lib*" and "*Programs*" contain information related to the software functioning; therefore, the authors highly recommend not modifying these folders.

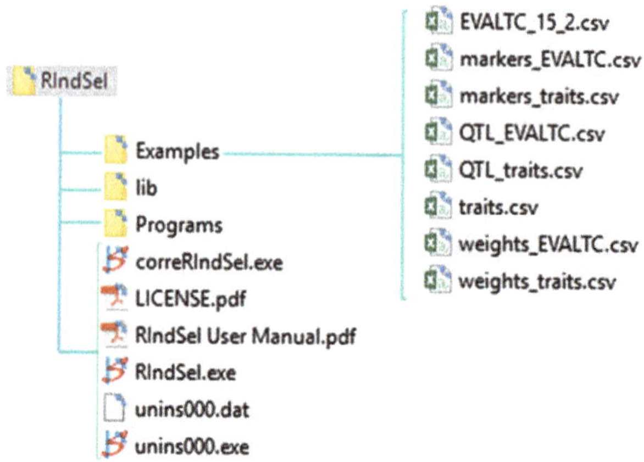

Fig. 11.3 Tree diagram of the RIndSel structure

11.4 Second Module: Capturing Parameters to Run

Once the data have been read (first module), RIndSel moves to the second module (Fig. 11.4), where some feedback is required:

1. To choose the selection index to calculate.
2. To select the experimental design.
3. To identify the variables of experimental design.
4. To choose the traits that will be used to calculate the selection index in the data file.

This module is structured in such a way that calculating any selection index is relatively easy. There are three other small buttons located upper left of the module: "*Back*," "*Analyze*," and "*Help*." *Back* returns to the previous module (Fig. 11.1), *Analyze* executes and calculates the selection index, and *Help* provides the same functions as described in the previous section. In addition, there are four windows, each of which must be filled with the correct parameters. The first one is related to the indices that RIndSel is able to calculate (Fig. 11.5).

11.5 Selection Index

In this menu, it is necessary to define the percentage of genotypes that will be selected. By default, it is 5%, but any other percentage can be chosen. RIndSel uses the correlation matrix or the variance–covariance matrix to obtain the index; however, by default, the variance–covariance matrix is used. To work with the correlation matrix box, "*Correlation*" should be checked. The sign for "*economic weights*"

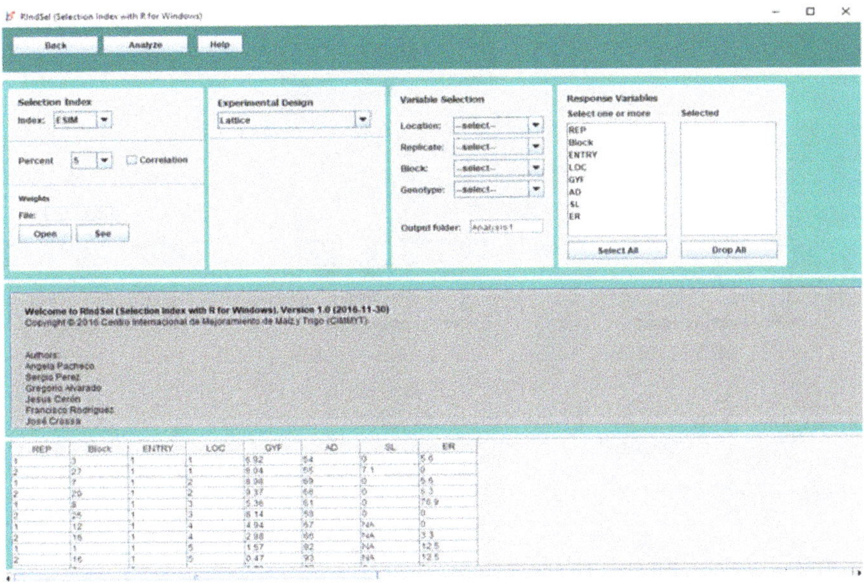

Fig. 11.4 RIndSel module of analysis

can be used to determine the behavior of the expected genetic gain of the traits. For example, with -1, the mean of the traits tends to decrease, whereas with 1, it increases. It is also possible to use the trait heritability. The economic weights can be assigned by creating a comma-delimited file with the name of the trait and economic weight sign (Fig. 11.6a). Once the file has been created, it can be browsed by pressing the open button and where the *.csv file is located (Fig. 11.6b).

To calculate the restricted linear phenotypic selection index (RLPSI or K&N, see Chap. 3 for details), it is necessary to create the same file and incorporate an additional column called "*Restrictions*." This last column must be filled with the number one for those traits that remain fixed (restricted) and zeros for those traits that change (Fig. 11.7). An additional option is to ignore the "*Weights*" box, which means that RIndSel automatically presents an Excel file covering the options for capturing economic weights; the only requirement is that the file must be saved as a comma delimited file.

11.6 Experimental Design

The menu allows the user to select the field array design to be used. There are two choices:

1. Lattice or alpha-lattice
2. Random complete block designs

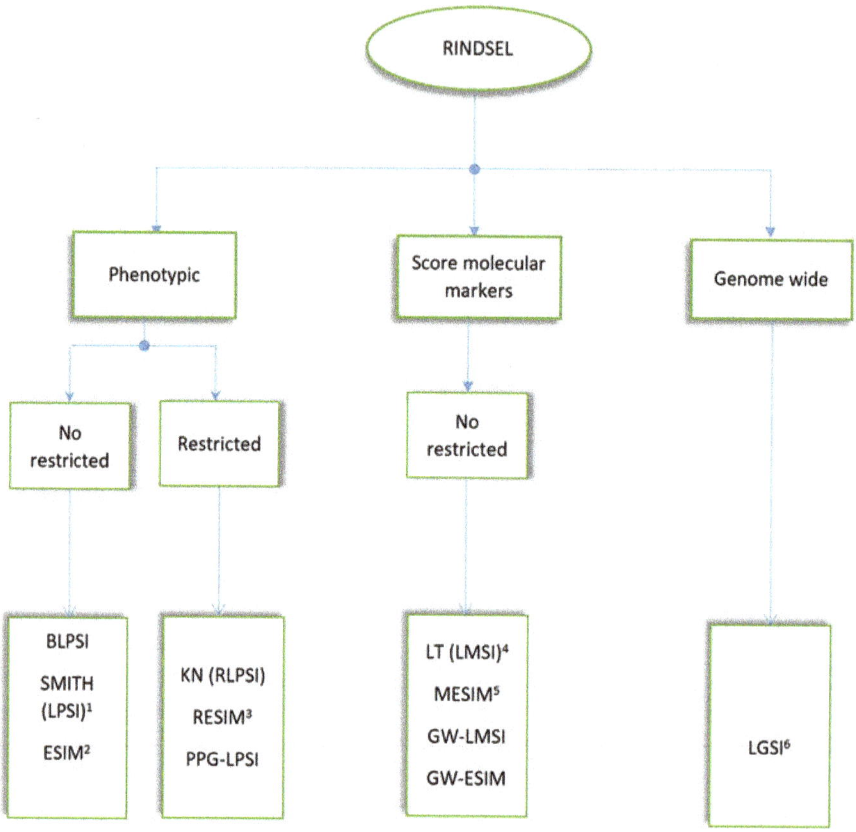

Fig. 11.5 Flow diagram of the selection indices that RIndSel is able to calculate; [1]Smith (1936), [2,3]Cerón-Rojas (2008a), [4]Lande R, Thompson R (1990), [5]Cerón-Rojas (2008b), [6]Cerón-Rojas (2015)

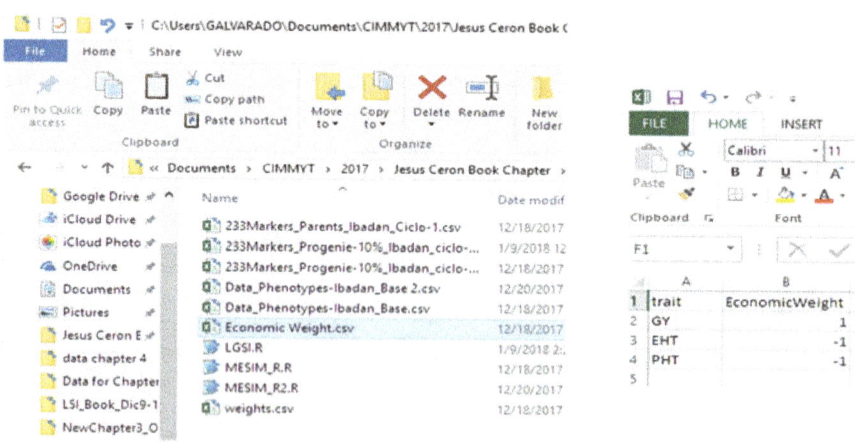

a b

Fig. 11.6 Example of content for (**a**) economic weights of (**b**) file location

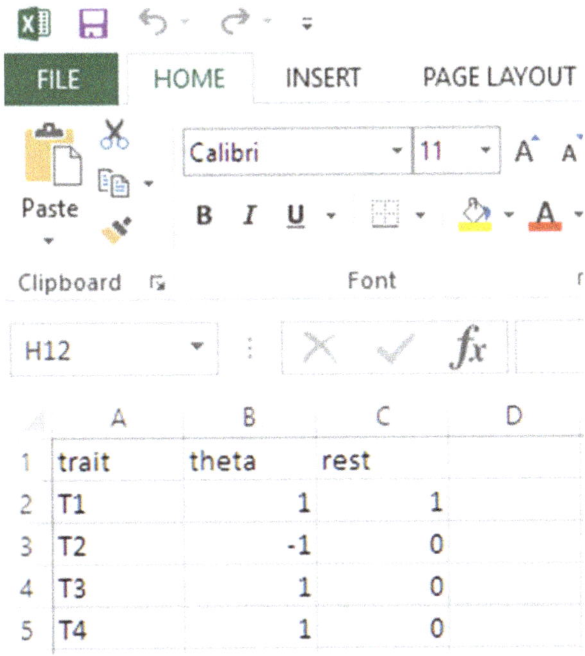

Fig. 11.7 Economic weights for restricted selection indices

11.7 Variable Selection

Experimental design is strongly related to the "*Variable Selection*" menu, where it is possible to identify the variables that constitute the experimental design. Thus, we can choose variables that match with the "*Location*," replicate for random complete block design and block, provided that we have a lattice or alpha-lattice experiment.

11.8 Response Variables

In this menu, the user can select traits to be used to calculate the selection index. It can be activated by clicking on the trait to be selected. Figure 11.8 shows an example of how this window must be filled when a Smith phenotypic selection index is calculated.

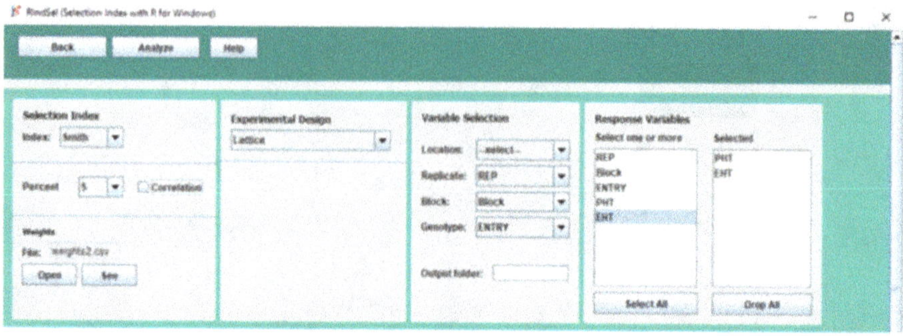

Fig. 11.8 Example of parameters that could be used to calculate a phenotypic selection index

11.9 Molecular Selection Indices

If the selection index to be calculated is molecular, such as the Lande and Thompson (1990) or the linear molecular selection index (Fig. 11.9, and see Table 1.1, Chap. 1, for details), two additional files are required:

1. Whole molecular markers matrix (green arrow).
2. Marker scores or estimated quantitative trait loci values (red arrow).

Marker scores can be obtained by making a regression of the phenotypic values on a codified molecular markers matrix (see Chap. 4 for details). The file can be created in Excel and must have the score with its respective marker for each trait; this file is saved with a .csv extension. An example of how these kinds of files must be generated is shown in Fig. 11.10a.

To calculate the scores in an F2 population, it is important for the molecular marker to have previously been codified as −1, 0, and 1 for genotypes aa, Aa, and AA respectively. When data come from an recombinant inbred line population, the molecular marker should be codified as −1 and 1 for homozygous genotype aa and AA respectively. In the genomic selection indices (LGSI) context (see Chap. 5 for details), it is only necessary to codify the molecular marker matrix (Fig. 11.10b), as these indices do not require a marker score.

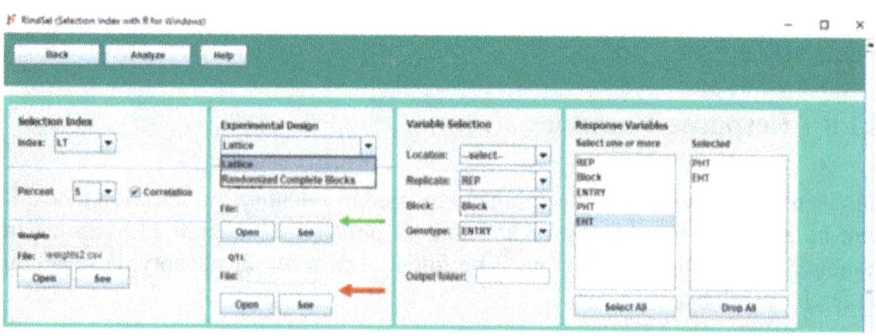

Fig. 11.9 Example of parameters that could be used to calculate a molecular selection index

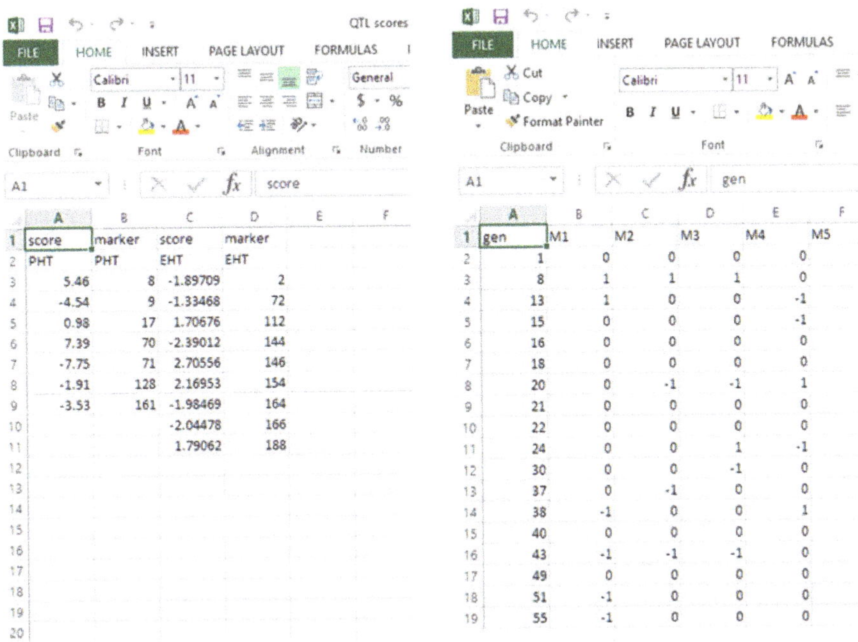

score	marker	score	marker
PHT	PHT	EHT	EHT
5.46	8	-1.89709	2
-4.54	9	-1.33468	72
0.98	17	1.70676	112
7.39	70	-2.39012	144
-7.75	71	2.70556	146
-1.91	128	2.16953	154
-3.53	161	-1.98469	164
		-2.04478	166
		1.79062	188

a

gen	M1	M2	M3	M4	M5
1	0	0	0	0	
8	1	1	1	0	
13	1	0	0	-1	
15	0	0	0	-1	
16	0	0	0	0	
18	0	0	0	0	
20	0	-1	-1	1	
21	0	0	0	0	
22	0	0	0	0	
24	0	0	1	-1	
30	0	0	-1	0	
37	0	-1	0	0	
38	-1	0	0	1	
40	0	0	0	0	
43	-1	-1	-1	0	
49	0	0	0	0	
51	-1	0	0	0	
55	-1	0	0	0	

b

Fig. 11.10 Comma delimited files read in Excel for (**a**) scores of markers for traits plant height (PHT) and ear height (EHT), (**b**) a codified molecular marker matrix

11.10 How to Use RIndSel

The use of RIndSel can be illustrated with an example from the Smith linear phenotypic selection index (LPSI) (Smith 1936, see Chap. 2 for details). Figure 11.11 shows the phenotypic data (Fig. 11.11a), together with the file of economic weights (Fig. 11.11b). Three simulated traits (T1, T2, and T3) described in Chap. 2 were used. T1 and T3 are positive (economic value = 1), whereas trait T2 is negative (economic value = −1). It is important to remember that all data files must be saved in comma delimited format (*.csv).

After the data and economic weights files have been generated, the data need to be loaded into RIndSel; thus, it is important to be able to find the pathway to where the files are located (e.g., "C://Book/datafile/C1_PSI_05_Phen.csv"). Once the data file has been located, it must be uploaded, which can be done by clicking on the file, causing it to automatically begin this process. It is then possible go to the second module (Fig. 11.12) and select subsequent parameters from the menus. In this case, **Selection Index**: *Smith*; **Percent**: *5;* **Weights**: *here we must look for where the economic weights are*, for example "C://Book/datafile/C1_PSI_05_Phen Weights. csv." Once this file has been located, it must be selected by clicking.

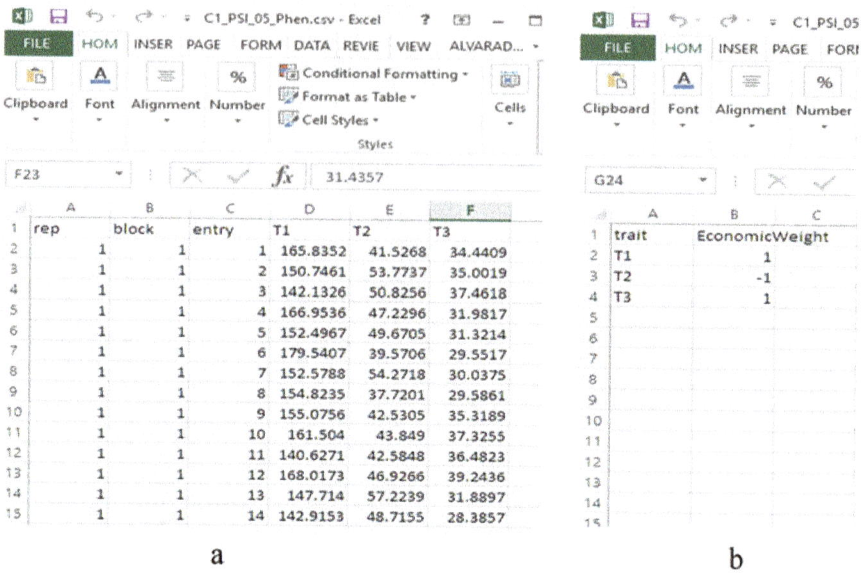

Fig. 11.11 Simulated data from Chap. 2 with (**a**) array in an alpha-lattice and (**b**) economic weights required to test the Smith linear phenotypic selection index (LPSI)

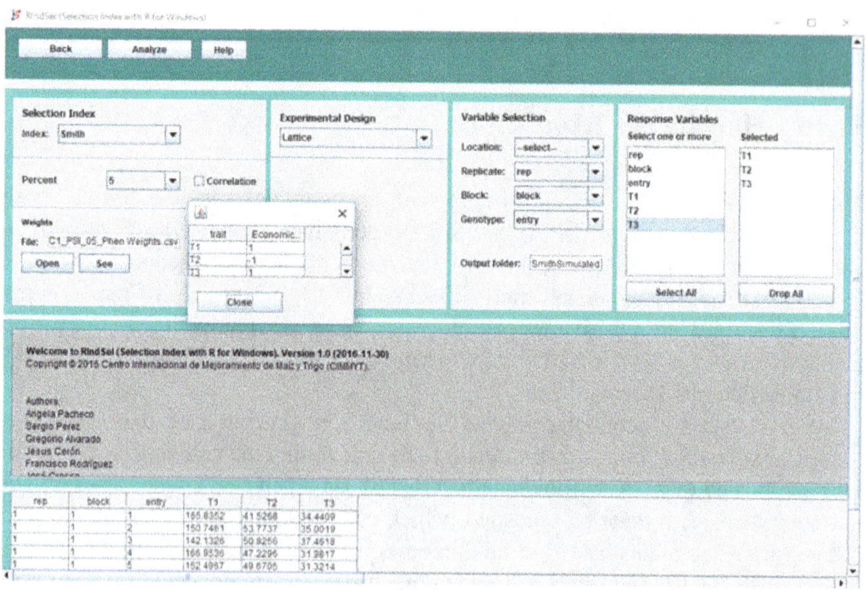

Fig. 11.12 Example of filling in a phenotypic selection index without restrictions

After the selection index windows are filled, the following menu is called: **Experimental design**, which allows the user to select the appropriate design – (for example, a lattice). To select the design variables, the user must navigate to the **Variable Selection**. In this example, the experiment has only one location, and the following should be selected: *rep* as **Replicate**, *block* as **Block** and *entry* as **Genotype**. An output name of the index must be assigned by writing its name in the **Box Output** folder, which is below the **Variable Selection** menu. For the Smith LPSI, the name chosen was *SmithSimulated*. Finally, the **Response Variables** menu should be filled by selecting the traits T1, T2, and T3.

11.11 RIndSel Output

This section explains the structure of the RIndSel output. First, RIndSel presents the genotypic variance–covariance matrix and the phenotypic variance–covariance matrix (Table 11.1). In addition, when the selection index involves molecular data, RIndSel presents an additional molecular variance–covariance matrix, which contains the additive variability associated with the markers (Table 11.2).

RIndSel also presents a table with the estimated values of the index parameters (Table 11.3). These estimates are the covariance of the selection index, the variance of the selection index, the net genetic merit (breeding value), the correlation between the selection index and the net genetic merit, the selection response, and the heritability of the index (see Chap. 2 for additional details).

Additional results are presented in Table 11.4, which show the ranked selected individuals; this ranking was done as a function of the estimated selection index values. Table 11.4 also presents the means of the traits of the selected individuals; the means of the traits of the total population; the selection differential (see Chap. 2),

Table 11.1 Matrices of variance–covariance deployed by RIndSel

rownames	T1	T2	T3
Genetic covariance matrix			
T1	36.21	−12.93	8.35
T2	−12.93	13.04	−3.40
T3	8.35	−3.40	9.96
Phenotypic covariance matrix			
T1	62.50	−12.74	8.53
T2	−12.74	17.52	−3.38
T3	8.53	−3.38	12.31

Table 11.2 Molecular covariance matrix

rownames	T1	T2	T3
T1	62.50	−12.74	8.53
T2	−12.74	17.52	−3.38
T3	8.53	−3.38	12.31

Table 11.3 Estimated selection index parameters given by the RIndSel output

Parameter	Output
Covariance between the selection index and the breeding value	86.7185
Variance of the selection index	86.7185
Variance of the breeding value	108.5746
Correlation between the selection index and the breeding value	0.8937
Response to selection	16.3431
Heritability	0.8168

Table 11.4 Values of the three traits for selected individuals and the values of the Smith linear phenotypic selection index, means and gains with $k = 5\%$

rownames	T1	T2	T3	Index
Entry 353	189.68	38.16	36.13	103.97
Entry 370	178.27	34.38	37.79	103.45
Entry 480	174.84	42.72	45.12	100.66
Entry 300	177.38	39.15	40.34	100.65
Entry 273	181.18	35.94	35.14	100.52
Entry 275	167.94	36.82	42.20	99.92
Entry 148	173.37	37.07	39.62	99.86
Entry 137	185.48	46.48	42.55	99.77
Entry 351	173.79	38.38	40.52	99.68
Entry 236	182.85	37.88	34.96	99.20
Entry 217	175.13	38.48	39.16	98.84
Entry 356	171.09	39.60	41.98	98.47
Entry 167	175.39	38.73	37.73	97.17
Entry 230	169.73	37.10	38.69	96.80
Entry 243	171.90	41.53	41.45	96.29
Entry 55	170.02	36.92	37.76	96.15
Entry 68	172.56	37.18	36.70	96.13
Entry 36	175.80	38.86	36.34	95.75
Entry 164	173.61	38.37	36.42	95.14
Entry 140	170.53	42.52	41.97	95.05
Entry 146	177.40	39.64	35.50	94.89
Entry 432	174.01	40.73	38.26	94.84
Entry 378	176.62	42.69	38.47	94.44
Entry 288	172.14	39.37	37.26	94.23
Entry 386	175.77	42.89	38.81	94.13
Mean of selected individuals	175.46	39.26	38.83	
Mean of all individuals	161.88	45.19	34.39	
Selection differential	13.58	−5.92	4.44	
Expected genetic gain 5%	9.51	−5.48	4.22	

Table 11.5 First 20 values of the entries and their corresponding selection index for all individuals when three traits are analyzed

rownames	T1	T2	T3	Index
Entry 1	164.46	39.63	34.66	86.81
Entry 2	144.39	50.77	34.65	63.82
Entry 3	157.48	48.04	37.90	77.52
Entry 4	167.30	47.98	30.49	74.97
Entry 5	164.11	49.89	32.03	72.85
Entry 6	166.26	40.44	29.93	81.81
Entry 7	154.59	52.22	30.31	63.22
Entry 8	160.00	42.91	31.23	77.12
Entry 9	158.51	46.32	34.52	76.25
Entry 10	163.63	45.43	35.73	81.35
Entry 11	156.16	46.75	35.58	75.62
Entry 12	171.38	41.17	35.13	89.52
Entry 13	153.17	54.18	36.23	66.79
Entry 14	149.89	52.33	31.13	61.39
Entry 15	159.63	49.01	31.72	70.96
Entry 16	160.70	42.51	32.99	79.85
Entry 17	157.07	45.49	28.40	69.68
Entry 18	167.50	41.69	36.73	88.55
Entry 19	159.17	50.60	36.25	73.93
Entry 20	161.80	46.58	37.33	80.85

and the expected genetic gain per trait. Selected individuals can be identified by the first column called "*rownames*," as columns 2 to 4 contain the best linear and unbiased estimator for each mean trait. Finally, column 5 presents the estimated selection index values.

Comparison between means of selected individuals and all individuals is done by selection differential, where in general traits whose economic weight was 1 are positive, whereas those traits whose economic weight was -1 are negative. The expected genetic gain is an inferential tool based on normal distribution that depends on the percentage of selected individuals and gives the estimated index expected genetic gain per trait.

Finally, Table 11.5 shows the best linear and unbiased estimators for all individuals accompanied by its respective selection index. In this case, only the first 20 individuals were included. This table output is important, because on some occasions, it is necessary to determine the specific behavior of a group of genotypes that may not have a good performance, even though they have shown a good general performance from previous analyses. Another possibility is that a group of individuals belongs to a specific population group; thus, it is possible to select the best individual for this population group.

References

Cerón-Rojas JJ, Sahagún-Castellanos J, Castillo-González F, Santacruz-Varela A, Crossa J (2008a) A restricted selection index method based on eigenanalysis. J Agric Biol Environ Stat 13(4):421–438

Cerón-Rojas JJ, Sahagún-Castellanos J, Castillo-González F, Santacruz-Varela A, Benítez-Riquelme I, Crossa J (2008b) A molecular selection index method based on eigenanalysis. Genetics 180:547–557

Cerón-Rojas JJ, Crossa J, Arief VN, Basford K, Rutkoski J, Jarquín D, Alvarado G, Beyene Y, Semagn K, DeLacy I (2015) A genomic selection index applied to simulated and real data. Genes/Genomes/Genetics 5:2155–2164

Lande R, Thompson R (1990) Efficiency of marker-assisted selection in the improvement of quantitative traits. Genetics 124:743–756

Pacheco A, Pérez S, Alvarado G, Ceron J, Rodríguez F, Crossa J, Burgueño J (2017) RIndSel: selection indices for plant breeding. hdl:11529/10854, CIMMYT Research Data & Software Repository Network, V1

SAS Institute (2017) SAS user's guide: statistics module. Version 9.4. Ed. Cary, NC

Smith HF (1936) A discriminant function for plant selection. In: Papers on quantitative genetics and related topics. Department of Genetics, North Carolina State College, Raleigh, NC, pp 466–476

Permissions

All chapters in this book were first published in LSIMPB, by Springer by J. Jesus Céron-Rojas & José Crossa; hereby published with permission under the Creative Commons Attribution License or equivalent. Every chapter published in this book has been scrutinized by our experts. Their significance has been extensively debated. The topics covered herein carry significant information for a comprehensive understanding. They may even be implemented as practical applications or may be referred to as a beginning point for further studies.

We would like to thank the editorial team for lending their expertise to make the book truly unique. They have played a crucial role in the development of this book. Without their invaluable contributions this book wouldn't have been possible. They have made vital efforts to compile up to date information on the varied aspects of this subject to make this book a valuable addition to the collection of many professionals and students.

This book was conceptualized with the vision of imparting up-to-date and integrated information in this field. To ensure the same, a matchless editorial board was set up. Every individual on the board went through rigorous rounds of assessment to prove their worth. After which they invested a large part of their time researching and compiling the most relevant data for our readers.

The editorial board has been involved in producing this book since its inception. They have spent rigorous hours researching and exploring the diverse topics which have resulted in the successful publishing of this book. They have passed on their knowledge of decades through this book. To expedite this challenging task, the publisher supported the team at every step. A small team of assistant editors was also appointed to further simplify the editing procedure and attain best results for the readers.

Apart from the editorial board, the designing team has also invested a significant amount of their time in understanding the subject and creating the most relevant covers. They scrutinized every image to scout for the most suitable representation of the subject and create an appropriate cover for the book.

The publishing team has been an ardent support to the editorial, designing and production team. Their endless efforts to recruit the best for this project, has resulted in the accomplishment of this book. They are a veteran in the field of academics and their pool of knowledge is as vast as their experience in printing. Their expertise and guidance has proved useful at every step. Their uncompromising quality standards have made this book an exceptional effort. Their encouragement from time to time has been an inspiration for everyone.

The publisher and the editorial board hope that this book will prove to be a valuable piece of knowledge for students, practitioners and scholars across the globe.

Index

Printed in the USA
CPSIA information can be obtained
at www.ICGtesting.com
LVHW010817221023
761596LV00005B/537

9 781647 403546